T0131148

THE EXTENDED SELFISH GENE

Richard Dawkins, FRS, was Charles Simonyi Professor for the Public Understanding of Science at Oxford University from 1995 to 2008. Born in Nairobi of British parents, he was educated at Oxford and did his doctorate under the Nobel Prize-winning ethologist Niko Tinbergen. From 1967 to 1969 he was an Assistant Professor at the University of California at Berkeley, returning as University Lecturer and later Reader in Zoology at New College, Oxford, before becoming the first holder of the Simonyi Chair. He is an Emeritus Fellow of New College.

The Selfish Gene (1976; second edition 1989) catapulted Richard Dawkins to fame, and remains his most famous and widely read work. It was followed by a string of bestselling books: *The Extended Phenotype* (1982), *The Blind Watchmaker* (1986), *River Out of Eden* (1995), *Climbing Mount Improbable* (1996), *Unweaving the Rainbow* (1998), *A Devil's Chaplain* (2003), *The Ancestor's Tale* (2004), *The God Delusion* (2006), and *The Greatest Show on Earth* (2009). He has also published a science book for children, *The Magic of Reality* (2011), and two volumes of memoirs, *An Appetite for Wonder* (2013) and *Brief Candle in the Dark* (2015). Dawkins is a Fellow of both the Royal Society and the Royal Society of Literature. He is the recipient of numerous honours and awards, including the 1987 Royal Society of Literature Award, the *Los Angeles Times* Literary Prize of the same year, the 1990 Michael Faraday Award of the Royal Society, the 1994 Nakayama Prize, the 1997 International Cosmos Prize for Achievement in Human Science, the Kistler Prize in 2001, and the Shakespeare Prize in 2005, the 2006 Lewis Thomas Prize for Writing About Science, and the Nierenberg Prize for Science in the Public Interest in 2009.

MARKING THE 40TH ANNIVERSARY

THE EXTENDED SELFISH GENE

RICHARD DAWKINS

OXFORD
UNIVERSITY PRESS

OXFORD

UNIVERSITY PRESS

Great Clarendon Street, Oxford, OX2 6DP,
United Kingdom

Oxford University Press is a department of the University of Oxford.
It furthers the University's objective of excellence in research, scholarship,
and education by publishing worldwide. Oxford is a registered trade mark of
Oxford University Press in the UK and in certain other countries

First published 1976
Second edition 1989
30th anniversary edition 2006
40th anniversary edition 2016

Chapters 2 and 3 from *The Extended Phenotype*
© Richard Dawkins 1982, 1999
First published 1982
Revised edition with new Afterword and Further Reading 1999
Reprinted with corrections 2008
Revised impression 2016

Published in the United States of America by Oxford University Press
198 Madison Avenue, New York, NY 10016, United States of America

British Library Cataloguing in Publication Data
Data available

Library of Congress Control Number: 2016935442

ISBN 978–0–19–878878–2

Printed in Great Britain by
Clays Ltd., St Ives plc

CONTENTS

TWO CHAPTERS FROM THE EXTENDED PHENOTYPE

NOTE ON THIS EXTENDED HARDBACK EDITION

The Extended Phenotype, my only book intended primarily for an audience of professional biologists, was published six years after *The Selfish Gene*. Its more innovative chapters, the ones that led me to say, 'If you never read anything else of mine, please at least read this', were later abbreviated and simplified as 'The long reach of the gene', Chapter 13 of the second edition of *The Selfish Gene*. Earlier chapters of *The Extended Phenotype* were devoted to preparing the ground in various ways. Among these, Chapters 2 and 3 specifically targeted two major misunderstandings of *The Selfish Gene*. Chapter 2 dealt with the Great Genetic Determinism Fallacy—the mistaken view that our behaviour is entirely determined by our genes, irrespective of environment or other factors. Chapter 3 tackled the Great Adaptationism Misunderstanding—the supposed 'adaptationist' view that *all* features and behaviours of organisms should be understood as adaptations. Because these chapters of *The Extended Phenotype* arose so directly from responses to *The Selfish Gene*, and because the confusions and misreadings they highlight persist in some corners to this day, Latha Menon of Oxford University Press (with whom I have worked so fruitfully on several books in the past) suggested that they should be included here at the end to produce this special 'extended' 40th anniversary hardback edition. We toyed with the idea of my rewriting them with a lay audience in mind, but eventually we thought it best to leave them more or less as they were originally published, with a selective glossary to help non-specialist readers. These two chapters give a flavour of some of the worse misunderstandings that met the first edition of *The Selfish Gene* and should serve to forestall any repetition of them forty years on.

RICHARD DAWKINS
February 2016

INTRODUCTION TO 30TH ANNIVERSARY EDITION

It is sobering to realize that I have lived nearly half my life with *The Selfish Gene*—for better, for worse. Over the years, as each of my seven subsequent books has appeared, publishers have sent me on tour to promote it. Audiences respond to the new book, whichever one it is, with gratifying enthusiasm, applaud politely and ask intelligent questions. Then they line up to buy, and have me sign...*The Selfish Gene*. That is a bit of an exaggeration. Some of them do buy the new book and, for the rest, my wife consoles me by arguing that people who newly discover an author will naturally tend to go back to his first book: having read *The Selfish Gene*, surely they'll work their way through to the latest and (to its fond parent) favourite baby?

I would mind more if I could claim that *The Selfish Gene* had become severely outmoded and superseded. Unfortunately (from one point of view) I cannot. Details have changed and factual examples burgeoned mightily. But, with an exception that I shall discuss in a moment, there is little in the book that I would rush to take back now, or apologize for. Arthur Cain, late Professor of Zoology at Liverpool and one of my inspiring tutors at Oxford in the sixties, described *The Selfish Gene* in 1976 as a 'young man's book'. He was deliberately quoting a commentator on A.J. Ayer's *Language Truth and Logic*. I was flattered by the comparison, although I knew that Ayer had recanted much of his first book and I could hardly miss Cain's pointed implication that I should, in the fullness of time, do the same.

Let me begin with some second thoughts about the title. In 1975, through the mediation of my friend Desmond Morris I showed the partially completed book to Tom Maschler, doyen of London publishers, and we discussed it in his room at Jonathan Cape. He liked the book but not the title. 'Selfish', he said, was a 'down word'. Why not call it *The Immortal Gene?* Immortal was an 'up' word, the immortality of genetic information was a central theme of the book, and 'immortal gene' had almost the same intriguing ring as 'selfish gene' (neither of us, I think, noticed the resonance with Oscar Wilde's *The Selfish Giant*). I now think Maschler may have been right. Many critics, especially vociferous ones learned in philosophy as I have discovered, prefer to read a book by title only. No doubt this works well enough for *The Tale of Benjamin Bunny* or *The Decline and Fall of the Roman Empire*, but I can readily see that 'The Selfish Gene' on its own, without the large footnote of the book itself, might give an inadequate impression of its contents. Nowadays, an American publisher would in any case have insisted on a subtitle.

The best way to explain the title is by locating the emphasis. Emphasize 'selfish' and you will think the book is about selfishness, whereas, if anything, it devotes more attention to altruism. The correct word of the title to stress is 'gene' and let me explain why. A central debate within Darwinism concerns the unit that is actually selected: what kind of entity is it that survives, or does not survive, as a consequence of natural selection. That unit will become, more or less by definition, 'selfish'. Altruism might well be favoured at other levels. Does natural selection choose between species? If so, we might expect individual organisms to behave altruistically 'for the good of the species'. They might limit their birth rates to avoid overpopulation, or restrain their hunting behaviour to conserve the species' future stocks of prey. It was such widely disseminated misunderstandings of Darwinism that originally provoked me to write the book.

Or does natural selection, as I urge instead here, choose between genes? In this case, we should not be surprised to find individual organisms behaving altruistically 'for the good of the genes', for example by feeding and protecting kin who are likely to share copies of the same genes. Such kin altruism is only one way in which gene selfishness can translate itself into individual altruism. This book explains how it works, together with reciprocation, Darwinian theory's other main generator of altruism. If I were ever to rewrite the book, as a late convert to the Zahavi/Grafen 'handicap principle' (see pages 406–12) I should also give some space to Amotz Zahavi's idea that altruistic donation might be a 'Potlatch' style of dominance signal: see how superior to you I am, I can afford to make a donation to you!

Let me repeat and expand the rationale for the word 'selfish' in the title. The critical question is: Which level in the hierarchy of life will turn out to be the inevitably 'selfish' level, at which natural selection acts? The Selfish Species? The Selfish Group? The Selfish Organism? The Selfish Ecosystem? Most of these could be argued, and most have been uncritically assumed by one or another author, but all of them are wrong. Given that the Darwinian message is going to be pithily encapsulated as The Selfish *Something*, that something turns out to be the gene, for cogent reasons which this book argues. Whether or not you end up buying the argument itself, that is the explanation for the title.

I hope that takes care of the more serious misunderstandings. Nevertheless, I do with hindsight notice lapses of my own on the very same subject. These are to be found especially in Chapter 1, epitomized by the sentence 'Let us try to teach generosity and altruism because we are born selfish'. There is nothing wrong with teaching generosity and altruism, but 'born selfish' is misleading. In partial explanation, it was not until 1978 that I began to think clearly about the distinction between 'vehicles' (usually organisms) and the 'replicators' that ride inside them (in practice genes: the whole matter is explained in Chapter 13, which was added in the second edition). Please mentally delete that rogue sentence and others like it, and substitute something along the lines of this paragraph.

Given the dangers of that style of error, I can readily see how the title could be misunderstood, and this is one reason why I should perhaps have gone for *The Immortal Gene*. *The Altruistic Vehicle* would have been another possibility. Perhaps it would have been too enigmatic but, at all events, the apparent dispute between the gene and the organism as rival units of natural selection (a dispute that exercised the late Ernst Mayr to the end) is resolved. There are two kinds of unit of natural selection, and there is no dispute between them. The gene is the unit in the sense of replicator. The organism is the unit in the sense of vehicle. Both are important. Neither should be denigrated. They represent two completely distinct kinds of unit and we shall be hopelessly confused unless we recognize the distinction.

Another good alternative to *The Selfish Gene* would have been *The Cooperative Gene*. It sounds paradoxically opposite, but a central part of the book argues for a form of cooperation among self-interested genes. This emphatically does not mean that groups of genes prosper at the expense of their members, or at the expense of other groups. Rather, each gene is seen as pursuing its own self-interested agenda against the background of the other genes in the gene pool—the set of candidates for sexual shuffling within a species. Those other genes are part of the environment in which each gene survives, in the same way as the weather, predators and prey, supporting vegetation and soil bacteria are parts of the environment. From each gene's point of view, the 'background' genes are those with which it shares bodies in its journey down the generations. In the short term, that means the other members of the genome. In the long term, it means the other genes in the gene pool of the species. Natural selection therefore sees to it that gangs of mutually compatible—which is almost to say cooperating—genes are favoured in the presence of each other. At no time does this evolution of the 'cooperative gene' violate the fundamental principle of the selfish gene. Chapter 5 develops the idea, using the analogy of a rowing crew, and Chapter 13 takes it further.

Now, given that natural selection for selfish genes tends to favour cooperation among genes, it has to be admitted that there are some genes that do no such thing and work against the interests of the rest of the genome. Some authors have called them outlaw genes, others ultra-selfish genes, yet others just 'selfish genes'—misunderstanding the subtle difference from genes that cooperate in self-interested cartels. Examples of ultra-selfish genes are the meiotic drive genes described on pages 304–6, and the 'parasitic DNA' originally proposed on pages 56–7 and developed further by various authors under the catchphrase 'Selfish DNA'. The uncovering of new and ever more bizarre examples of ultra-selfish genes has become a feature of the years since this book was first published.*

The Selfish Gene has been criticized for anthropomorphic personification and this too needs an explanation, if not an apology. I employ two levels of personification: of genes, and of organisms. Personification of genes really ought not to be a problem, because no sane person thinks DNA molecules have conscious personalities, and no sensible reader would impute such a delusion to an author. I once had the honour of hearing the great molecular biologist Jacques Monod talking about creativity in science.

---

* Austin Burt and Robert Trivers (2006), *Genes in Conflict: the biology of selfish genetic elements* (Harvard University Press) arrived too late for inclusion in the first printing of this edition. It will undoubtedly become the definitive reference work on this important subject.

I have forgotten his exact words, but he said approximately that, when trying to think through a chemical problem, he would ask himself what he would do if he were an electron. Peter Atkins, in his wonderful book *Creation Revisited*, uses a similar personification when considering the refraction of a light beam, passing into a medium of higher refractive index which slows it down. The beam behaves as if trying to minimize the time taken to travel to an end point. Atkins imagines it as a lifeguard on a beach racing to rescue a drowning swimmer. Should he head straight for the swimmer? No, because he can run faster than he can swim and would be wise to increase the dry-land proportion of his travel time. Should he run to a point on the beach directly opposite his target, thereby minimizing his swimming time? Better, but still not the best. Calculation (if he had time to do it) would disclose to the lifeguard an optimum intermediate angle, yielding the ideal combination of fast running followed by inevitably slower swimming. Atkins concludes:

> That is exactly the behaviour of light passing into a denser medium. But how does light know, apparently in advance, which is the briefest path? And, anyway, why should it care?

He develops these questions in a fascinating exposition, inspired by quantum theory.

Personification of this kind is not just a quaint didactic device. It can also help a professional scientist to get the right answer, in the face of tricky temptations to error. Such is the case with Darwinian calculations of altruism and selfishness, cooperation and spite. It is very easy to get the wrong answer. Personifying genes, if done with due care and caution, often turns out to be the shortest route to rescuing a Darwinian theorist drowning in muddle. While trying to exercise that caution, I was encouraged by the masterful precedent of W. D. Hamilton, one of the four named heroes of the book. In a paper of 1972 (the year in which I began to write *The Selfish Gene*) Hamilton wrote:

> A gene is being favoured in natural selection if the aggregate of its replicas forms an increasing fraction of the total gene pool. We are going to be concerned with genes supposed to affect the social behaviour of their bearers, so let us try to make the argument more vivid by attributing to the genes, temporarily, intelligence and a certain freedom of choice. Imagine that a gene is considering the problem of increasing the number of its replicas, and imagine that it can choose between ...

That is exactly the right spirit in which to read much of *The Selfish Gene*.

Personifying an organism could be more problematical. This is because organisms, unlike genes, have brains and therefore really might have selfish or altruistic motives in something like the subjective sense we would recognize. A book called *The Selfish Lion* might actually confuse, in a way that *The Selfish Gene* should not. Just as one can put oneself in the position of an imaginary light beam, intelligently choosing the optimal route through a cascade of lenses and prisms, or an imaginary gene choosing an optimal route through the generations, so one can postulate an individual lioness, calculating an optimal behavioural strategy for the long term future survival of her genes. Hamilton's first gift to biology was the precise mathematics that a truly Darwinian individual such as a lion would, in effect, have to employ, when taking decisions calculated to maximize the long term survival of its genes. In this book I used informal verbal equivalents of such calculations—on the two levels.

On page 168 we switch rapidly from one level to the other:

> We have considered the conditions under which it would actually pay a mother to let a runt die. We might suppose intuitively that

the runt himself should go on struggling to the last, but the theory does not necessarily preict this. As soon as a runt becomes so small and weak that his expectation of life is reduced to the point where benefit to him due to parental investment is less than half the benefit that the same investment could potentially confer on the other babies, the runt should die gracefully and willingly. He can benefit his genes most by doing so.

That is all individual-level introspection. The assumption is not that the runt chooses what gives him pleasure, or what feels good. Rather, individuals in a Darwinian world are assumed to be making an *as-if* calculation of what would be best for their genes. This particular paragraph goes on to make it explicit by a quick change to gene-level personification:

> That is to say, a gene that gives the instruction 'Body, if you are very much smaller than your litter-mates, give up the struggle and die' could be successful in the gene pool, because it has a 50 per cent chance of being in the body of each brother and sister saved, and its chances of surviving in the body of the runt are very small anyway.

And then the paragraph immediately switches back to the introspective runt:

> There should be a point of no return in the career of a runt. Before he reaches this point he should go on struggling. As soon as he reaches it he should give up and preferably let himself be eaten by his litter-mates or his parents.

I really believe that these two levels of personification are not confusing if read in context and in full. The two levels of 'as-if calculation' come to exactly the same conclusion if done correctly: that, indeed, is the criterion for judging their correctness. So, I don't think personification is something I would undo if I were to write the book again today.

Unwriting a book is one thing. Unreading it is something else. What are we to make of the following verdict, from a reader in Australia?

> Fascinating, but at times I wish I could unread it…On one level, I can share in the sense of wonder Dawkins so evidently sees in the workings-out of such complex processes…But at the same time, I largely blame *The Selfish Gene* for a series of bouts of depression I suffered from for more than a decade…Never sure of my spiritual outlook on life, but trying to find something deeper—trying to believe, but not quite being able to—I found that this book just about blew away any vague ideas I had along these lines, and prevented them from coalescing any further. This created quite a strong personal crisis for me some years ago.

I have previously described a pair of similar responses from readers:

> A foreign publisher of my first book confessed that he could not sleep for three nights after reading it, so troubled was he by what he saw as its cold, bleak message. Others have asked me how I can bear to get up in the mornings. A teacher from a distant country wrote to me reproachfully that a pupil had come to him in tears after reading the same book, because it had persuaded her that life was empty and purposeless. He advised her not to show the book to any of her friends, for fear of contaminating them with the same nihilistic pessimism (*Unweaving the Rainbow*).

If something is true, no amount of wishful thinking can undo it. That is the first thing to say, but the second is almost as important. As I went on to write,

> Presumably there is indeed no purpose in the ultimate fate of the cosmos, but do any of us really tie our life's hopes to the ultimate fate of the cosmos anyway? Of course we don't; not if we are sane. Our lives are ruled by all sorts of closer, warmer,

human ambitions and perceptions. To accuse science of robbing life of the warmth that makes it worth living is so preposterously mistaken, so diametrically opposite to my own feelings and those of most working scientists, I am almost driven to the despair of which I am wrongly suspected.

A similar tendency to shoot the messenger is displayed by other critics who have objected to what they see as the disagreeable social, political or economic implications of *The Selfish Gene*. Soon after Mrs Thatcher won her first election victory in 1979, my friend Steven Rose wrote the following in *New Scientist*:

> I am not implying that Saatchi and Saatchi engaged a team of sociobiologists to write the Thatcher scripts, nor even that certain Oxford and Sussex dons are beginning to rejoice at this practical expression of the simple truths of selfish genery they have been struggling to convey to us. The coincidence of fashionable theory with political events is messier than that. I do believe though, that when the history of the move to the right of the late 1970s comes to be written, from law and order to monetarism and to the (more contradictory) attack on statism, then the switch in scientific fashion, if only from group to kin selection models in evolutionary theory, will come to be seen as part of the tide which has rolled the Thatcherites and their concept of a fixed, 19th century competitive and xenophobic human nature into power.

The 'Sussex don' was the late John Maynard Smith, admired by Steven Rose and me alike, and he replied characteristically in a letter to *New Scientist*: 'What should we have done, fiddled the equations?' One of the dominant messages of *The Selfish Gene* (reinforced by the title essay of *A Devil's Chaplain*) is that we should not derive our values from Darwinism, unless it is with a negative sign. Our brains have evolved to the point where we are capable of rebelling against our selfish genes. The fact that we can do so is made obvious by our use of contraceptives. The same principle can and should work on a wider scale.

Unlike the second edition of 1989, this anniversary edition adds no new material except this Introduction, and some extracts from reviews chosen by my three-times Editor and champion, Latha Menon. Nobody but Latha could have filled the shoes of Michael Rodgers, K-selected Editor Extraordinary, whose indomitable belief in this book was the booster rocket of its first edition's trajectory.

This edition does, however—and it is a source of particular joy to me—restore the original Foreword by Robert Trivers. I have mentioned Bill Hamilton as one of the four intellectual heroes of the book. Bob Trivers is another. His ideas dominate large parts of Chapters 9, 10 and 12, and the whole of Chapter 8. Not only is his Foreword a beautifully crafted introduction to the book: unusually, he chose the medium to announce to the world a brilliant new idea, his theory of the evolution of self-deception. I am most grateful to him for giving permission for the original Foreword to grace this anniversary edition.

RICHARD DAWKINS
Oxford, October 2005

PREFACE TO SECOND EDITION

In the dozen years since *The Selfish Gene* was published its central message has become textbook orthodoxy. This is paradoxical, but not in the obvious way. It is not one of those books that was reviled as revolutionary when published, then steadily won converts until it ended up so orthodox that we now wonder what the fuss was about. Quite the contrary. From the outset the reviews were gratifyingly favourable and it was not seen, initially, as a controversial book. Its reputation for contentiousness took years to grow until, by now, it is widely regarded as a work of radical extremism. But over the very same years as the book's *reputation* for extremism has escalated, its actual *content* has seemed less and less extreme, more and more the common currency.

The selfish gene theory is Darwin's theory, expressed in a way that Darwin did not choose but whose aptness, I should like to think, he would instantly have recognized and delighted in. It is in fact a logical outgrowth of orthodox neo-Darwinism, but expressed as a novel image. Rather than focus on the individual organism, it takes a gene's eye view of nature. It is a different way of seeing, not a different theory. In the opening pages of *The Extended Phenotype* I explained this using the metaphor of the Necker cube.

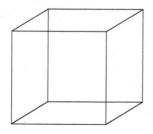

This is a two-dimensional pattern of ink on paper, but it is perceived as a transparent, three-dimensional cube. Stare at it for a few seconds and it will change to face in a different direction. Carry on staring and it will flip back to the original cube. Both cubes are equally compatible with the two-dimensional data on the retina, so the brain happily alternates between them. Neither is more correct than the other. My point was that there are two ways of looking at natural selection, the gene's angle and that of the individual. If properly understood they are equivalent; two views of the same truth. You can flip from one to the other and it will still be the same neo-Darwinism.

I now think that this metaphor was too cautious. Rather than propose a new theory or unearth a new fact, often the most important contribution a scientist can make is to discover a new way of seeing old theories or facts. The Necker cube model is misleading because it suggests that the two ways of seeing are equally good. To be sure, the metaphor gets it partly right: 'angles', unlike theories, cannot be judged by experiment; we cannot resort to our familiar criteria of verification and falsification. But a change of vision can, at its best, achieve something loftier than a theory. It can usher in a whole climate of thinking, in which many exciting and testable theories are born, and unimagined facts laid bare. The Necker cube metaphor misses this completely. It captures the idea of a flip in vision, but fails to do justice to its value. What we are talking about is not a flip to an equivalent view but, in extreme cases, a transfiguration.

I hasten to disclaim any such status for my own modest contributions. Nevertheless, it is for this kind of reason that I prefer not to make a clear separation between science and its 'popularization'. Expounding ideas that have hitherto appeared only in the technical literature is a difficult art. It requires insightful new twists of language and revealing metaphors. If you push novelty of language and metaphor far enough, you can end up with a new way of seeing. And a new way of seeing, as I have just argued, can in its own right make an original contribution to science. Einstein himself was no mean popularizer, and I've often suspected that his vivid metaphors did more than just help the rest of us. Didn't they also fuel his creative genius?

The gene's eye view of Darwinism is implicit in the writings of R. A. Fisher and the other great pioneers of neo-Darwinism in the early thirties, but was made explicit by W. D. Hamilton and G. C. Williams in the sixties. For me their insight had a visionary quality. But I found their expressions of it too laconic, not full-throated enough. I was convinced that an amplified and developed version could make everything about life fall into place, in the heart as well as in the brain. I would write a book extolling the gene's eye view of evolution. It should concentrate its examples on social behaviour, to help correct the unconscious group selectionism that then pervaded popular Darwinism. I began the book in 1972 when powers-cuts resulting from industrial strife interrupted my laboratory research. The blackouts unfortunately (from one point of view) ended after a mere two chapters, and I shelved the project until I had a sabbatical leave in 1975. Meanwhile the theory had been extended, notably by John Maynard Smith and Robert Trivers. I now see that it was one of those mysterious periods in which new ideas are hovering in the air. I wrote *The Selfish Gene* in something resembling a fever of excitement.

When Oxford University Press approached me for a second edition they insisted that a conventional, comprehensive, page by page revision was inappropriate. There are some books that, from their conception, are obviously destined for a string of editions, and *The Selfish Gene* was not one of them. The first edition borrowed a youthful quality from the times in which it was written.

There was a whiff of revolution abroad, a streak of Wordsworth's blissful dawn. A pity to change a child of those times, fatten it with new facts or wrinkle it with complications and cautions. So, the original text should stand, warts, sexist pronouns and all. Notes at the end would cover corrections, responses and developments. And there should be entirely new chapters, on subjects whose novelty in their own time would carry forward the mood of revolutionary dawn. The result was Chapters 12 and 13. For these I took my inspiration from the two books in the field that have most excited me during the intervening years: Robert Axelrod's *The Evolution of Cooperation*, because it seems to offer some sort of hope for our future; and my own *The Extended Phenotype* because for me it dominated those years and because—for what that is worth—it is probably the finest thing I shall ever write.

The title 'Nice guys finish first' is borrowed from the BBC *Horizon* television programme that I presented in 1985. This was a fifty-minute documentary on game-theoretic approaches to the evolution of cooperation, produced by Jeremy Taylor. The making of this film, and another, *The Blind Watchmaker*, by the same producer, gave me a new respect for his profession. At their best, *Horizon* producers (some of their programmes can be seen in America, often repackaged under the name *Nova*) turn themselves into advanced scholarly experts on the subject in hand. Chapter 12 owes more than just its title to my experience of working closely with Jeremy Taylor and the *Horizon* team, and I am grateful.

I recently learned a disagreeable fact: there are influential scientists in the habit of putting their names to publications in whose composition they have played no part. Apparently some senior scientists claim joint authorship of a paper when all that they have contributed is bench space, grant money and an editorial read-through of the manuscript. For all I know, entire scientific reputations may have been built on the work of students and colleagues!

I don't know what can be done to combat this dishonesty. Perhaps journal editors should require signed testimony of what each author contributed. But that is by the way. My reason for raising the matter here is to make a contrast. Helena Cronin has done so much to improve every line—every word—that she should, but for her adamant refusal, be named as joint author of all the new portions of this book. I am deeply grateful to her, and sorry that my acknowledgment must be limited to this. I also thank Mark Ridley, Marian Dawkins and Alan Grafen for advice and for constructive criticism of particular sections. Thomas Webster, Hilary McGlynn and others at Oxford University Press cheerfully tolerated my whims and procrastinations.

RICHARD DAWKINS

1989

FOREWORD TO
FIRST EDITION

The chimpanzee and the human share about 99.5 per cent of their evolutionary history, yet most human thinkers regard the chimp as a malformed, irrelevant oddity while seeing themselves as stepping-stones to the Almighty. To an evolutionist this cannot be so. There exists no objective basis on which to elevate one species above another. Chimp and human, lizard and fungus, we have all evolved over some three billion years by a process known as natural selection. Within each species some individuals leave more surviving offspring than others, so that the inheritable traits (genes) of the reproductively successful become more numerous in the next generation. This is natural selection: the non-random differential reproduction of genes. Natural selection has built us, and it is natural selection we must understand if we are to comprehend our own identities.

Although Darwin's theory of evolution through natural selection is central to the study of social behavior (especially when wedded to Mendel's genetics), it has been very widely neglected. Whole industries have grown up in the social sciences dedicated to the construction of a pre-Darwinian and pre-Mendelian view of the social and psychological world. Even within biology the neglect and misuse of Darwinian theory has been astonishing. Whatever the reasons for this strange development, there is evidence that it is coming to an end. The great work of Darwin and Mendel has been extended by a growing number of workers, most notably by R. A. Fisher, W. D. Hamilton, G. C. Williams, and J. Maynard Smith. Now, for the first time, this important body of social theory based on natural selection is presented in a simple and popular form by Richard Dawkins.

One by one Dawkins takes up the major themes of the new work in social theory: the concepts of altruistic and selfish behavior, the genetical definition of self-interest, the evolution of aggressive behavior, kinship theory (including parent-offspring relations and the evolution of the social insects), sex ratio theory, reciprocal altruism, deceit, and the natural selection of sex differences. With a confidence that comes from mastering the underlying theory, Dawkins unfolds the new work with admirable clarity and style. Broadly educated in biology, he gives the reader a taste of its rich and fascinating literature. Where he differs from published work (as he does in criticizing a fallacy of my own), he is almost invariably exactly on target. Dawkins also takes pains to make clear the logic of his arguments, so that the reader, by applying the logic given, can extend the arguments (and even take on Dawkins himself). The arguments themselves extend in many directions. For example, if (as Dawkins argues) deceit is fundamental in animal communication, then there must be strong selection to spot deception and this ought, in turn, to select for a degree of self-deception, rendering some facts and motives unconscious so as not to betray—by the subtle signs of self-knowledge—the deception being practiced. Thus, the conventional view that natural selection favors nervous systems which produce ever more accurate images of the world must be a very naïve view of mental evolution.

The recent progress in social theory has been substantial enough to have generated a minor flurry of counter-revolutionary activity. It has been alleged, for example, that the recent progress is, in fact, part of a cyclical conspiracy to impede social advancement by making such advancement appear to be genetically impossible. Similar feeble thoughts have been strung together to produce the impression that Darwinian social theory is reactionary in its political implications. This is very far from the truth. The genetic equality of the sexes is, for the first time, clearly established by

Fisher and Hamilton. Theory and quantitative data from the social insects demonstrate that there is no inherent tendency for parents to dominate their offspring (or vice versa). And the concepts of parental investment and female choice provide an objective and unbiased basis for viewing sex differences, a considerable advance over popular efforts to root women's powers and rights in the functionless swamp of biological identity. In short, Darwinian social theory gives us a glimpse of an underlying symmetry and logic in social relationships which, when more fully comprehended by ourselves, should revitalize our political understanding and provide the intellectual support for a science and medicine of psychology. In the process it should also give us a deeper understanding of the many roots of our suffering.

ROBERT L. TRIVERS
Harvard University, July, 1976

PREFACE TO FIRST EDITION

This book should be read almost as though it were science fiction. It is designed to appeal to the imagination. But it is not science fiction: it is science. Cliché or not, 'stranger than fiction' expresses exactly how I feel about the truth. We are survival machines—robot vehicles blindly programmed to preserve the selfish molecules known as genes. This is a truth which still fills me with astonishment. Though I have known it for years, I never seem to get fully used to it. One of my hopes is that I may have some success in astonishing others.

Three imaginary readers looked over my shoulder while I was writing, and I now dedicate the book to them. First the general reader, the layman. For him I have avoided technical jargon almost totally, and where I have had to use specialized words I have defined them. I now wonder why we don't censor most of our jargon from learned journals too. I have assumed that the layman has no special knowledge, but I have not assumed that he is stupid. Anyone can popularize science if he oversimplifies. I have worked hard to try to popularize some subtle and complicated ideas in non-mathematical language, without losing their essence. I do not know how far I have succeeded in this, nor how far I have succeeded in another of my ambitions: to try to make the book as entertaining and gripping as its subject matter deserves. I have long felt that biology ought to seem as exciting as a mystery story, for a mystery story is exactly what biology is. I do not dare to hope that I have conveyed more than a tiny fraction of the excitement which the subject has to offer.

My second imaginary reader was the expert. He has been a harsh critic, sharply drawing in his breath at some of my analogies and figures of speech. His favourite phrases are 'with the exception of'; 'but on the other hand'; and 'ugh'. I listened to him attentively, and even completely rewrote one chapter entirely for his benefit, but in the end I have had to tell the story my way. The expert will still not be totally happy with the way I put things. Yet my greatest hope is that even he will find something new here; a new way of looking at familiar ideas perhaps; even stimulation of new ideas of his own. If this is too high an aspiration, may I at least hope that the book will entertain him on a train?

The third reader I had in mind was the student, making the transition from layman to expert. If he still has not made up his mind what field he wants to be an expert in, I hope to encourage him to give my own field of zoology a second glance. There is a better reason for studying zoology than its possible 'usefulness', and the general likeableness of animals. This reason is that we animals are the most complicated and perfectly-designed pieces of machinery in the known universe. Put it like that, and it is hard to see why anybody studies anything else! For the student who has already committed himself to zoology, I hope my book may have some educational value. He is having to work through the original papers and technical books on which my treatment is based. If he finds the original sources hard to digest, perhaps my non-mathematical interpretation may help, as an introduction and adjunct.

There are obvious dangers in trying to appeal to three different kinds of reader. I can only say that I have been very conscious of these dangers, but that they seemed to be outweighed by the advantages of the attempt.

I am an ethologist, and this is a book about animal behaviour. My debt to the ethological tradition in which I was trained will be obvious. In particular, Niko Tinbergen does not realize the extent of his influence on me during the twelve years I worked under him at Oxford. The phrase 'survival machine', though not actually his own, might well be. But ethology has recently been invigorated by an invasion of fresh ideas from sources not conventionally regarded as ethological. This book is largely based on these new ideas. Their originators are acknowledged in the appropriate places in the text; the dominant figures are G. C. Williams, J. Maynard Smith, W. D. Hamilton, and R. L. Trivers.

Various people suggested titles for the book, which I have gratefully used as chapter titles: 'Immortal Coils', John Krebs; 'The Gene Machine', Desmond Morris; 'Genesmanship', Tim Clutton-Brock and Jean Dawkins, independently with apologies to Stephen Potter.

Imaginary readers may serve as targets for pious hopes and aspirations, but they are of less practical use than real readers and critics. I am addicted to revising, and Marian Dawkins has been subjected to countless drafts and redrafts of every page. Her considerable knowledge of the biological literature and her understanding of theoretical issues, together with her ceaseless encouragement and moral support, have been essential to me. John Krebs too read the whole book in draft. He knows more about the subject than I do, and he has been generous and unstinting with his advice and suggestions. Glenys Thomson and Walter Bodmer criticized my handling of genetic topics kindly but firmly. I fear that my revision may still not fully satisfy them, but I hope they will find it somewhat improved. I am most grateful for their time and patience. John Dawkins exercised an unerring eye for misleading phraseology, and made excellent constructive suggestions for re-wording. I could not have wished for a more suitable 'intelligent layman' than Maxwell Stamp. His perceptive spotting of an important general flaw in the style of the first draft did much for the final version. Others who constructively criticized particular chapters, or otherwise gave expert advice, were John Maynard Smith, Desmond

Morris, Tom Maschler, Nick Blurton Jones, Sarah Kettlewell, Nick Humphrey, Tim Clutton-Brock, Louise Johnson, Christopher Graham, Geoff Parker, and Robert Trivers. Pat Searle and Stephanie Verhoeven not only typed with skill, but encouraged me by seeming to do so with enjoyment. Finally, I wish to thank Michael Rodgers of Oxford University Press who, in addition to helpfully criticizing the manuscript, worked far beyond the call of duty in attending to all aspects of the production of this book.

RICHARD DAWKINS
1976

WHY ARE PEOPLE?

Intelligent life on a planet comes of age when it first works out the reason for its own existence. If superior creatures from space ever visit earth, the first question they will ask, in order to assess the level of our civilization, is: 'Have they discovered evolution yet?' Living organisms had existed on earth, without ever knowing why, for over three thousand million years before the truth finally dawned on one of them. His name was Charles Darwin. To be fair, others had had inklings of the truth, but it was Darwin who first put together a coherent and tenable account of why we exist. Darwin made it possible for us to give a sensible answer to the curious child whose question heads this chapter. We no longer have to resort to superstition when faced with the deep problems: Is there a meaning to life? What are we for? What is man? After posing the last of these questions, the eminent zoologist G. G. Simpson put it thus: 'The point I want to make now is that all attempts to answer that question before 1859 are worthless and that we will be better off if we ignore them completely.'*

Today the theory of evolution is about as much open to doubt as the theory that the earth goes round the sun, but the full implications of Darwin's revolution have yet to be widely realized. Zoology is still a minority subject in universities, and even those who choose to study it often make their decision without appreciating its profound philosophical significance. Philosophy and the subjects known as 'humanities' are still taught almost as if Darwin had never lived. No doubt this will change in time. In any case, this book is not intended as a general advocacy of Darwinism. Instead, it will explore the consequences of the evolution theory for a particular issue. My purpose is to examine the biology of selfishness and altruism.

Apart from its academic interest, the human importance of this subject is obvious. It touches every aspect of our social lives, our loving and hating, fighting and cooperating, giving and stealing, our greed and our generosity. These are claims that could have been made for Lorenz's *On Aggression*, Ardrey's *The Social Contract*, and Eibl-Eibesfeldt's *Love and Hate*. The trouble with these books is that their authors got it totally and utterly wrong. They got it wrong because they misunderstood how evolution works. They made the erroneous assumption that the important thing in evolution is the good of the *species* (or the group) rather than the good of the individual (or the gene). It is ironic that Ashley Montagu should criticize Lorenz as a 'direct descendant of the "nature red in tooth and claw" thinkers of the nineteenth century...'. As I understand Lorenz's view of evolution, he would be very much at one with Montagu in rejecting the implications of Tennyson's famous phrase. Unlike both of them, I think 'nature red in tooth and claw' sums up our modern understanding of natural selection admirably.

Before beginning on my argument itself, I want to explain briefly what sort of an argument it is, and what sort of an argument it is not. If we were told that a man had lived a long and prosperous life in the world of Chicago gangsters, we would be entitled to make some guesses as to the sort of man he was. We might expect that he would have qualities such as toughness, a quick trigger finger, and the ability to attract loyal friends. These would not be infallible deductions, but you can make some inferences about a man's character if you know something about the conditions in which he has survived and prospered. The argument of this book is that we, and all other animals, are machines created by our genes. Like successful Chicago gangsters, our genes have survived, in some cases for millions of years, in a highly competitive world. This entitles us to expect certain qualities in our genes. I shall argue that a predominant quality to be expected in a successful gene is ruthless selfishness. This gene selfishness will usually give rise to selfishness in individual behaviour. However, as we shall see, there are special circumstances in which a gene can achieve its own selfish goals best by fostering a limited form of altruism at the level of individual animals. 'Special' and 'limited' are important words in the last sentence. Much as we might wish to believe otherwise, universal love and the welfare of the species as a whole are concepts that simply do not make evolutionary sense.

This brings me to the first point I want to make about what this book is *not*. I am not advocating a morality based on evolution.* I am saying how things have evolved. I am not saying how we humans morally ought to behave. I stress this, because I know I am in danger of being misunderstood by those people, all too numerous, who cannot distinguish a statement of belief in what is the case from an advocacy of what ought to be the case. My own feeling is that a human society based simply on the gene's law of universal ruthless selfishness would be a very nasty society in which to live. But unfortunately, however much we may deplore something, it does not stop it being true. This book is mainly intended to be interesting, but if you would extract a moral from it, read it as a warning. Be warned that if you wish, as I do, to build a society in which individuals cooperate generously and unselfishly towards a common good, you can expect little help from biological nature. Let us try to *teach* generosity and altruism, because we are born selfish. Let us understand what our own selfish genes are up to, because we may then at least have the chance to upset their designs, something that no other species has ever aspired to.

As a corollary to these remarks about teaching, it is a fallacy— incidentally a very common one—to suppose that genetically inherited traits are by definition fixed and unmodifiable. Our genes may instruct us to be selfish, but we are not necessarily compelled to obey them all our lives. It may just be more difficult to learn altruism than it would be if we were genetically pro- grammed to be altruistic. Among animals, man is uniquely dom- inated by culture, by influences learned and handed down. Some would say that culture is so important that genes, whether selfish or not, are virtually irrelevant to the understanding of human nature. Others would disagree. It all depends where you stand in the debate over 'nature versus nurture' as determinants of human attributes. This brings me to the second thing this book is not: it is not an advocacy of one position or another in the nature/nur- ture controversy. Naturally I have an opinion on this, but I am not going to express it, except insofar as it is implicit in the view of culture that I shall present in the final chapter. If genes really turn out to be totally irrelevant to the determination of modern human behaviour, if we really are unique among animals in this respect, it is, at the very least, still interesting to inquire about the rule to which we have so recently become the exception. And if our species is not so exceptional as we might like to think, it is even more important that we should study the rule.

The third thing this book is not is a descriptive account of the detailed behaviour of man or of any other particular animal species. I shall use factual details only as illustrative examples. I shall not be saying: 'If you look at the behaviour of baboons you will find it to be selfish; therefore the chances are that human behaviour is selfish also'. The logic of my 'Chicago gangster' argument is quite different. It is this. Humans and baboons have evolved by natural selection. If you look at the way natural selection works, it seems to follow that anything that has evolved by natural selection should be selfish. Therefore we must expect that when we go and look at the behaviour of baboons, humans, and all other living creatures, we shall find it to be selfish. If we find that our expectation is wrong, if we observe that human behaviour is truly altruistic, then we shall be faced with something puzzling, something that needs explaining.

Before going any further, we need a definition. An entity, such as a baboon, is said to be altruistic if it behaves in such a way as to increase another such entity's welfare at the expense of its own. Selfish behaviour has exactly the opposite effect. 'Welfare' is defined as 'chances of survival', even if the effect on actual life and death prospects is so small as to *seem* negligible. One of the surprising consequences of the modern version of the Darwinian theory is that apparently trivial tiny influences on survival probability can have a major impact on evolution. This is because of the enormous time available for such influences to make themselves felt.

It is important to realize that the above definitions of altruism and selfishness are *behavioural*, not subjective. I am not concerned here with the psychology of motives. I am not going to argue about whether people who behave altruistically are 'really' doing it for secret or subconscious selfish motives. Maybe they are and maybe they aren't, and maybe we can never know, but in any case that is not what this book is about. My definition is concerned only with whether the *effect* of an act is to lower or raise the survival prospects of the presumed altruist and the survival prospects of the presumed beneficiary.

It is a very complicated business to demonstrate the effects of behaviour on long-term survival prospects. In practice, when we apply the definition to real behaviour, we must qualify it with the word 'apparently'. An apparently altruistic act is one that looks, superficially, as if it must tend to make the altruist more likely (however slightly) to die, and the recipient more likely to survive. It often turns out on closer inspection that acts of apparent altruism are really selfishness in disguise. Once again, I do not mean that the underlying motives are secretly selfish, but that the real effects of the act on survival prospects are the reverse of what we originally thought.

I am going to give some examples of apparently selfish and apparently altruistic behaviour. It is difficult to suppress subjective habits of thought when we are dealing with our own species, so I shall choose examples from other animals instead. First some miscellaneous examples of selfish behaviour by individual animals.

Blackheaded gulls nest in large colonies, the nests being only a few feet apart. When the chicks first hatch out they are small and defenceless and easy to swallow. It is quite common for a gull to wait until a neighbour's back is turned, perhaps while it is away fishing, and then pounce on one of the neighbour's chicks and swallow it whole. It thereby obtains a good nutritious meal, without having to go to the trouble of catching a fish, and without having to leave its own nest unprotected.

More well known is the macabre cannibalism of female praying mantises. Mantises are large carnivorous insects. They normally eat smaller insects such as flies, but they will attack almost anything that moves. When they mate, the male cautiously creeps up on the female, mounts her, and copulates. If the female gets the chance, she will eat him, beginning by biting his head off, either as the male is approaching, or immediately after he mounts, or after they separate. It might seem most sensible for her to wait until copulation is over before she starts to eat him. But the loss of the head does not seem to throw the rest of the male's body off its sexual stride. Indeed, since the insect head is the seat of some inhibitory nerve centres, it is possible that the female improves the male's sexual performance by eating his head.* If so, this is an added benefit. The primary one is that she obtains a good meal.

The word 'selfish' may seem an understatement for such extreme cases as cannibalism, although these fit well with our definition. Perhaps we can sympathize more directly with the reported cowardly behaviour of emperor penguins in the Antarctic. They have been seen standing on the brink of the water, hesitating before diving in, because of the danger of being eaten by seals. If only one of them would dive in, the rest would know whether there was a seal there or not. Naturally nobody wants to be the guinea pig, so they wait, and sometimes even try to push each other in.

More ordinarily, selfish behaviour may simply consist of refusing to share some valued resource such as food, territory, or sexual partners. Now for some examples of apparently altruistic behaviour.

The stinging behaviour of worker bees is a very effective defence against honey robbers. But the bees who do the stinging are kamikaze fighters. In the act of stinging, vital internal organs are usually torn out of the body, and the bee dies soon afterwards. Her suicide mission may have saved the colony's vital food stocks, but she herself is not around to reap the benefits. By our definition this is an altruistic behavioural act. Remember that we are not talking about conscious motives. They may or may not be present, both here and in the selfishness examples, but they are irrelevant to our definition.

Laying down one's life for one's friends is obviously altruistic, but so also is taking a slight risk for them. Many small birds, when they see a flying predator such as a hawk, give a charac-teristic 'alarm call', upon which the whole flock takes appropriate evasive action. There is indirect evidence that the bird who gives the alarm call puts itself in special danger, because it attracts the predator's attention particularly to itself. This is only a slight additional risk, but it nevertheless seems, at least at first sight, to qualify as an altruistic act by our definition.

The commonest and most conspicuous acts of animal altruism are done by parents, especially mothers, towards their children. They may incubate them, either in nests or in their own bodies, feed them at enormous cost to themselves, and take great risks in protecting them from predators. To take just one particular example, many ground-nesting birds perform a so-called 'distraction display' when a predator such as a fox approaches. The parent bird limps away from the nest, holding out one wing as though it were broken. The predator, sensing easy prey, is lured away from the nest containing the chicks. Finally the parent bird gives up its pretence and leaps into the air just in time to escape the fox's jaws. It has probably saved the life of its nestlings, but at some risk to itself.

I am not trying to make a point by telling stories. Chosen examples are never serious evidence for any worthwhile generalization. These stories are simply intended as illustrations of what I mean by altruistic and selfish behaviour at the level of individuals. This book will show how both individual selfishness and individual altruism are explained by the fundamental law that I am calling *gene selfishness*. But first I must deal with a particular erroneous explanation for altruism, because it is widely known, and even widely taught in schools.

This explanation is based on the misconception that I have already mentioned, that living creatures evolve to do things 'for the good of the species' or 'for the good of the group'. It is easy to see how this idea got its start in biology. Much of an animal's life is devoted to reproduction, and most of the acts of altruistic self-sacrifice that are observed in nature are performed by parents towards their young. 'Perpetuation of the species' is a common euphemism for reproduction, and it is undeniably a *consequence* of reproduction. It requires only a slight over-stretching of logic to deduce that the 'function' of reproduction is 'to' perpetuate the species. From this it is but a further short false step to conclude that animals will in general behave in such a way as to favour the perpetuation of the species. Altruism towards fellow members of the species seems to follow.

This line of thought can be put into vaguely Darwinian terms. Evolution works by natural selection, and natural selection means the differential survival of the 'fittest'. But are we talking about the fittest individuals, the fittest races, the fittest species, or what? For some purposes this does not greatly matter, but when we are talking about altruism it is obviously crucial. If it is species that are competing in what Darwin called the struggle for existence, the individual seems best regarded as a pawn in the game, to be sacrificed when the greater interest of the species as a whole requires it. To put it in a slightly more respectable way, a group, such as a species or a population within a species, whose individual members are prepared to sacrifice themselves for the welfare of the group, may be less likely to go extinct than a rival group whose individual members place their own selfish interests first. Therefore the world becomes populated mainly by groups consisting of self-sacrificing individuals. This is the theory of 'group selection', long assumed to be true by biologists not familiar with the details of evolutionary theory, brought out into the open in a famous book by V. C. Wynne-Edwards, and popularized by Robert Ardrey in *The Social Contract*. The orthodox alternative is normally called 'individual selection', although I personally prefer to speak of gene selection.

The quick answer of the 'individual selectionist' to the argument just put might go something like this. Even in the group of altruists, there will almost certainly be a dissenting minority who refuse to make any sacrifice. If there is just one selfish rebel, prepared to exploit the altruism of the rest, then he, by definition, is more likely than they are to survive and have children. Each of these children will tend to inherit his selfish traits. After several generations of this natural selection, the 'altruistic group' will be over-run by selfish individuals, and will be indistinguishable from the selfish group. Even if we grant the improbable chance existence initially of pure altruistic groups without any rebels, it is very difficult to see what is to stop selfish individuals migrating in from neighbouring selfish groups, and, by inter-marriage, contaminating the purity of the altruistic groups.

The individual selectionist would admit that groups do indeed die out, and that whether or not a group goes extinct may be influenced by the behaviour of the individuals in that group. He might even admit that *if only* the individuals in a group had the gift of foresight they could see that in the long run their own best interests lay in restraining their selfish greed, to prevent the destruction of the whole group. How many times must this have been said in recent years to the working people of Britain? But group extinction is a slow process compared with the rapid cut and thrust of individual competition. Even while the group is going slowly and inexorably downhill, selfish individuals prosper in the short term at the expense of altruists. The citizens of Britain may or may not be blessed with foresight, but evolution is blind to the future.

Although the group selection theory now commands little support within the ranks of those professional biologists who understand evolution, it does have great intuitive appeal. Successive generations of zoology students are surprised, when they come up from school, to find that it is not the orthodox point of view. For this they are hardly to be blamed, for in the *Nuffield Biology Teachers' Guide*, written for advanced level biology schoolteachers in Britain, we find the following: 'In higher animals, behaviour may take the form of individual suicide to ensure the survival of the species.' The anonymous author of this guide is blissfully ignorant of the fact that he has said something controversial. In this respect he is in Nobel Prize-winning company. Konrad Lorenz, in *On Aggression*, speaks of the 'species preserving' functions of aggressive behaviour, one of these functions being to make sure that only the fittest individuals are allowed to breed. This is a gem of a circular argument, but the point I am making here is that the group selection idea is so deeply ingrained that Lorenz, like the author of the *Nuffield Guide*, evidently did not realize that his statements contravened orthodox Darwinian theory.

I recently heard a delightful example of the same thing on an otherwise excellent B.B.C. television programme about Australian spiders. The 'expert' on the programme observed that the vast majority of baby spiders end up as prey for other species, and she then went on to say: 'Perhaps this is the real purpose of their existence, as only a few need to survive in order for the species to be preserved'!

Robert Ardrey, in *The Social Contract*, used the group selection theory to account for the whole of social order in general. He clearly sees man as a species that has strayed from the path of animal righteousness. Ardrey at least did his homework. His decision to disagree with orthodox theory was a conscious one, and for this he deserves credit.

Perhaps one reason for the great appeal of the group selection theory is that it is thoroughly in tune with the moral and political ideals that most of us share. We may frequently behave selfishly as individuals, but in our more idealistic moments we honour and admire those who put the welfare of others first. We get a bit muddled over how widely we want to interpret the word 'others', though. Often altruism within a group goes with selfishness between groups. This is a basis of trade unionism. At another level the nation is a major beneficiary of our altruistic self-sacrifice, and young men are expected to die as individuals for the greater glory of their country as a whole. Moreover, they are encouraged to kill other individuals about whom nothing is known except that they belong to a different nation. (Curiously, peace-time appeals for individuals to make some small sacrifice in the rate at which they increase their standard of living seem to be less effective than war-time appeals for individuals to lay down their lives.)

Recently there has been a reaction against racialism and patriotism, and a tendency to substitute the whole human species as the object of our fellow feeling. This humanist broadening of the target of our altruism has an interesting corollary, which again seems to buttress the 'good of the species' idea in evolution. The politically liberal, who are normally the most convinced spokesmen of the species ethic, now often have the greatest scorn for those who have gone a little further in widening their altruism, so that it includes other species. If I say that I am more interested in preventing the slaughter of large whales than I am in improving housing conditions for people, I am likely to shock some of my friends.

The feeling that members of one's own species deserve special moral consideration as compared with members of other species is old and deep. Killing people outside war is the most serious-ly-regarded crime ordinarily committed. The only thing more strongly forbidden by our culture is eating people (even if they are already dead). We enjoy eating members of other species, however. Many of us shrink from judicial execution of even the

most horrible human criminals, while we cheerfully counte-
nance the shooting without trial of fairly mild animal pests.
Indeed we kill members of other harmless species as a means
of recreation and amusement. A human foetus, with no more
human feeling than an amoeba, enjoys a reverence and legal protection far in excess of those granted to an adult chimpanzee.
Yet the chimp feels and thinks and—according to recent experimental evidence—may even be capable of learning a form of
human language. The foetus belongs to our own species, and
is instantly accorded special privileges and rights because of it.
Whether the ethic of 'speciesism', to use Richard Ryder's term,
can be put on a logical footing any more sound than that of
'racism', I do not know. What I do know is that it has no proper
basis in evolutionary biology.

The muddle in human ethics over the level at which altruism is
desirable—family, nation, race, species, or all living things—is
mirrored by a parallel muddle in biology over the level at which
altruism is to be expected according to the theory of evolution.
Even the group selectionist would not be surprised to find members
of rival groups being nasty to each other: in this way, like trade
unionists or soldiers, they are favouring their own group in the
struggle for limited resources. But then it is worth asking how
the group selectionist decides *which* level is the important one. If
selection goes on between groups within a species, and between
species, why should it not also go on between larger groupings?
Species are grouped together into genera, genera into orders, and
orders into classes. Lions and antelopes are both members of the
class Mammalia, as are we. Should we then not expect lions to
refrain from killing antelopes, 'for the good of the mammals'?
Surely they should hunt birds or reptiles instead, in order to
prevent the extinction of the class. But then, what of the need to
perpetuate the whole phylum of vertebrates?

It is all very well for me to argue by *reductio ad absurdum*, and to point to the difficulties of the group selection theory, but the apparent existence of individual altruism still has to be explained. Ardrey goes so far as to say that group selection is the only possible explanation for behaviour such as 'stotting' in Thomson's gazelles. This vigorous and conspicuous leaping in front of a predator is analogous to bird alarm calls, in that it seems to warn companions of danger while apparently calling the predator's attention to the stotter himself. We have a responsibility to explain stotting Tommies and all similar phenomena, and this is something I am going to face in later chapters.

Before that I must argue for my belief that the best way to look at evolution is in terms of selection occurring at the lowest level of all. In this belief I am heavily influenced by G. C. Williams's great book *Adaptation and Natural Selection*. The central idea I shall make use of was foreshadowed by A. Weismann in pre-gene days at the turn of the century—his doctrine of the 'continuity of the germ-plasm'. I shall argue that the fundamental unit of selection, and therefore of self-interest, is not the species, nor the group, nor even, strictly, the individual. It is the gene, the unit of heredity.* To some biologists this may sound at first like an extreme view. I hope when they see in what sense I mean it they will agree that it is, in substance, orthodox, even if it is expressed in an unfamiliar way. The argument takes time to develop, and we must begin at the beginning, with the very origin of life itself.

THE REPLICATORS

In the beginning was simplicity. It is difficult enough explaining how even a simple universe began. I take it as agreed that it would be even harder to explain the sudden springing up, fully armed, of complex order—life, or a being capable of creating life. Darwin's theory of evolution by natural selection is satisfying because it shows us a way in which simplicity could change into complexity, how unordered atoms could group themselves into ever more complex patterns until they ended up manufacturing people. Darwin provides a solution, the only feasible one so far suggested, to the deep problem of our existence. I will try to explain the great theory in a more general way than is customary, beginning with the time before evolution itself began.

Darwin's 'survival of the fittest' is really a special case of a more general law of *survival of the stable*. The universe is populated by stable things. A stable thing is a collection of atoms that is permanent enough or common enough to deserve a name. It may be a unique collection of atoms, such as the Matterhorn, that lasts long enough to be worth naming. Or it may be a *class* of entities, such as rain drops, that come into existence at a sufficiently high rate to deserve a collective name, even if any one of them is short-lived. The things that we see around us, and which we think of as needing explanation—rocks, galaxies, ocean waves—are all, to a greater or lesser extent, stable patterns of atoms. Soap bubbles tend to be spherical because this is a stable configuration for thin films filled with gas. In a spacecraft, water is also stable in spherical globules, but on earth, where there is gravity, the stable surface for standing water is flat and horizontal. Salt crystals tend to be cubes because this is a stable way of packing sodium and chloride ions together. In the sun the simplest atoms of all, hydrogen atoms, are fusing to form helium atoms, because in the conditions that prevail there the helium configuration is more stable. Other even more complex atoms are being formed in stars all over the universe, ever since soon after the 'big bang' which, according to the prevailing theory, initiated the universe. This is originally where the elements on our world came from.

Sometimes when atoms meet they link up together in chemical reaction to form molecules, which may be more or less stable. Such molecules can be very large. A crystal such as a diamond can be regarded as a single molecule, a proverbially stable one in this case, but also a very simple one since its internal atomic structure is endlessly repeated. In modern living organisms there are other large molecules which are highly complex, and their complexity shows itself on several levels. The haemoglobin of our blood is a typical protein molecule. It is built up from chains of smaller molecules, amino acids, each containing a few dozen atoms arranged in a precise pattern. In the haemoglobin molecule there are 574 amino acid molecules. These are arranged in four chains, which twist around each other to form a globular three-dimensional structure of bewildering complexity. A model of a haemoglobin molecule looks rather like a dense thornbush. But unlike a real thornbush it is not a haphazard approximate pattern but a definite invariant structure, identically repeated, with not a twig nor a twist out of place, over six thousand million million

million times in an average human body. The precise thornbush shape of a protein molecule such as haemoglobin is stable in the sense that two chains consisting of the same sequences of amino acids will tend, like two springs, to come to rest in exactly the same three-dimensional coiled pattern. Haemoglobin thornbushes are springing into their 'preferred' shape in your body at a rate of about four hundred million million per second, and others are being destroyed at the same rate.

Haemoglobin is a modern molecule, used to illustrate the principle that atoms tend to fall into stable patterns. The point that is relevant here is that, before the coming of life on earth, some rudimentary evolution of molecules could have occurred by ordinary processes of physics and chemistry. There is no need to think of design or purpose or directedness. If a group of atoms in the presence of energy falls into a stable pattern it will tend to stay that way. The earliest form of natural selection was simply a selection of stable forms and a rejection of unstable ones. There is no mystery about this. It had to happen by definition.

From this, of course, it does not follow that you can explain the existence of entities as complex as man by exactly the same principles on their own. It is no good taking the right number of atoms and shaking them together with some external energy till they happen to fall into the right pattern, and out drops Adam! You may make a molecule consisting of a few dozen atoms like that, but a man consists of over a thousand million million million million atoms. To try to make a man, you would have to work at your biochemical cocktail-shaker for a period so long that the entire age of the universe would seem like an eye-blink, and even then you would not succeed. This is where Darwin's theory, in its most general form, comes to the rescue. Darwin's theory takes over from where the story of the slow building up of molecules leaves off.

The account of the origin of life that I shall give is necessarily speculative; by definition, nobody was around to see what happened. There are a number of rival theories, but they all have certain features in common. The simplified account I shall give is probably not too far from the truth.*

We do not know what chemical raw materials were abundant on earth before the coming of life, but among the plausible possibilities are water, carbon dioxide, methane, and ammonia: all simple compounds known to be present on at least some of the other planets in our solar system. Chemists have tried to imitate the chemical conditions of the young earth. They have put these simple substances in a flask and supplied a source of energy such as ultraviolet light or electric sparks—artificial simulation of primordial lightning. After a few weeks of this, something interesting is usually found inside the flask: a weak brown soup containing a large number of molecules more complex than the ones originally put in. In particular, amino acids have been found—the building blocks of proteins, one of the two great classes of biological molecules. Before these experiments were done, naturally-occurring amino acids would have been thought of as diagnostic of the presence of life. If they had been detected on, say Mars, life on that planet would have seemed a near certainty. Now, however, their existence need imply only the presence of a few simple gases in the atmosphere and some volcanoes, sunlight, or thundery weather. More recently, laboratory simulations of the chemical conditions of earth before the coming of life have yielded organic substances called purines and pyrimidines. These are building blocks of the genetic molecule, DNA itself.

Processes analogous to these must have given rise to the 'primeval soup' which biologists and chemists believe constituted the seas some three to four thousand million years ago. The organic substances became locally concentrated, perhaps in drying scum round the shores, or in tiny suspended droplets. Under the further influence of energy such as ultraviolet light from the sun, they combined into larger molecules. Nowadays large organic molecules would not last long enough to be noticed: they would be quickly absorbed and broken down by bacteria or other living creatures. But bacteria and the rest of us are late-comers, and in those days large organic molecules could drift unmolested through the thickening broth.

At some point a particularly remarkable molecule was formed by accident. We will call it the *replicator*. It may not necessarily have been the biggest or the most complex molecule around, but it had the extraordinary property of being able to create copies of itself. This may seem a very unlikely sort of accident to happen. So it was. It was exceedingly improbable. In the lifetime of a man, things that are that improbable can be treated for practical purposes as impossible. That is why you will never win a big prize on the football pools. But in our human estimates of what is probable and what is not, we are not used to dealing in hundreds of millions of years. If you filled in pools coupons every week for a hundred million years you would very likely win several jackpots.

Actually a molecule that makes copies of itself is not as difficult to imagine as it seems at first, and it only had to arise once. Think of the replicator as a mould or template. Imagine it as a large molecule consisting of a complex chain of various sorts of building block molecules. The small building blocks were abundantly available in the soup surrounding the replicator. Now suppose that each building block has an affinity for its own kind. Then whenever a building block from out in the soup lands up next to a part of the replicator for which it has an affinity, it will tend to stick there. The building blocks that attach themselves in this way will automatically be arranged in a sequence that mimics that of the replicator itself. It is easy then to think of them joining up to form a stable chain just as in the formation of the original replicator. This process could continue as a progressive stacking up, layer upon layer. This is how crystals are formed. On the other hand, the two chains might split apart, in which case we have two replicators, each of which can go on to make further copies.

A more complex possibility is that each building block has affinity not for its own kind, but reciprocally for one particular other kind. Then the replicator would act as a template not for an identical copy, but for a kind of 'negative', which would in its turn re-make an exact copy of the original positive. For our purposes it does not matter whether the original replication process was positive-negative or positive-positive, though it is worth remarking that the modern equivalents of the first replicator, the DNA molecules, use positive-negative replication. What does matter is that suddenly a new kind of 'stability' came into the world. Previously it is probable that no particular kind of complex molecule was very abundant in the soup, because each was dependent on building blocks happening to fall by luck into a particular stable configuration. As soon as the replicator was born it must have spread its copies rapidly throughout the seas, until the smaller building block molecules became a scarce resource, and other larger molecules were formed more and more rarely.

So we seem to arrive at a large population of identical replicas. But now we must mention an important property of any copying process: it is not perfect. Mistakes will happen. I hope there are no misprints in this book, but if you look carefully you may find one or two. They will probably not seriously distort the meaning of the sentences, because they will be 'first generation' errors. But imagine the days before printing, when books such as the Gospels were copied by hand. All scribes, however careful, are bound to make a few errors, and some are not above a little wilful 'improvement'. If they all copied from a single master original, meaning would not be greatly perverted. But let copies be made from other copies, which in their turn were made from other copies, and errors will start to become cumulative and serious. We tend to regard erratic copying as a bad thing, and in the case of human documents it is hard to think of examples where errors can be described as improvements. I suppose the scholars of the Septuagint could at least be said to have started something big when they mistranslated the Hebrew word for 'young woman' into the Greek word for 'virgin', coming up with the prophecy: 'Behold a virgin shall conceive and bear a son ...'* Anyway, as we shall see, erratic copying in biological replicators can in a real sense give rise to improvement, and it was essential for the progressive evolution of life that some errors were made. We do not know how accurately the original replicator molecules made their copies. Their modern descendants, the DNA molecules, are astonishingly faithful compared with the most high-fidelity human copying process, but even they occasionally make mistakes, and it is ultimately these mistakes that make evolution possible. Probably the original replicators were far more erratic, but in any case we may be sure that mistakes were made, and these mistakes were cumulative.

As mis-copyings were made and propagated, the primeval soup became filled by a population not of identical replicas, but of several varieties of replicating molecules, all 'descended' from the same ancestor. Would some varieties have been more numerous than others? Almost certainly yes. Some varieties would have been inherently more stable than others. Certain molecules, once formed, would be less likely than others to break up again. These types would become relatively numerous in the soup, not only as a direct logical consequence of their 'longevity', but also because they would have a long time available for making copies of themselves. Replicators of high longevity would therefore tend to become more numerous and, other things being equal, there would have been an 'evolutionary trend' towards greater longevity in the population of molecules.

But other things were probably not equal, and another property of a replicator variety that must have had even more importance in spreading it through the population was speed of replication or 'fecundity'. If replicator molecules of type A make copies of themselves on average once a week while those of type B make copies of themselves once an hour, it is not difficult to see that pretty soon type A molecules are going to be far outnumbered, even if they 'live' much longer than B molecules. There would therefore probably have been an 'evolutionary trend' towards higher 'fecundity' of molecules in the soup. A third characteristic of replicator molecules which would have been positively selected is accuracy of replication. If molecules of type X and type Y last the same length of time and replicate at the same rate, but X makes a mistake on average every tenth replication while Y makes a mistake only every hundredth replication, Y will obviously become more numerous. The X contingent in the population loses not only the errant 'children' themselves, but also all their descendants, actual or potential.

If you already know something about evolution, you may find something slightly paradoxical about the last point. Can we reconcile the idea that copying errors are an essential prerequisite for evolution to occur, with the statement that natural selection favours high copying-fidelity? The answer is that although evolution may seem, in some vague sense, a 'good thing', especially since we are the product of it, nothing actually 'wants' to evolve. Evolution is something that happens, willy-nilly, in spite of all the efforts of the replicators (and nowadays of the genes) to prevent it happening. Jacques Monod made this point very well in his Herbert Spencer lecture, after wryly remarking: 'Another curious aspect of the theory of evolution is that everybody thinks he understands it!'

To return to the primeval soup, it must have become populated by stable varieties of molecule; stable in that either the individual molecules lasted a long time, or they replicated rapidly, or they replicated accurately. Evolutionary trends toward these three kinds of stability took place in the following sense: if you had sampled the soup at two different times, the later sample would have contained a higher proportion of varieties with high longevity/fecundity/copying-fidelity. This is essentially what a biologist means by evolution when he is speaking of living creatures, and the mechanism is the same—natural selection.

Should we then call the original replicator molecules 'living'? Who cares? I might say to you 'Darwin was the greatest man who has ever lived', and you might say 'No, Newton was', but I hope we would not prolong the argument. The point is that no conclusion of substance would be affected whichever way our argument was resolved. The facts of the lives and achievements of Newton and Darwin remain totally unchanged whether we label them 'great' or not. Similarly, the story of the replicator molecules probably happened something like the way I am telling it, regardless of whether we choose to call them 'living'. Human suffering has been caused because too many of us cannot grasp that words are only tools for our use, and that the mere presence in the dictionary of a word like 'living' does not mean it necessarily has to refer to something definite in the real world. Whether we call the early replicators living or not, they were the ancestors of life; they were our founding fathers.

The next important link in the argument, one that Darwin himself laid stress on (although he was talking about animals and plants, not molecules), is *competition*. The primeval soup was not capable of supporting an infinite number of replicator molecules.

For one thing, the earth's size is finite, but other limiting factors must also have been important. In our picture of the replicator acting as a template or mould, we supposed it to be bathed in a soup rich in the small building block molecules necessary to make copies. But when the replicators became numerous, building blocks must have been used up at such a rate that they became a scarce and precious resource. Different varieties or strains of replicator must have competed for them. We have considered the factors that would have increased the numbers of favoured kinds of replicator. We can now see that less-favoured varieties must actually have become *less* numerous because of competition, and ultimately many of their lines must have gone extinct. There was a struggle for existence among replicator varieties. They did not know they were struggling, or worry about it; the struggle was conducted without any hard feelings, indeed without feelings of any kind. But they were struggling, in the sense that any mis-copying that resulted in a new higher level of stability, or a new way of reducing the stability of rivals, was automatically preserved and multiplied. The process of improvement was cumulative. Ways of increasing stability and of decreasing rivals' stability became more elaborate and more efficient. Some of them may even have 'discovered' how to break up molecules of rival varieties chemically, and to use the building blocks so released for making their own copies. These proto-carnivores simultaneously obtained food and removed competing rivals. Other replicators perhaps discovered how to protect themselves, either chemically, or by building a physical wall of protein around themselves. This may have been how the first living cells appeared. Replicators began not merely to exist, but to construct for themselves containers, vehicles for their continued existence. The replicators that survived were the ones that built *survival machines* for themselves to live in. The first survival machines probably consisted of nothing more than a protective coat. But making a living got steadily harder as new rivals arose with better and more effective survival machines. Survival machines got bigger and more elaborate, and the process was cumulative and progressive.

Was there to be any end to the gradual improvement in the techniques and artifices used by the replicators to ensure their own continuation in the world? There would be plenty of time for improvement. What weird engines of self-preservation would the millennia bring forth? Four thousand million years on, what was to be the fate of the ancient replicators? They did not die out, for they are past masters of the survival arts. But do not look for them floating loose in the sea; they gave up that cavalier freedom long ago. Now they swarm in huge colonies, safe inside gigantic lumbering robots,* sealed off from the outside world, communicating with it by tortuous indirect routes, manipulating it by remote control. They are in you and in me; they created us, body and mind; and their preservation is the ultimate rationale for our existence. They have come a long way, those replicators. Now they go by the name of genes, and we are their survival machines.

IMMORTAL COILS

W e are survival machines, but 'we' does not mean just people. It embraces all animals, plants, bacteria, and viruses. The total number of survival machines on earth is very difficult to count and even the total number of species is unknown. Taking just insects alone, the number of living species has been estimated at around three million, and the number of individual insects may be a million million million.

Different sorts of survival machine appear very varied on the outside and in their internal organs. An octopus is nothing like a mouse, and both are quite different from an oak tree. Yet in their fundamental chemistry they are rather uniform, and, in particular, the replicators that they bear, the genes, are basically the same kind of molecule in all of us—from bacteria to elephants. We are all survival machines for the same kind of replicator—molecules called DNA—but there are many different ways of making a living in the world, and the replicators have built a vast range of machines to exploit them. A monkey is a machine that preserves genes up trees, a fish is a machine that preserves genes in the water; there is even a small worm that preserves genes in German beer mats. DNA works in mysterious ways.

For simplicity I have given the impression that modern genes, made of DNA, are much the same as the first replicators in the primeval soup. It does not matter for the argument, but this may not really be true. The original replicators may have been a related kind of molecule to DNA, or they may have been totally different. In the latter case we might say that their survival machines must have been seized at a later stage by DNA. If so, the original replicators were utterly destroyed, for no trace of them remains in modern survival machines. Along these lines, A. G. Cairns-Smith has made the intriguing suggestion that our ancestors, the first replicators, may have been not organic molecules at all, but inorganic crystals—minerals, little bits of clay. Usurper or not, DNA is in undisputed charge today, unless, as I tentatively suggest in Chapter 11, a new seizure of power is now just beginning.

A DNA molecule is a long chain of building blocks, small molecules called nucleotides. Just as protein molecules are chains of amino acids, so DNA molecules are chains of nucleotides. A DNA molecule is too small to be seen, but its exact shape has been ingeniously worked out by indirect means. It consists of a pair of nucleotide chains twisted together in an elegant spiral; the 'double helix'; the 'immortal coil'. The nucleotide building blocks come in only four different kinds, whose names may be shortened to A, T, C, and G. These are the same in all animals and plants. What differs is the order in which they are strung together. A G building block from a man is identical in every particular to a G building block from a snail. But the *sequence* of building blocks in a man is not only different from that in a snail. It is also different—though less so—from the sequence in every other man (except in the special case of identical twins).

Our DNA lives inside our bodies. It is not concentrated in a particular part of the body, but is distributed among the cells. There are about a thousand million million cells making up an average human body, and, with some exceptions which we can ignore, every one of those cells contains a complete copy of that body's DNA. This DNA can be regarded as a set of instructions for how to make a body, written in the A, T, C, G alphabet of the nucleotides. It is as though, in every room of a gigantic building, there was a book-case containing the architect's plans for the entire building. The 'book-case' in a cell is called the nucleus. The architect's plans run to 46 volumes in man—the number is different in other species. The 'volumes' are called chromosomes. They are visible under a microscope as long threads, and the genes are strung out along them in order. It is not easy, indeed it may not even be meaningful, to decide where one gene ends and the next one begins. Fortunately, as this chapter will show, this does not matter for our purposes.

I shall make use of the metaphor of the architect's plans, freely mixing the language of the metaphor with the language of the real thing. 'Volume' will be used interchangeably with chromosome. 'Page' will provisionally be used interchangeably with gene, although the division between genes is less clear-cut than the division between the pages of a book. This metaphor will take us quite a long way. When it finally breaks down I shall introduce other metaphors. Incidentally, there is of course no 'architect'. The DNA instructions have been assembled by natural selection.

DNA molecules do two important things. Firstly, they replicate, that is to say they make copies of themselves. This has gone on non-stop ever since the beginning of life, and the DNA molecules are now very good at it indeed. As an adult, you consist of a thousand million million cells, but when you were first conceived you were just a single cell, endowed with one master copy of the architect's plans. This cell divided into two, and each of the two cells received its own copy of the plans. Successive divisions took the number of cells up to 4, 8, 16, 32, and so on into the billions. At every division the DNA plans were faithfully copied, with scarcely any mistakes.

It is one thing to speak of the duplication of DNA. But if the DNA is really a set of plans for building a body, how are the plans put into practice? How are they translated into the fabric of the body? This brings me to the second important thing DNA does. It indirectly supervises the manufacture of a different kind of molecule—protein. The haemoglobin which was mentioned in the last chapter is just one example of the enormous range of protein molecules. The coded message of the DNA, written in the four-letter nucleotide alphabet, is translated in a simple mechanical way into another alphabet. This is the alphabet of amino acids which spells out protein molecules.

Making proteins may seem a far cry from making a body, but it is the first small step in that direction. Proteins not only constitute much of the physical fabric of the body; they also exert sensitive control over all the chemical processes inside the cell, selectively turning them on and off at precise times and in precise places. Exactly how this eventually leads to the development of a baby is a story which it will take decades, perhaps centuries, for embryologists to work out. But it is a fact that it does. Genes do indirectly control the manufacture of bodies, and the influence is strictly one way: acquired characteristics are not inherited. No matter how much knowledge and wisdom you acquire during your life, not one jot will be passed on to your children by genetic means. Each new generation starts from scratch. A body is the genes' way of preserving the genes unaltered.

The evolutionary importance of the fact that genes control embryonic development is this: it means that genes are at least partly responsible for their own survival in the future, because their survival depends on the efficiency of the bodies in which they live and which they helped to build. Once upon a time, natural selection consisted of the differential survival of replicators floating free in the primeval soup. Now, natural selection favours replicators that are good at building survival machines, genes that are skilled in the art of controlling embryonic development. In this, the replicators are no more conscious or purposeful than they ever were. The same old processes of automatic selection between rival molecules by reason of their longevity, fecundity, and copying-fidelity still go on as blindly and as inevitably as they did in the far-off days. Genes have no foresight. They do not plan ahead. Genes just *are*, some genes more so than others, and that is all there is to it. But the qualities that determine a gene's longevity and fecundity are not so simple as they were. Not by a long way.

In recent years—the last six hundred million or so—the replicators have achieved notable triumphs of survival-machine technology such as the muscle, the heart, and the eye (evolved several times independently). Before that, they radically altered fundamental features of their way of life as replicators, which must be understood if we are to proceed with the argument.

The first thing to grasp about a modern replicator is that it is highly gregarious. A survival machine is a vehicle containing not just one gene but many thousands. The manufacture of a body is a cooperative venture of such intricacy that it is almost impossible to disentangle the contribution of one gene from that of another.* A given gene will have many different effects on quite different parts of the body. A given part of the body will be influenced by many genes, and the effect of any one gene depends on interaction with many others. Some genes act as master genes controlling the operation of a cluster of other genes. In terms of the analogy, any given page of the plans makes reference to many different parts of the building; and each page makes sense only in terms of cross-references to numerous other pages.

This intricate inter-dependence of genes may make you wonder why we use the word 'gene' at all. Why not use a collective

noun like 'gene complex'? The answer is that for many purposes that is indeed quite a good idea. But if we look at things
in another way, it does make sense too to think of the gene complex as being divided up into discrete replicators or genes. This
arises because of the phenomenon of sex. Sexual reproduction
has the effect of mixing and shuffling genes. This means that
any one individual body is just a temporary vehicle for a short-
lived combination of genes. The *combination* of genes that is any
one individual may be short-lived, but the genes themselves are
potentially very long-lived. Their paths constantly cross and
recross down the generations. One gene may be regarded as a
unit that survives through a large number of successive individual bodies. This is the central argument that will be developed
in this chapter. It is an argument that some of my most respected
colleagues obstinately refuse to agree with, so you must forgive
me if I seem to labour it! First I must briefly explain the facts
of sex.

I said that the plans for building a human body are spelt out in
46 volumes. In fact this was an over-simplification. The truth is
rather bizarre. The 46 chromosomes consist of 23 *pairs* of chromosomes. We might say that filed away in the nucleus of every
cell are two alternative sets of 23 volumes of plans. Call them
Volume 1a and Volume 1b, Volume 2a and Volume 2b, etc., down
to Volume 23a and Volume 23b. Of course the identifying numbers I use for volumes and, later, pages, are purely arbitrary.

We receive each chromosome intact from one of our two
parents, in whose testis or ovary it was assembled. Volumes 1a,
2a, 3a, ... came, say, from the father. Volumes 1b, 2b, 3b, ... came
from the mother. It is very difficult in practice, but in theory you
could look with a microscope at the 46 chromosomes in any one
of your cells, and pick out the 23 that came from your father and
the 23 that came from your mother.

The paired chromosomes do not spend all their lives physically in contact with each other, or even near each other. In what sense then are they 'paired'? In the sense that each volume coming originally from the father can be regarded, page for page, as a direct alternative to one particular volume coming originally from the mother. For instance, Page 6 of Volume 13a and Page 6 of Volume 13b might both be 'about' eye colour; perhaps one says 'blue' while the other says 'brown'.

Sometimes the two alternative pages are identical, but in other cases, as in our example of eye colour, they differ. If they make contradictory 'recommendations', what does the body do? The answer varies. Sometimes one reading prevails over the other. In the eye colour example just given, the person would actually have brown eyes: the instructions for making blue eyes would be ignored in the building of the body, though this does not stop them being passed on to future generations. A gene that is ignored in this way is called *recessive*. The opposite of a recessive gene is a *dominant* gene. The gene for brown eyes is dominant to the gene for blue eyes. A person has blue eyes only if both copies of the relevant page are unanimous in recommending blue eyes. More usually when two alternative genes are not identical, the result is some kind of compromise—the body is built to an intermediate design or something completely different.

When two genes, like the brown eye and the blue eye gene, are rivals for the same slot on a chromosome, they are called *alleles* of each other. For our purposes, the word allele is synonymous with rival. Imagine the volumes of architects' plans as being loose-leaf binders, whose pages can be detached and interchanged. Every Volume 13 must have a Page 6, but there are several possible Page 6s which could go in the binder between Page 5 and Page 7. One version says 'blue eyes', another possible version says 'brown eyes'; there may be yet other versions in the population at large which spell out other colours like green. Perhaps there are half a dozen alternative alleles sitting in the Page 6 position on the 13th chromosomes scattered around the population as a whole. Any given person only has two Volume 13 chromosomes. Therefore he can have a maximum of two alleles in the Page 6 slot. He may, like a blue-eyed person, have two copies of the same allele, or he may have any two alleles chosen from the half dozen alternatives available in the population at large.

You cannot, of course, literally go and choose your genes from a pool of genes available to the whole population. At any given time all the genes are tied up inside individual survival machines. Our genes are doled out to us at conception, and there is nothing we can do about this. Nevertheless, there is a sense in which, in the long term, the genes of the population in general can be regarded as a *gene pool*. This phrase is in fact a technical term used by geneticists. The gene pool is a worthwhile abstraction because sex mixes genes up, albeit in a carefully organized way. In particular, something like the detaching and interchanging of pages and wads of pages from loose-leaf binders really does go on, as we shall presently see.

I have described the normal division of a cell into two new cells, each one receiving a complete copy of all 46 chromosomes. This normal cell division is called *mitosis*. But there is another kind of cell division called *meiosis*. This occurs only in the production of the sex cells; the sperms or eggs. Sperms and eggs are unique among our cells in that, instead of containing 46 chromosomes, they contain only 23. This is, of course, exactly half of 46— convenient when they fuse in sexual fertilization to make a new individual! Meiosis is a special kind of cell division, taking place only in testicles and ovaries, in which a cell with the full double set of 46 chromosomes divides to form sex cells with the single set of 23 (all the time using the human numbers for illustration).

A sperm, with its 23 chromosomes, is made by the meiotic division of one of the ordinary 46-chromosome cells in the testicle. Which 23 are put into any given sperm cell? It is clearly important that a sperm should not get just any old 23 chromosomes: it mustn't end up with two copies of Volume 13 and none of Volume 17. It would theoretically be possible for an individual to endow one of his sperms with chromosomes which came, say, entirely from his mother; that is Volume 1b, 2b, 3b,..., 23b. In this unlikely event, a child conceived by the sperm would inherit half her genes from her paternal grandmother, and none from her paternal grandfather. But in fact this kind of gross, whole-chromosome distribution does not happen. The truth is rather more complex. Remember that the volumes (chromosomes) are to be thought of as loose-leaf binders. What happens is that, during the manufacture of the sperm, single pages, or rather multi-page chunks, are detached and swapped with the corresponding chunks from the alternative volume. So, one particular sperm cell might make up its Volume 1 by taking the first 65 pages from Volume 1a, and pages 66 to the end from Volume 1b. This sperm cell's other 22 volumes would be made up in a similar way. Therefore every sperm cell made by an individual is unique, even though all his sperms assembled their 23 chromosomes from bits of the same set of 46 chromosomes. Eggs are made in a similar way in ovaries, and they too are all unique.

The real-life mechanics of this mixing are fairly well understood. During the manufacture of a sperm (or egg), bits of each paternal chromosome physically detach themselves and change places with exactly corresponding bits of maternal chromosome. (Remember that we are talking about chromosomes that came originally from the parents of the individual making the sperm, i.e., from the paternal grandparents of the child who is eventually conceived by the sperm.) The process of swapping bits of chromosome is called *crossing over*. It is very important for the whole argument of this book. It means that if you got out your microscope and looked at the chromosomes in one of your own sperms (or eggs if you are female) it would be a waste of time trying to identify chromosomes that originally came from your father and chromosomes that originally came from your mother. (This is in marked contrast to the case of ordinary body cells (see page 31).) Any one chromosome in a sperm would be a patchwork, a mosaic of maternal genes and paternal genes.

The metaphor of the page for the gene starts to break down here. In a loose-leaf binder a whole page may be inserted, removed or exchanged, but not a fraction of a page. But the gene complex is just a long string of nucleotide letters, not divided into discrete pages in an obvious way at all. To be sure, there are special symbols for END OF PROTEIN CHAIN MESSAGE and START OF PROTEIN CHAIN MESSAGE written in the same four-letter alphabet as the protein messages themselves. In between these two punctuation marks are the coded instructions for making one protein. If we wish, we can define a single gene as a sequence of nucleotide letters lying between a START and an END symbol, and coding for one protein chain. The word *cistron* has been used for a unit defined in this way, and some people use the word gene interchangeably with cistron. But crossing-over does not respect boundaries between cistrons. Splits may occur within cistrons as well as between them. It is as though the architect's plans were written out, not on discrete pages, but on 46 rolls of ticker tape. Cistrons are not of fixed length. The only way to tell where one cistron ends and the next begins would be to read the symbols on the tape, looking for END OF MESSAGE and START OF MESSAGE symbols. Crossing-over is represented by taking matching paternal and maternal tapes, and cutting and exchanging matching portions, regardless of what is written on them.

In the title of this book the word gene means not a single cistron but something more subtle. My definition will not be to everyone's taste, but there is no universally agreed definition of a gene. Even if there were, there is nothing sacred about definitions. We can define a word how we like for our own purposes, provided we do so clearly and unambiguously. The definition I want to use comes from G. C. Williams.* A gene is defined as any portion of chromosomal material that potentially lasts for enough generations to serve as a unit of natural selection. In the words of the previous chapter, a gene is a replicator with high copying-fidelity. Copying-fidelity is another way of saying longevity-in-the-form-of-copies and I shall abbreviate this simply to longevity. The definition will take some justifying.

On any definition, a gene has to be a portion of a chromosome. The question is: How big a portion—how much of the ticker tape? Imagine any sequence of adjacent code-letters on the tape. Call the sequence a *genetic unit*. It might be a sequence of only ten letters within one cistron; it might be a sequence of eight cistrons; it might start and end in mid-cistron. It will overlap with other genetic units. It will include smaller units, and it will form part of larger units. No matter how long or short it is, for the purposes of the present argument, this is what we are calling a genetic unit. It is just a length of chromosome, not physically differentiated from the rest of the chromosome in any way.

Now comes the important point. The shorter a genetic unit is, the longer—in generations—it is likely to live. In particular, the less likely it is to be split by any one crossing-over. Suppose a whole chromosome is, on average, likely to undergo one cross-over every time a sperm or egg is made by meiotic division, and this cross-over can happen anywhere along its length. If we consider a very large genetic unit, say half the length of the chromosome, there is a 50 per cent chance that the unit will be split at each meiosis. If the genetic unit we are considering is only 1 per cent of the length of the chromosome, we can assume that it has only a 1 per cent chance of being split in any one meiotic division. This means that the unit can expect to survive for a large number of generations in the individual's descendants. A single cistron is likely to be much less than 1 per cent of the length of a chromosome. Even a group of several neighbouring cistrons can expect to live many generations before being broken up by crossing-over.

The average life-expectancy of a genetic unit can conveniently be expressed in generations, which can in turn be translated into years. If we take a whole chromosome as our presumptive genetic unit, its life story lasts for only one generation. Suppose it is your chromosome number 8a, inherited from your father. It was created inside one of your father's testicles, shortly before you were conceived. It had never existed before in the whole history of the world. It was created by the meiotic shuffling process, forged by the coming together of pieces of chromosome from your paternal grandmother and your paternal grandfather. It was placed inside one particular sperm, and it was unique. The sperm was one of several millions, a vast armada of tiny vessels, and together they sailed into your mother. This particular sperm (unless you are a non-identical twin) was the only one of the flotilla which found harbour in one of your mother's eggs—that is why you exist. The genetic unit we are considering, your chromosome number 8a, set about replicating itself along with all the rest of your genetic material. Now it exists, in duplicate form, all over your body. But when you in your turn come to have children, this chromosome will be destroyed when you manufacture eggs (or sperms). Bits of it will be interchanged with bits of your maternal chromosome number 8b. In any one sex cell, a new chromosome number 8 will be created, perhaps 'better' than the old one, perhaps 'worse', but, barring a rather improbable coincidence, definitely different, definitely unique. The life-span of a chromosome is one generation.

What about the life-span of a smaller genetic unit, say 1/100 of the length of your chromosome 8a? This unit too came from your father, but it very probably was not originally assembled in him. Following the earlier reasoning, there is a 99 per cent chance that he received it intact from one of his two parents. Suppose it was from his mother, your paternal grandmother. Again, there is a 99 per cent chance that she inherited it intact from one of her parents. Eventually, if we trace the ancestry of a small genetic unit back far enough, we will come to its original creator. At some stage it must have been created for the first time inside a testicle or an ovary of one of your ancestors.

Let me repeat the rather special sense in which I am using the word 'create'. The smaller sub-units which make up the genetic unit we are considering may well have existed long before. Our genetic unit was created at a particular moment only in the sense that the particular *arrangement* of sub-units by which it is defined did not exist before that moment. The moment of creation may have occurred quite recently, say in one of your grandparents. But if we consider a very small genetic unit, it may have been first assembled in a much more distant ancestor, perhaps an ape-like pre-human ancestor. Moreover, a small genetic unit inside you may go on just as far into the future, passing intact through a long line of your descendants.

Remember too that an individual's descendants constitute not a single line but a branching line. Whichever of your ancestors it was who 'created' a particular short length of your chromosome 8a, he or she very likely has many other descendants besides you. One of your genetic units may also be present in your second cousin. It may be present in me, and in the Prime Minister, and in your dog, for we all share ancestors if we go back far enough.

Also the same small unit might be assembled several times independently by chance: if the unit is small, the coincidence is not too improbable. But even a close relative is unlikely to share a whole chromosome with you. The smaller a genetic unit is, the more likely it is that another individual shares it—the more likely it is to be represented many times over in the world, in the form of copies.

The chance coming together, through crossing-over, of previously existing sub-units is the usual way for a new genetic unit to be formed. Another way—of great evolutionary importance even though it is rare—is called *point mutation*. A point mutation is an error corresponding to a single misprinted letter in a book. It is rare, but clearly the longer a genetic unit is, the more likely it is to be altered by a mutation somewhere along its length.

Another rare kind of mistake or mutation which has important long-term consequences is called *inversion*. A piece of chromosome detaches itself at both ends, turns head over heels, and reattaches itself in the inverted position. In terms of the earlier analogy, this would necessitate some renumbering of pages. Sometimes portions of chromosomes do not simply invert, but become reattached in a completely different part of the chromosome, or even join up with a different chromosome altogether. This corresponds to the transfer of a wad of pages from one volume to another. The importance of this kind of mistake is that, though usually disastrous, it can occasionally lead to the close *linkage* of pieces of genetic material which happen to work well together. Perhaps two cistrons which have a beneficial effect only when they are both present—they complement or reinforce each other in some way—will be brought close to each other by means of inversion. Then natural selection may tend to favour the new 'genetic unit' so formed, and it will spread through the future population. It is possible that gene complexes have,

over the years, been extensively rearranged or 'edited' in this kind
of way.

One of the neatest examples of this concerns the phenomenon known as *mimicry*. Some butterflies taste nasty. They are
usually brightly and distinctively coloured, and birds learn to
avoid them by their 'warning' marks. Now other species of butterfly that do not taste nasty cash in. They *mimic* the nasty ones.
They are born looking like them in colour and shape (but not
taste). They frequently fool human naturalists, and they also fool
birds. A bird who has once tasted a genuinely nasty butterfly
tends to avoid all butterflies that look the same. This includes the
mimics, and so genes for mimicry are favoured by natural selection. That is how mimicry evolves.

There are many different species of 'nasty' butterfly and they
do not all look alike. A mimic cannot resemble all of them: it has
to commit itself to one particular nasty species. In general, any
particular species of mimic is a specialist at mimicking one particular nasty species. But there are species of mimic that do
something very strange. Some individuals of the species mimic
one nasty species; other individuals mimic another. Any individual who was intermediate or who tried to mimic both would
soon be eaten; but such intermediates are not born. Just as an
individual is either definitely male or definitely female, so an individual butterfly mimics either one nasty species or the other.
One butterfly may mimic species A while his brother mimics
species B.

It looks as though a single gene determines whether an individual will mimic species A or species B. But how can a single
gene determine all the multifarious aspects of mimicry—colour,
shape, spot pattern, rhythm of flight? The answer is that one
gene in the sense of a *cistron* probably cannot. But by the unconscious and automatic 'editing' achieved by inversions and other

accidental rearrangements of genetic material, a large cluster of formerly separate genes has come together in a tight linkage group on a chromosome. The whole cluster behaves like a single gene—indeed, by our definition it now *is* a single gene—and it has an 'allele' which is really another cluster. One cluster contains the cistrons concerned with mimicking species A; the other those concerned with mimicking species B. Each cluster is so rarely split up by crossing-over that an intermediate butterfly is never seen in nature, but they do very occasionally turn up if large numbers of butterflies are bred in the laboratory.

I am using the word gene to mean a genetic unit that is small enough to last for a large number of generations and to be distributed around in the form of many copies. This is not a rigid all-or-nothing definition, but a kind of fading-out definition, like the definition of 'big' or 'old'. The more likely a length of chromosome is to be split by crossing-over, or altered by mutations of various kinds, the less it qualifies to be called a gene in the sense in which I am using the term. A cistron presumably qualifies, but so also do larger units. A dozen cistrons may be so close to each other on a chromosome that for our purposes they constitute a single long-lived genetic unit. The butterfly mimicry cluster is a good example. As the cistrons leave one body and enter the next, as they board sperm or egg for the journey into the next generation, they are likely to find that the little vessel contains their close neighbours of the previous voyage, old shipmates with whom they sailed on the long odyssey from the bodies of distant ancestors. Neighbouring cistrons on the same chromosome form a tightly-knit troupe of travelling companions who seldom fail to get on board the same vessel when meiosis time comes around.

To be strict, this book should be called not *The Selfish Cistron* nor *The Selfish Chromosome*, but *The slightly selfish big bit of chromosome and the even more selfish little bit of chromosome*. To say the least this is not a catchy title so, defining a gene as a little bit of chromosome which potentially lasts for many generations, I call the book *The Selfish Gene*.

We have now arrived back at the point we left at the end of Chapter 1. There we saw that selfishness is to be expected in any entity that deserves the title of a basic unit of natural selection. We saw that some people regard the species as the unit of natural selection, others the population or group within the species, and yet others the individual. I said that I preferred to think of the gene as the fundamental unit of natural selection, and therefore the fundamental unit of self-interest. What I have now done is to *define* the gene in such a way that I cannot really help being right!

Natural selection in its most general form means the differential survival of entities. Some entities live and others die but, in order for this selective death to have any impact on the world, an additional condition must be met. Each entity must exist in the form of lots of copies, and at least some of the entities must be *potentially* capable of surviving—in the form of copies—for a significant period of evolutionary time. Small genetic units have these properties: individuals, groups, and species do not. It was the great achievement of Gregor Mendel to show that hereditary units can be treated in practice as indivisible and independent particles. Nowadays we know that this is a little too simple. Even a cistron is occasionally divisible and any two genes on the same chromosome are not wholly independent. What I have done is to define a gene as a unit which, to a high degree, *approaches* the ideal of indivisible particulateness. A gene is not indivisible, but it is seldom divided. It is either definitely present or definitely absent in the body of any given individual. A gene travels intact from grandparent to grandchild, passing straight through the intermediate generation without being merged with other genes. If genes continually blended with each other, natural selection as we now understand it would be impossible. Incidentally, this was proved in Darwin's lifetime, and it caused Darwin great worry since in those days it was assumed that heredity was a blending process. Mendel's discovery had already been published, and it could have rescued Darwin, but alas he never knew about it: nobody seems to have read it until years after Darwin and Mendel had both died. Mendel perhaps did not realize the significance of his findings; otherwise, he might have written to Darwin.

Another aspect of the particulateness of the gene is that it does not grow senile; it is no more likely to die when it is a million years old than when it is only a hundred. It leaps from body to body down the generations, manipulating body after body in its own way and for its own ends, abandoning a succession of mortal bodies before they sink in senility and death.

The genes are the immortals, or rather, they are defined as genetic entities that come close to deserving the title. We, the individual survival machines in the world, can expect to live a few more decades. But the genes in the world have an expectation of life that must be measured not in decades but in thousands and millions of years.

In sexually reproducing species, the individual is too large and too temporary a genetic unit to qualify as a significant unit of natural selection.* The group of individuals is an even larger unit. Genetically speaking, individuals and groups are like clouds in the sky or dust-storms in the desert. They are temporary aggregations or federations. They are not stable through evolutionary time. Populations may last a long while, but they are constantly blending with other populations and so losing their identity. They are also subject to evolutionary change from within. A population is not a discrete enough entity to be a unit of natural selection, not stable and unitary enough to be 'selected' in preference to another population.

An individual body seems discrete enough while it lasts, but alas, how long is that? Each individual is unique. You cannot get evolution by selecting between entities when there is only one copy of each entity! Sexual reproduction is not replication. Just as a population is contaminated by other populations, so an individual's posterity is contaminated by that of his sexual partner. Your children are only half you, your grandchildren only a quarter you. In a few generations the most you can hope for is a large number of descendants, each of whom bears only a tiny portion of you—a few genes—even if a few do bear your surname as well.

Individuals are not stable things, they are fleeting. Chromosomes too are shuffled into oblivion, like hands of cards soon after they are dealt. But the cards themselves survive the shuffling. The cards are the genes. The genes are not destroyed by crossing-over, they merely change partners and march on. Of course they march on. That is their business. They are the replicators and we are their survival machines. When we have served our purpose we are cast aside. But genes are denizens of geological time: genes are forever.

Genes, like diamonds, are forever, but not quite in the same way as diamonds. It is an individual diamond crystal that lasts, as an unaltered pattern of atoms. DNA molecules don't have that kind of permanence. The life of any one physical DNA molecule is quite short—perhaps a matter of months, certainly not more than one lifetime. But a DNA molecule could theoretically live on in the form of *copies* of itself for a hundred million years. Moreover, just like the ancient replicators in the primeval soup, copies of a particular gene may be distributed all over the world. The difference is that the modern versions are all neatly packaged inside the bodies of survival machines.

What I am doing is emphasizing the potential near-immortality of a gene, in the form of copies, as its defining property. To define a gene as a single cistron is good for some purposes, but for the purposes of evolutionary theory it needs to be enlarged. The extent of the enlargement is determined by the purpose of the definition. We want to find the practical unit of natural selection. To do this we begin by identifying the properties that a successful unit of natural selection must have. In the terms of the last chapter, these are longevity, fecundity, and copying-fidelity. We then simply define a 'gene' as the largest entity which, at least potentially, has these properties. The gene is a long-lived replicator, existing in the form of many duplicate copies. It is not infinitely long-lived. Even a diamond is not literally everlasting, and even a cistron can be cut in two by crossing-over. The gene is defined as a piece of chromosome which is sufficiently short for it to last, potentially, for *long enough* for it to function as a significant unit of natural selection.

Exactly how long is 'long enough'? There is no hard and fast answer. It will depend on how severe the natural selection 'pressure' is. That is, on how much more likely a 'bad' genetic unit is to die than its 'good' allele. This is a matter of quantitative detail which will vary from example to example. The largest practical unit of natural selection—the gene—will usually be found to lie somewhere on the scale between cistron and chromosome.

It is its potential immortality that makes a gene a good candidate as the basic unit of natural selection. But now the time has come to stress the word 'potential'. A gene *can* live for a million years, but many new genes do not even make it past their first generation. The few new ones that succeed do so partly because they are lucky, but mainly because they have what it takes, and that means they are good at making survival machines. They have an effect on the embryonic development of each successive body in which they find themselves, such that that body is a little bit more likely to live and reproduce than it would have been under the influence of the rival gene or allele. For example, a 'good' gene might ensure its survival by tending to endow the successive bodies in which it finds itself with long legs, which help those bodies to escape from predators. This is a particular example, not a universal one. Long legs, after all, are not always an asset. To a mole they would be a handicap. Rather than bog ourselves down in details, can we think of any *universal* qualities that we would expect to find in all good (i.e. long-lived) genes? Conversely, what are the properties that instantly mark a gene out as a 'bad', short-lived one? There might be several such universal properties, but there is one that is particularly relevant to this book: at the gene level, altruism must be bad and selfishness good. This follows inexorably from our definitions of altruism and selfishness. Genes are competing directly with their alleles for survival, since their alleles in the gene pool are rivals for their slot on the chromosomes of future generations. Any gene that behaves in such a way as to increase its own survival chances in the gene pool at the expense of its alleles will, by definition, tautologously, tend to survive. The gene is the basic unit of selfishness.

The main message of this chapter has now been stated. But I have glossed over some complications and hidden assumptions. The first complication has already been briefly mentioned. However independent and free genes may be in their journey through the generations, they are very much *not* free and independent agents in their control of embryonic development. They collaborate and interact in inextricably complex ways, both with each other, and with their external environment. Expressions like 'gene for long legs' or 'gene for altruistic behaviour' are convenient figures of speech, but it is important to understand what they mean. There is no gene which single-handedly builds a leg, long or short. Building a leg is a multigene cooperative enterprise.

Influences from the external environment too are indispensable: after all, legs are actually made of food! But there may well be a single gene which, *other things being equal*, tends to make legs longer than they would have been under the influence of the gene's allele.

As an analogy, think of the influence of a fertilizer, say nitrate, on the growth of wheat. Everybody knows that wheat plants grow bigger in the presence of nitrate than in its absence. But nobody would be so foolish as to claim that, on its own, nitrate can make a wheat plant. Seed, soil, sun, water, and various minerals are obviously all necessary as well. But if all these other factors are held constant, and even if they are allowed to vary within limits, addition of nitrate will make the wheat plants grow bigger. So it is with single genes in the development of an embryo. Embryonic development is controlled by an interlocking web of relationships so complex that we had best not contemplate it. No one factor, genetic or environmental, can be considered as the single 'cause' of any part of a baby. All parts of a baby have a near infinite number of antecedent causes. But a *difference* between one baby and another, for example a difference in length of leg, might easily be traced to one or a few simple antecedent differences, either in environment or in genes. It is *differences* that matter in the competitive struggle to survive; and it is genetically-controlled differences that matter in evolution.

As far as a gene is concerned, its alleles are its deadly rivals, but other genes are just a part of its environment, comparable to temperature, food, predators, or companions. The effect of the gene depends on its environment, and this includes other genes. Sometimes a gene has one effect in the presence of a particular other gene, and a completely different effect in the presence of another set of companion genes. The whole set of genes in a body constitutes a kind of genetic climate or background, modifying and influencing the effects of any particular gene.

But now we seem to have a paradox. If building a baby is such an intricate cooperative venture, and if every gene needs several thousands of fellow genes to complete its task, how can we reconcile this with my picture of indivisible genes, springing like immortal chamois from body to body down the ages: the free, untrammelled, and self-seeking agents of life? Was that all nonsense? Not at all. I may have got a bit carried away with the purple passages, but I was not talking nonsense, and there is no real paradox. We can explain this by means of another analogy.

One oarsman on his own cannot win the Oxford and Cambridge boat race. He needs eight colleagues. Each one is a specialist who always sits in a particular part of the boat—bow or stroke or cox, etc. Rowing the boat is a cooperative venture, but some men are nevertheless better at it than others. Suppose a coach has to choose his ideal crew from a pool of candidates, some specializing in the bow position, others specializing as cox, and so on. Suppose that he makes his selection as follows. Every day he puts together three new trial crews, by random shuffling of the candidates for each position, and he makes the three crews race against each other. After some weeks of this it will start to emerge that the winning boat often tends to contain the same individual men. These are marked up as good oarsmen. Other individuals seem consistently to be found in slower crews, and these are eventually rejected. But even an outstandingly good oarsman might sometimes be a member of a slow crew, either because of the inferiority of the other members, or because of bad luck— say a strong adverse wind. It is only *on average* that the best men tend to be in the winning boat.

The oarsmen are genes. The rivals for each seat in the boat are alleles potentially capable of occupying the same slot along the length of a chromosome. Rowing fast corresponds to building a body which is successful at surviving. The wind is the external environment. The pool of alternative candidates is the gene pool. As far as the survival of any one body is concerned, all its genes are in the same boat. Many a good gene gets into bad company, and finds itself sharing a body with a lethal gene, which kills the body off in childhood. Then the good gene is destroyed along with the rest. But this is only one body, and replicas of the same good gene live on in other bodies which lack the lethal gene. Many copies of good genes are dragged under because they happen to share a body with bad genes, and many perish through other forms of ill luck, say when their body is struck by lightning. But by definition luck, good and bad, strikes at random, and a gene that is *consistently* on the losing side is not unlucky; it is a bad gene.

One of the qualities of a good oarsman is teamwork, the ability to fit in and cooperate with the rest of a crew. This may be just as important as strong muscles. As we saw in the case of the butterflies, natural selection may unconsciously 'edit' a gene complex by means of inversions and other gross movements of bits of chromosome, thereby bringing genes that cooperate well together into closely linked groups. But there is also a sense in which genes which are in no way linked to each other physically can be selected for their mutual compatibility. A gene that cooperates well with most of the other genes that it is likely to meet in successive bodies, i.e. the genes in the whole of the rest of the gene pool, will tend to have an advantage.

For example, a number of attributes are desirable in an efficient carnivore's body, among them sharp cutting teeth, the right kind of intestine for digesting meat, and many other things. An efficient herbivore, on the other hand, needs flat grinding teeth, and a much longer intestine with a different kind of digestive chemistry. In a herbivore gene pool, any new gene that conferred on its possessors sharp meat-eating teeth would not be very successful. This is not because meat-eating is universally a bad idea, but because you cannot efficiently eat meat unless you also have the right sort of intestine, and all the other attributes of a meat-eating way of life. Genes for sharp, meat-eating teeth are not inherently bad genes. They are only bad genes in a gene pool that is dominated by genes for herbivorous qualities.

This is a subtle, complicated idea. It is complicated because the 'environment' of a gene consists largely of other genes, each of which is itself being selected for its ability to cooperate with *its* environment of other genes. An analogy adequate to cope with this subtle point does exist, but it is not from everyday experience. It is the analogy with human 'game theory', which will be introduced in Chapter 5 in connection with aggressive contests between individual animals. I therefore postpone further discussion of this point until the end of that chapter, and return to the central message of this one. This is that the basic unit of natural selection is best regarded not as the species, nor as the population, nor even as the individual, but as some small unit of genetic material which it is convenient to label the gene. The cornerstone of the argument, as given earlier, was the assumption that genes are potentially immortal, while bodies and all other higher units are temporary. This assumption rests upon two facts: the fact of sexual reproduction and crossing-over, and the fact of individual mortality. These facts are undeniably true. But this does not stop us asking why they are true. Why do we and most other survival machines practise sexual reproduction? Why do our chromosomes cross over? And why do we not live for ever?

The question of why we die of old age is a complex one, and the details are beyond the scope of this book. In addition to particular reasons, some more general ones have been proposed. For example, one theory is that senility represents an accumulation of deleterious copying errors and other kinds of gene damage which occur during the individual's lifetime. Another theory, due to Sir Peter Medawar, is a good example of evolutionary thinking in terms of gene selection.* Medawar first dismisses traditional arguments such as: 'Old individuals die as an act of altruism to the rest of the species, because if they stayed around when they were too decrepit to reproduce, they would clutter up the world to no good purpose.' As Medawar points out, this is a circular argument, assuming what it sets out to prove, namely that old animals are too decrepit to reproduce. It is also a naïve group selection or species selection kind of explanation, although that part of it could be rephrased more respectably. Medawar's own theory has a beautiful logic. We can build up to it as follows.

We have already asked what are the most general attributes of a 'good' gene, and we decided that 'selfishness' was one of them. But another general quality that successful genes will have is a tendency to postpone the death of their survival machines at least until after reproduction. No doubt some of your cousins and great-uncles died in childhood, but not a single one of your ancestors did. Ancestors just don't die young!

A gene that makes its possessors die is called a lethal gene. A semilethal gene has some debilitating effect, such that it makes death from other causes more probable. Any gene exerts its maximum effect on bodies at some particular stage of life, and lethals and semilethals are not exceptions. Most genes exert their influence during foetal life, others during childhood, other during young adulthood, others in middle age, and yet others in old age. (Reflect that a caterpillar and the butterfly it turns into have exactly the same set of genes.) Obviously lethal genes will tend to be removed from the gene pool. But equally obviously a late-acting lethal will be more stable in the gene pool than an early-acting lethal. A gene that is lethal in an older body may still be successful in the gene pool, provided its lethal effect does not show itself until after the body has had time to do at least some reproducing. For instance, a gene that made old bodies develop cancer could be passed on to numerous offspring because the individuals would reproduce before they got cancer. On the other hand, a gene that made young adult bodies develop cancer would not be passed on to very many offspring, and a gene that made young children develop fatal cancer would not be passed on to any offspring at all. According to this theory then, senile decay is simply a by-product of the accumulation in the gene pool of late-acting lethal and semi-lethal genes, which have been allowed to slip through the net of natural selection simply because they are late-acting.

The aspect that Medawar himself emphasizes is that selection will favour genes that have the effect of postponing the operation of other, lethal genes, and it will also favour genes that have the effect of hastening the effect of good genes. It may be that a great deal of evolution consists of genetically-controlled changes in the time of onset of gene activity.

It is important to notice that this theory does not need to make any prior assumptions about reproduction occurring only at certain ages. Taking as a starting assumption that all individuals were equally likely to have a child at any age, the Medawar theory would quickly predict the accumulation in the gene pool of late-acting deleterious genes, and the tendency to reproduce less in old age would follow as a secondary consequence.

As an aside, one of the good features of this theory is that it leads us to some rather interesting speculations. For instance it follows from it that if we wanted to increase the human life span, there are two general ways in which we could do it. Firstly, we could ban reproduction before a certain age, say forty. After some centuries of this the minimum age limit would be raised to fifty, and so on. It is conceivable that human longevity could be pushed up to several centuries by this means. I cannot imagine that anyone would seriously want to institute such a policy.

Secondly we could try to 'fool' genes into thinking that the body they are sitting in is younger than it really is. In practice this would mean identifying changes in the internal chemical environment of a body that take place during ageing. Any of these could be the 'cues' that 'turn on' late-acting lethal genes. By simulating the superficial chemical properties of a young body it might be possible to prevent the turning on of late-acting deleterious genes. The interesting point is that chemical signals of old age need not in any normal sense be deleterious in themselves. For instance, suppose that it incidentally happens to be a fact that a substance S is more concentrated in the bodies of old individuals than of young individuals. S in itself might be quite harmless, perhaps some substance in the food which accumulates in the body over time. But automatically, any gene that just happened to exert a deleterious effect in the presence of S, but which otherwise had a good effect, would be positively selected in the gene pool, and would in effect be a gene 'for' dying of old age. The cure would simply be to remove S from the body.

What is revolutionary about this idea is that S itself is only a 'label' for old age. Any doctor who noticed that high concentrations of S tended to lead to death, would probably think of S as a kind of poison, and would rack his brains to find a direct causal link between S and bodily malfunctioning. But in the case of our hypothetical example, he might be wasting his time!

There might also be a substance Y, a 'label' for youth in the sense that it was more concentrated in young bodies than in old ones. Once again, genes might be selected that would have good effects in the presence of Y, but which would be deleterious in its absence. Without having any way of knowing what S or Y are— there could be many such substances—we can simply make the general prediction that the more you can simulate or mimic the properties of a young body in an old one, however superficial these properties may seem, the longer should that old body live.

I must emphasize that these are just speculations based on the Medawar theory. Although there is a sense in which the Medawar theory logically must have some truth in it, this does not mean necessarily that it is the right explanation for any given practical example of senile decay. What matters for present purposes is that the gene selection view of evolution has no difficulty in accounting for the tendency of individuals to die when they get old. The assumption of individual mortality, which lay at the heart of our argument in this chapter, is justifiable within the framework of the theory.

The other assumption I have glossed over, that of the existence of sexual reproduction and crossing-over, is more difficult to justify. Crossing-over does not always have to happen. Male fruit-flies do not do it. There is a gene that has the effect of suppressing crossing-over in females as well. If we were to breed a population of flies in which this gene was universal, the *chromosome* in a 'chromosome pool' would become the basic indivisible unit of natural selection. In fact, if we followed our definition to its logical conclusion, a whole chromosome would have to be regarded as one 'gene'.

Then again, alternatives to sex do exist. Female greenflies can bear live, fatherless, female offspring, each one containing all the genes of its mother. (Incidentally, an embryo in her mother's 'womb' may have an even smaller embryo inside her own womb. So a greenfly female may give birth to a daughter and a grand-daughter simultaneously, both of them being equivalent to her own identical twins.) Many plants propagate vegetatively by sending out suckers. In this case we might prefer to speak of *growth* rather than of reproduction; but then, if you think about it, there

is rather little distinction between growth and non-sexual reproduction anyway, since both occur by simple mitotic cell division.
Sometimes the plants produced by vegetative reproduction become
detached from the 'parent'. In other cases, for instance elm trees,
the connecting suckers remain intact. In fact an entire elm wood
might be regarded as a single individual.

So, the question is: If greenflies and elm trees don't do it, why
do the rest of us go to such lengths to mix our genes up with
somebody else's before we make a baby? It does seem an odd way
to proceed. Why did sex, that bizarre perversion of straightforward
replication, ever arise in the first place? What is the good of sex?*

This is an extremely difficult question for the evolutionist to
answer. Most serious attempts to answer it involve sophisticated
mathematical reasoning. I am frankly going to evade it except to
say one thing. This is that at least some of the difficulty that theorists have with explaining the evolution of sex results from the
fact that they habitually think of the individual as trying to maximize the number of his genes that survive. In these terms, sex
appears paradoxical because it is an 'inefficient' way for an individual to propagate her genes: each child has only 50 per cent of
the individual's genes, the other 50 per cent being provided by
the sexual partner. If only, like a greenfly, she would bud-off children who were exact replicas of herself, she would pass 100 per
cent of her genes on to the next generation in the body of every
child. This apparent paradox has driven some theorists to embrace
group selectionism, since it is relatively easy to think of group-
level advantages for sex. As W. F. Bodmer has succinctly put it,
sex 'facilitates the accumulation in a single individual of advantageous mutations which arose separately in different individuals.'

But the paradox seems less paradoxical if we follow the argument of this book, and treat the individual as a survival machine
built by a short-lived confederation of long-lived genes. 'Efficiency'

from the whole individual's point of view is then seen to be irrelevant. Sexuality versus non-sexuality will be regarded as an attribute under single-gene control, just like blue eyes versus brown eyes. A gene 'for' sexuality manipulates all the other genes for its own selfish ends. So does a gene for crossing-over. There are even genes—called mutators—that manipulate the rates of copying errors in other genes. By definition, a copying error is to the disadvantage of the gene which is miscopied. But if it is to the advantage of the selfish mutator gene that induces it, the mutator can spread through the gene pool. Similarly, if crossing-over benefits a gene for crossing-over, that is a sufficient explanation for the existence of crossing-over. And if sexual, as opposed to non-sexual, reproduction benefits a gene for sexual reproduction, that is a sufficient explanation for the existence of sexual reproduction. Whether or not it benefits all the rest of an individual's genes is comparatively irrelevant. Seen from the selfish gene's point of view, sex is not so bizarre after all.

This comes perilously close to being a circular argument, since the existence of sexuality is a precondition for the whole chain of reasoning that leads to the gene being regarded as the unit of selection. I believe there are ways of escaping from the circularity, but this book is not the place to pursue the question. Sex exists. That much is true. It is a consequence of sex and crossing-over that the small genetic unit or gene can be regarded as the nearest thing we have to a fundamental, independent agent of evolution.

Sex is not the only apparent paradox that becomes less puzzling the moment we learn to think in selfish gene terms. For instance, it appears that the amount of DNA in organisms is more than is strictly necessary for building them: a large fraction of the DNA is never translated into protein. From the point of view of the individual organism this seems paradoxical. If the 'purpose' of DNA is to supervise the building of bodies, it is surprising to find a large quantity of DNA which does no such thing. Biologists are racking their brains trying to think what useful task this apparently surplus DNA is doing. But from the point of view of the selfish genes themselves, there is no paradox. The true 'purpose' of DNA is to survive, no more and no less. The simplest way to explain the surplus DNA is to suppose that it is a parasite, or at best a harmless but useless passenger, hitching a ride in the survival machines created by the other DNA.*

Some people object to what they see as an excessively gene-centred view of evolution. After all, they argue, it is whole individuals with all their genes who actually live or die. I hope I have said enough in this chapter to show that there is really no disagreement here. Just as whole boats win or lose races, it is indeed individuals who live or die, and the *immediate* manifestation of natural selection is nearly always at the individual level. But the long-term consequences of non-random individual death and reproductive success are manifested in the form of changing gene frequencies in the gene pool. With reservations, the gene pool plays the same role for the modern replicators as the primeval soup did for the original ones. Sex and chromosomal crossing-over have the effect of preserving the liquidity of the modern equivalent of the soup. Because of sex and crossing-over the gene pool is kept well stirred, and the genes partially shuffled. Evolution is the process by which some genes become more numerous and others less numerous in the gene pool. It is good to get into the habit, whenever we are trying to explain the evolution of some characteristic, such as altruistic behaviour, of asking ourselves simply: What effect will this characteristic have on frequencies of genes in the gene pool? At times, gene language gets a bit tedious, and for brevity and vividness we shall lapse into metaphor. But we shall always keep a sceptical eye on our metaphors, to make sure they can be translated back into gene language if necessary.

As far as the gene is concerned, the gene pool is just the new sort of soup where it makes its living. All that has changed is that nowadays it makes its living by cooperating with successive groups of companions drawn from the gene pool in building one mortal survival machine after another. It is to survival machines themselves, and the sense in which genes may be said to control their behaviour, that we turn in the next chapter.

THE GENE MACHINE

S urvival machines began as passive receptacles for the genes, providing little more than walls to protect them from the chemical warfare of their rivals and the ravages of accidental molecular bombardment. In the early days they 'fed' on organic molecules freely available in the soup. This easy life came to an end when the organic food in the soup, which had been slowly built up under the energetic influence of centuries of sunlight, was all used up. A major branch of survival machines, now called plants, started to use sunlight directly themselves to build up complex molecules from simple ones, re-enacting at much higher speed the synthetic processes of the original soup. Another branch, now known as animals, 'discovered' how to exploit the chemical labours of the plants, either by eating them or by eating other animals. Both main branches of survival machines evolved more and more ingenious tricks to increase their efficiency in their various ways of life, and new ways of life were continually being opened up. Sub-branches and sub-sub-branches evolved, each one excelling in a particular specialized way of making a living: in the sea, on the ground, in the air, underground, up trees, inside other living bodies. This sub-branching has given rise to the immense diversity of animals and plants which so impresses us today.

Both animals and plants evolved into many-celled bodies, complete copies of all the genes being distributed to every cell. We do not know when, why, or how many times independently, this happened. Some people use the metaphor of a colony, describing a body as a colony of cells. I prefer to think of the body as a colony of *genes*, and of the cell as a convenient working unit for the chemical industries of the genes.

Colonies of genes they may be but, in their behaviour, bodies have undeniably acquired an individuality of their own. An animal moves as a coordinated whole, as a unit. Subjectively I feel like a unit, not a colony. This is to be expected. Selection has favoured genes that cooperate with others. In the fierce competition for scarce resources, in the relentless struggle to eat other survival machines, and to avoid being eaten, there must have been a premium on central coordination rather than anarchy within the communal body. Nowadays the intricate mutual co-evolution of genes has proceeded to such an extent that the communal nature of an individual survival machine is virtually unrecognizable. Indeed many biologists do not recognize it, and will disagree with me.

Fortunately for what journalists would call the 'credibility' of the rest of this book, the disagreement is largely academic. Just as it is not convenient to talk about quanta and fundamental particles when we discuss the workings of a car, so it is often tedious and unnecessary to keep dragging genes in when we discuss the behaviour of survival machines. In practice it is usually convenient, as an approximation, to regard the individual body as an agent 'trying' to increase the numbers of all its genes in future generations. I shall use the language of convenience. Unless otherwise stated, 'altruistic behaviour' and 'selfish behaviour' will mean behaviour directed by one animal body toward another.

This chapter is about *behaviour*—the trick of rapid movement which has been largely exploited by the animal branch of survival machines. Animals became active go-getting gene vehicles: gene machines. The characteristic of behaviour, as biologists use the term, is that it is fast. Plants move, but very slowly. When seen in highly speeded-up film, climbing plants look like active animals. But most plant movement is really irreversible growth. Animals, on the other hand, have evolved ways of moving hundreds of thousands of times faster. Moreover, the movements they make are reversible, and repeatable an indefinite number of times.

The gadget that animals evolved to achieve rapid movement was the muscle. Muscles are engines which, like the steam engine and the internal combustion engine, use energy stored in chemical fuel to generate mechanical movement. The difference is that the immediate mechanical force of a muscle is generated in the form of tension, rather than gas pressure as in the case of the steam and internal combustion engines. But muscles are like engines in that they often exert their force on cords, and levers with hinges. In us the levers are known as bones, the cords as tendons, and the hinges as joints. Quite a lot is known about the exact molecular ways in which muscles work, but I find more interesting the question of how muscle contractions are *timed*.

Have you ever watched an artificial machine of some complexity, a knitting or sewing machine, a loom, an automatic bottling factory, or a hay baler? Motive power comes from somewhere, an electric motor say, or a tractor. But much more baffling is the intricate timing of the operations. Valves open and shut in the right order, steel fingers deftly tie a knot round a hay bale, and then at just the right moment a knife shoots out and cuts the string. In many artificial machines timing is achieved by that brilliant invention the cam. This translates simple rotary motion into a complex rhythmic pattern of operations by means of an eccentric or specially shaped wheel. The principle of the musical box is similar. Other machines such as the steam organ and the pianola use paper rolls or cards with holes punched in a pattern. Recently there has been a trend towards replacing such simple mechanical timers with electronic ones. Digital computers are examples of large and versatile electronic devices which can be used for generating complex timed patterns of movements. The basic component of a modern electronic machine like a computer is the semiconductor, of which a familiar form is the transistor.

Survival machines seem to have bypassed the cam and the punched card altogether. The apparatus they use for timing their movements has more in common with an electronic computer, although it is strictly different in fundamental operation. The basic unit of biological computers, the nerve cell or neurone, is really nothing like a transistor in its internal workings. Certainly the code in which neurones communicate with each other seems to be a little bit like the pulse codes of digital computers, but the individual neurone is a much more sophisticated data-processing unit than the transistor. Instead of just three connections with other components, a single neurone may have tens of thousands. The neurone is slower than the transistor, but it has gone much further in the direction of miniaturization, a trend which has dominated the electronics industry over the past two decades. This is brought home by the fact that there are some ten thousand million neurones in the human brain: you could pack only a few hundred transistors into a skull.

Plants have no need of the neurone, because they get their living without moving around, but it is found in the great majority of animal groups. It may have been 'discovered' early in animal evolution, and inherited by all groups, or it may have been rediscovered several times independently.

Neurones are basically just cells, with a nucleus and chromosomes like other cells. But their cell walls are drawn out in long, thin, wire-like projections. Often a neurone has one particularly long 'wire' called the axon. Although the width of an axon is microscopic, its length may be many feet: there are single axons which run the whole length of a giraffe's neck. The axons are usually bundled together in thick multi-stranded cables called nerves. These lead from one part of the body to another carrying messages, rather like trunk telephone cables. Other neurones have short axons, and are confined to dense concentrations of nervous tissue called ganglia, or, when they are very large, brains. Brains may be regarded as analogous in function to computers.* They are analogous in that both types of machine generate complex patterns of output, after analysis of complex patterns of input, and after reference to stored information.

The main way in which brains actually contribute to the success of survival machines is by controlling and coordinating the contractions of muscles. To do this they need cables leading to the muscles, and these are called motor nerves. But this leads to efficient preservation of genes only if the timing of muscle contractions bears some relation to the timing of events in the outside world. It is important to contract the jaw muscles only when the jaws contain something worth biting, and to contract the leg muscles in running patterns only when there is something worth running towards or away from. For this reason, natural selection favoured animals that became equipped with sense organs, devices which translate patterns of physical events in the outside world into the pulse code of the neurones. The brain is connected to the sense organs—eyes, ears, taste-buds, etc.—by means of cables called sensory nerves. The workings of the sensory systems are particularly baffling, because they can achieve far more sophisticated feats of pattern-recognition than the best and most expensive man-made machines; if this were not so, all typists would be redundant, superseded by speech-recognizing machines, or machines for reading handwriting. Human typists will be needed for many decades yet.

There may have been a time when sense organs communicated more or less directly with muscles; indeed, sea anemones are not far from this state today, since for their way of life it is efficient. But to achieve more complex and indirect relationships between the timing of events in the outside world and the timing of muscular contractions, some kind of brain was needed as an intermediary. A notable advance was the evolutionary 'invention' of memory. By this device, the timing of muscle contractions could be influenced not only by events in the immediate past, but by events in the distant past as well. The memory, or store, is an essential part of a digital computer too. Computer memories are more reliable than human ones, but they are less capacious, and enormously less sophisticated in their techniques of information-retrieval.

One of the most striking properties of survival-machine behaviour is its apparent purposiveness. By this I do not just mean that it seems to be well calculated to help the animal's genes to survive, although of course it is. I am talking about a closer analogy to human purposeful behaviour. When we watch an animal 'searching' for food, or for a mate, or for a lost child, we can hardly help imputing to it some of the subjective feelings we ourselves experience when we search. These may include 'desire' for some object, a 'mental picture' of the desired object, an 'aim' or 'end in view'. Each one of us knows, from the evidence of our own introspection, that, at least in one modern survival machine, this purposiveness has evolved the property we call 'consciousness'. I am not philosopher enough to discuss what this means, but fortunately it does not matter for our present purposes because it is easy to talk about machines that behave *as if* motivated by a purpose, and to leave open the question whether they actually are conscious. These machines are basically very simple, and the principles of unconscious purposive behaviour are among the commonplaces of engineering science. The classic example is the Watt steam governor.

The fundamental principle involved is called negative feedback, of which there are various different forms. In general what happens is this. The 'purpose machine', the machine or thing that behaves as if it had a conscious purpose, is equipped with some kind of measuring device which measures the discrepancy between the current state of things, and the 'desired' state. It is built in such a way that the larger this discrepancy is, the harder the machine works. In this way the machine will automatically tend to reduce the discrepancy—this is why it is called *negative* feedback—and it may actually come to rest if the 'desired' state is reached. The Watt governor consists of a pair of balls which are whirled round by a steam engine. Each ball is on the end of a hinged arm. The faster the balls fly round, the more does centrifugal force push the arms towards a horizontal position, this tendency being resisted by gravity. The arms are connected to the steam valve feeding the engine, in such a way that the steam tends to be shut off when the arms approach the horizontal position. So, if the engine goes too fast, some of its steam will be shut off, and it will tend to slow down. If it slows down too much, more steam will automatically be fed to it by the valve, and it will speed up again. Such purpose machines often oscillate due to over-shooting and time-lags, and it is part of the engineer's art to build in supplementary devices to reduce the oscillations.

The 'desired' state of the Watt governor is a particular speed of rotation. Obviously it does not consciously desire it. The 'goal' of a machine is simply defined as that state to which it tends to return. Modern purpose machines use extensions of basic principles like negative feedback to achieve much more complex 'lifelike' behaviour. Guided missiles, for example, appear to search actively for their target, and when they have it in range they seem to pursue it, taking account of its evasive twists and turns, and sometimes even 'predicting' or 'anticipating' them. The details of how this is done are not worth going into. They involve negative feedback of various kinds, 'feed-forward', and other principles well understood by engineers and now known to be extensively involved in the working of living bodies. Nothing remotely approaching consciousness needs to be postulated, even though a layman, watching its apparently deliberate and purposeful behaviour, finds it hard to believe that the missile is not under the direct control of a human pilot.

It is a common misconception that because a machine such as a guided missile was originally designed and built by conscious man, then it must be truly under the immediate control of conscious man. Another variant of this fallacy is 'computers do not really play chess, because they can only do what a human operator tells them'. It is important that we understood why this is fallacious, because it affects our understanding of the sense in which genes can be said to 'control' behaviour. Computer chess is quite a good example for making the point, so I will discuss it briefly.

Computers do not yet play chess as well as human grand masters, but they have reached the standard of a good amateur. More strictly, one should say *programs* have reached the standard of a good amateur, for a chess-playing program is not fussy which physical computer it uses to act out its skills. Now, what is the role of the human programmer? First, he is definitely not manipulating the computer from moment to moment, like a puppeteer pulling strings. That would be just cheating. He writes the program, puts it in the computer, and then the computer is on its own: there is no further human intervention, except for the opponent typing in his moves. Does the programmer perhaps anticipate all possible chess positions, and provide the computer with a long list of good moves, one for each possible contingency? Most certainly not, because the number of possible positions in chess is so great that the world would come to an end before the list had been completed. For the same reason, the computer cannot possibly be programmed to try out 'in its head' all possible moves, and all possible follow-ups, until it finds a winning strategy. There are more possible games of chess than there are atoms in the galaxy. So much for the trivial non-solutions to the problem of programming a computer to play chess. It is in fact an exceedingly difficult problem, and it is hardly surprising that the best programs have still not achieved grand master status.

The programmer's actual role is rather more like that of a father teaching his son to play chess. He tells the computer the basic moves of the game, not separately for every possible starting position, but in terms of more economically expressed rules. He does not literally say in plain English 'bishops move in a diagonal', but he does say something mathematically equivalent, such as, though more briefly: 'New coordinates of bishop are obtained from old coordinates, by adding the same constant, though not necessarily with the same sign, to both old x coordinate and old y coordinate.' Then he might program in some 'advice', written in the same sort of mathematical or logical language, but amounting in human terms to hints such as 'don't leave your king unguarded', or useful tricks such as 'forking' with the knight. The details are intriguing, but they would take us too far afield. The important point is this. When it is actually playing, the computer is on its own, and can expect no help from its master. All the programmer can do is to set the computer up *beforehand* in the best way possible, with a proper balance between lists of specific knowledge and hints about strategies and techniques.

The genes too control the behaviour of their survival machines, not directly with their fingers on puppet strings, but indirectly like the computer programmer. All they can do is to set it up beforehand; then the survival machine is on its own, and the genes can only sit passively inside. Why are they so passive? Why don't they grab the reins and take charge from moment to moment? The answer is that they cannot because of time-lag problems. This is best shown by another analogy, taken from science fiction. *A for Andromeda* by Fred Hoyle and John Elliot is an exciting story, and, like all good science fiction, it has some interesting scientific points lying behind it. Strangely, the book seems to lack explicit mention of the most important of these underlying points. It is left to the reader's imagination. I hope the authors will not mind if I spell it out here.

There is a civilization 200 light-years away, in the constellation of Andromeda.* They want to spread their culture to distant worlds. How best to do it? Direct travel is out of the question. The speed of light imposes a theoretical upper limit to the rate at which you can get from one place to another in the universe, and mechanical considerations impose a much lower limit in practice. Besides, there may not be all that many worlds worth going to, and how do you know which direction to go in? Radio is a better way of communicating with the rest of the universe, since, if you have enough power to broadcast your signals in all directions rather than beam them in one direction, you can reach a very large number of worlds (the number increasing as the square of the distance the signal travels). Radio waves travel at the speed of light, which means the signal takes 200 years to reach earth from Andromeda. The trouble with this sort of distance is that you can never hold a conversation. Even if you discount the fact that each successive message from earth would be transmitted by people separated from each other by twelve generations, it would be just plain wasteful to attempt to converse over such distances.

This problem will soon arise in earnest for us: it takes about four minutes for radio waves to travel between earth and Mars. There can be no doubt that spacemen will have to get out of the habit of conversing in short alternating sentences, and will have to use long soliloquies or monologues, more like letters than conversations. As another example, Roger Payne has pointed out that the acoustics of the sea have certain peculiar properties, which mean that the exceedingly loud 'song' of some whales could theoretically be heard all the way round the world, provided the whales swim at a certain depth. It is not known whether they actually do communicate with each other over very great distances, but if they do they must be in much the same predicament as an astronaut on Mars. The speed of sound in water is such that it would take nearly two hours for the song to travel across the Atlantic Ocean and for a reply to return. I suggest this as an explanation for the fact that some whales deliver a continuous soliloquy, without repeating themselves, for a full eight minutes. They then go back to the beginning of the song and repeat it all over again, many times over, each complete cycle lasting about eight minutes.

The Andromedans of the story did the same thing. Since there was no point in waiting for a reply, they assembled everything they wanted to say into one huge unbroken message, and then they broadcast it out into space, over and over again, with a cycle time of several months. Their message was very different from that of the whales, however. It consisted of coded instructions for the building and programming of a giant computer. Of course the instructions were in no human language, but almost any code

can be broken by a skilled cryptographer, especially if the designers of the code intended it to be easily broken. Picked up by the
Jodrell Bank radio telescope, the message was eventually decoded,
the computer built, and the program run. The results were nearly
disastrous for mankind, for the intentions of the Andromedans
were not universally altruistic, and the computer was well on the
way to dictatorship over the world before the hero eventually finished it off with an axe.

From our point of view, the interesting question is in what
sense the Andromedans could be said to be manipulating events
on Earth. They had no direct control over what the computer did
from moment to moment; indeed they had no possible way of
even knowing the computer had been built, since the information would have taken 200 years to get back to them. The decisions and actions of the computer were entirely its own. It could
not even refer back to its masters for general policy instructions.
All its instructions had to be built-in in advance, because of the
inviolable 200 year barrier. In principle, it must have been programmed very much like a chess-playing computer, but with
greater flexibility and capacity for absorbing local information.
This was because the program had to be designed to work not
just on earth, but on any world possessing an advanced technology, any of a set of worlds whose detailed conditions the
Andromedans had no way of knowing.

Just as the Andromedans had to have a computer on earth
to take day-to-day decisions for them, our genes have to build a
brain. But the genes are not only the Andromedans who sent the
coded instructions; they are also the instructions themselves.
The reason why they cannot manipulate our puppet strings directly
is the same: time-lags. Genes work by controlling protein synthesis. This is a powerful way of manipulating the world, but it is
slow. It takes months of patiently pulling protein strings to build

an embryo. The whole point about behaviour, on the other hand, is that it is fast. It works on a time-scale not of months but of seconds and fractions of seconds. Something happens in the world, an owl flashes overhead, a rustle in the long grass betrays prey, and in milliseconds nervous systems crackle into action, muscles leap, and someone's life is saved—or lost. Genes don't have reaction-times like that. Like the Andromedans, the genes can only do their best *in advance* by building a fast executive computer for themselves, and programming it in advance with rules and 'advice' to cope with as many eventualities as they can 'anticipate'. But life, like the game of chess, offers too many different possible eventualities for all of them to be anticipated. Like the chess programmer, the genes have to 'instruct' their survival machines not in specifics, but in the general strategies and tricks of the living trade.*

As J. Z. Young has pointed out, the genes have to perform a task analogous to prediction. When an embryo survival machine is being built, the dangers and problems of its life lie in the future. Who can say what carnivores crouch waiting for it behind what bushes, or what fleet-footed prey will dart and zig-zag across its path? No human prophet, nor any gene. But some general predictions can be made. Polar bear genes can safely predict that the future of their unborn survival machine is going to be a cold one. They do not think of it as a prophecy, they do not think at all: they just build in a thick coat of hair, because that is what they have always done before in previous bodies, and that is why they still exist in the gene pool. They also predict that the ground is going to be snowy, and their prediction takes the form of making the coat of hair white and therefore camouflaged. If the climate of the Arctic changed so rapidly that the baby bear found itself born into a tropical desert, the predictions of the genes would be wrong, and they would pay the penalty. The young bear would die, and they inside it.

Prediction in a complex world is a chancy business. Every decision that a survival machine takes is a gamble, and it is the business of genes to program brains in advance so that on average they take decisions that pay off. The currency used in the casino of evolution is survival, strictly gene survival, but for many purposes individual survival is a reasonable approximation. If you go down to the water-hole to drink, you increase your risk of being eaten by predators who make their living lurking for prey by water-holes. If you do not go down to the water-hole you will eventually die of thirst. There are risks whichever way you turn, and you must take the decision that maximizes the long-term survival chances of your genes. Perhaps the best policy is to postpone drinking until you are very thirsty, then go and have one good long drink to last you a long time. That way you reduce the number of separate visits to the water-hole, but you have to spend a long time with your head down when you finally do drink. Alternatively the best gamble might be to drink little and often, snatching quick gulps of water while running past the water-hole. Which is the best gambling strategy depends on all sorts of complex things, not least the hunting habit of the predators, which itself is evolved to be maximally efficient from their point of view. Some form of weighing up of the odds has to be done. But of course we do not have to think of the animals as making the calculations consciously. All we have to believe is that those individuals whose genes build brains in such a way that they tend to gamble correctly are as a direct result more likely to survive, and therefore to propagate those same genes.

We can carry the metaphor of gambling a little further. A gambler must think of three main quantities, stake, odds, and prize. If the prize is very large, a gambler is prepared to risk a big stake. A gambler who risks his all on a single throw stands to gain a great deal. He also stands to lose a great deal, but on average high-stake gamblers are no better and no worse off than other players who play for low winnings with low stakes. An analogous comparison is that between speculative and safe investors on the stock market. In some ways the stock market is a better analogy than a casino, because casinos are deliberately rigged in the bank's favour (which means, strictly, that high-stake players will on average end up poorer than low-stake players; and low-stake players poorer than those who do not gamble at all. But this is for a reason not germane to our discussion). Ignoring this, both high-stake play and low-stake play seem reasonable. Are there animal gamblers who play for high stakes, and others with a more conservative game? In Chapter 9 we shall see that it is often possible to picture males as high-stake, high-risk gamblers, and females as safe investors, especially in polygamous species in which males compete for females. Naturalists who read this book may be able to think of species that can be described as high-stake, high-risk players, and other species that play a more conservative game. I now return to the more general theme of how genes make 'predictions' about the future.

One way for genes to solve the problem of making predictions in rather unpredictable environments is to build in a capacity for learning. Here the program may take the form of the following instructions to the survival machine: 'Here is a list of things defined as rewarding: sweet taste in the mouth, orgasm, mild temperature, smiling child. And here is a list of nasty things: various sorts of pain, nausea, empty stomach, screaming child. If you should happen to do something that is followed by one of the nasty things, don't do it again, but on the other hand repeat anything that is followed by one of the nice things.' The advantage of this sort of programming is that it greatly cuts down the number of detailed rules that have to be built into the original program; and it is also capable of coping with changes in the environment that could not have been predicted in detail. On the other hand, certain predictions have to be made still. In our example the genes are predicting that sweet taste in the mouth and orgasm are going to be 'good' in the sense that eating sugar and copulating are likely to be beneficial to gene survival. The possibilities of saccharine and masturbation are not anticipated according to this example; nor are the dangers of over-eating sugar in our environment where it exists in unnatural plenty.

Learning-strategies have been used in some chess-playing computer programs. These programs actually get better as they play against human opponents or against other computers. Although they are equipped with a repertoire of rules and tactics, they also have a small random tendency built into their decision procedure. They record past decisions, and whenever they win a game they slightly increase the weighting given to the tactics that preceded the victory, so that next time they are a little bit more likely to choose those same tactics again.

One of the most interesting methods of predicting the future is simulation. If a general wishes to know whether a particular military plan will be better than alternatives, he has a problem in prediction. There are unknown quantities in the weather, in the morale of his own troops, and in the possible countermeasures of the enemy. One way of discovering whether it is a good plan is to try and see, but it is undesirable to use this test for all the tentative plans dreamed up, if only because the supply of young men prepared to die 'for their country' is exhaustible, and the supply of possible plans is very large. It is better to try the various plans out in dummy runs rather than in deadly earnest. This may take the form of full-scale exercises with 'Northland' fighting 'Southland' using blank ammunition, but even this is expensive in time and materials. Less wastefully, war games may be played, with tin soldiers and little toy tanks being shuffled around a large map.

Recently, computers have taken over large parts of the simulation function, not only in military strategy, but in all fields where prediction of the future is necessary, fields like economics, ecology, sociology, and many others. The technique works like this. A model of some aspect of the world is set up in the computer. This does not mean that if you unscrewed the lid you would see a little miniature dummy inside with the same shape as the object simulated. In the chess-playing computer there is no 'mental picture' inside the memory banks recognizable as a chess board with knights and pawns sitting on it. The chess board and its current position would be represented by lists of electronically coded numbers. To us a map is a miniature scale model of a part of the world, compressed into two dimensions. In a computer, a map might alternatively be represented as a list of towns and other spots, each with two numbers—its latitude and longitude. But it does not matter how the computer actually holds its model of the world in its head, provided that it holds it in a form in which it can operate on it, manipulate it, do experiments with it, and report back to the human operators in terms which they can understand. Through the technique of simulation, model battles can be won or lost, simulated airliners fly or crash, economic policies lead to prosperity or to ruin. In each case the whole process goes on inside the computer in a tiny fraction of the time it would take in real life. Of course there are good models of the world and bad ones, and even the good ones are only approximations. No amount of simulation can predict exactly what will happen in reality, but a good simulation is enormously preferable to blind trial and error. Simulation could be called vicarious trial and error, a term unfortunately pre-empted long ago by rat psychologists.

If simulation is such a good idea, we might expect that survival machines would have discovered it first. After all, they invented many of the other techniques of human engineering long before we came on the scene: the focusing lens and the parabolic reflector, frequency analysis of sound waves, servo-control, sonar, buffer storage of incoming information, and countless others with long names, whose details don't matter. What about simulation? Well, when you yourself have a difficult decision to make involving unknown quantities in the future, you do go in for a form of simulation. You *imagine* what would happen if you did each of the alternatives open to you. You set up a model in your head, not of everything in the world, but of the restricted set of entities which you think may be relevant. You may see them vividly in your mind's eye, or you may see and manipulate stylized abstractions of them. In either case it is unlikely that somewhere laid out in your brain is an actual spatial model of the events you are imagining. But, just as in the computer, the details of how your brain represents its model of the world are less important than the fact that it is able to use it to predict possible events. Survival machines that can simulate the future are one jump ahead of survival machines who can only learn on the basis of overt trial and error. The trouble with overt trial is that it takes time and energy. The trouble with overt error is that it is often fatal. Simulation is both safer and faster.

The evolution of the capacity to simulate seems to have culminated in subjective consciousness. Why this should have happened is, to me, the most profound mystery facing modern biology. There is no reason to suppose that electronic computers are conscious when they simulate, although we have to admit that in the future they may become so. Perhaps consciousness arises when the brain's simulation of the world becomes so complete that it must include a model of itself.* Obviously the limbs and body of a survival machine must constitute an important part of its simulated world; presumably for the same kind of reason, the simulation itself could be regarded as part of the world to be simulated. Another word for this might indeed be 'self-awareness', but I don't find this a fully satisfying explanation of the evolution of consciousness, and this is only partly because it involves an infinite regress—if there is a model of the model, why not a model of the model of the model...?

Whatever the philosophical problems raised by consciousness, for the purpose of this story it can be thought of as the culmination of an evolutionary trend towards the emancipation of survival machines as executive decision-takers from their ultimate masters, the genes. Not only are brains in charge of the day-to-day running of survival-machine affairs, they have also acquired the ability to predict the future and act accordingly. They even have the power to rebel against the dictates of the genes, for instance in refusing to have as many children as they are able to. But in this respect man is a very special case, as we shall see.

What has all this to do with altruism and selfishness? I am trying to build up the idea that animal behaviour, altruistic or selfish, is under the control of genes in only an indirect, but still very powerful, sense. By dictating the way survival machines and their nervous systems are built, genes exert ultimate power over behaviour. But the moment-to-moment decisions about what to do next are taken by the nervous system. Genes are the primary policy-makers; brains are the executives. But as brains became more highly developed, they took over more and more of the actual policy decisions, using tricks like learning and simulation in doing so. The logical conclusion to this trend, not yet reached in any species, would be for the genes to give the survival machine a single overall policy instruction: do whatever you think best to keep us alive.

Analogies with computers and with human decision-taking are all very well. But now we must come down to earth and remember that evolution in fact occurs step-by-step, through the differential survival of genes in the gene pool. Therefore, in order for a behaviour pattern—altruistic or selfish—to evolve, it is necessary that a gene 'for' that behaviour should survive in the gene pool more successfully than a rival gene or allele 'for' some different behaviour. A gene for altruistic behaviour means any gene that influences the development of nervous systems in such a way as to make them likely to behave altruistically.* Is there any experimental evidence for the genetic inheritance of altruistic behaviour? No, but that is hardly surprising, since little work has been done on the genetics of any behaviour. Instead, let me tell you about one study of a behaviour pattern which does not happen to be obviously altruistic, but which is complex enough to be interesting. It serves as a model for how altruistic behaviour might be inherited.

Honey bees suffer from an infectious disease called foul brood. This attacks the grubs in their cells. Of the domestic breeds used by beekeepers, some are more at risk from foul brood than others, and it turns out that the difference between strains is, at least in some cases, a behavioural one. There are so-called hygienic strains which quickly stamp out epidemics by locating infected grubs, pulling them from their cells and throwing them out of the hive. The susceptible strains are susceptible because they do not practise this hygienic infanticide. The behaviour actually involved in hygiene is quite complicated. The workers have to locate the cell of each diseased grub, remove the wax cap from the cell, pull out the larva, drag it through the door of the hive, and throw it on the rubbish tip.

Doing genetic experiments with bees is quite a complicated business for various reasons. Worker bees themselves do not ordinarily reproduce, and so you have to cross a queen of one strain with a drone (= male) of the other, and then look at the behaviour of the daughter workers. This is what W. C. Rothenbuhler did. He found that all first-generation hybrid daughter hives were non-hygienic: the behaviour of their hygienic parent seemed to have been lost, although as things turned out the hygienic genes were still there but were recessive, like human genes for blue eyes. When Rothenbuhler 'back-crossed' first-generation hybrids with a pure hygienic strain (again of course using queens and drones), he obtained a most beautiful result. The daughter hives fell into three groups. One group showed perfect hygienic behaviour, a second showed no hygienic behaviour at all, and the third went half way. This last group uncapped the wax cells of diseased grubs, but they did not follow through and throw out the larvae. Rothenbuhler surmised that there might be two separate genes, one gene for uncapping, and one gene for throwing-out. Normal hygienic strains possess both genes, susceptible strains possess the alleles—rivals—of both genes instead. The hybrids who only went half way presumably possessed the uncapping gene (in double dose) but not the throwing-out gene. Rothenbuhler guessed that his experimental group of apparently totally non-hygienic bees might conceal a sub-group possessing the throwing-out gene, but unable to show it because they lacked the uncapping gene. He confirmed this most elegantly by removing caps himself. Sure enough, half of the apparently non-hygienic bees thereupon showed perfectly normal throwing-out behaviour.*

This story illustrates a number of important points which came up in the previous chapter. It shows that it can be perfectly proper to speak of a 'gene for behaviour so-and-so' even if we haven't the faintest idea of the chemical chain of embryonic causes leading from gene to behaviour. The chain of causes could even turn out to involve learning. For example, it could be that the uncapping gene exerts its effect by giving bees a taste for infected wax. This means they will find the eating of the wax caps covering disease-victims rewarding, and will therefore tend to repeat it. Even if this is how the gene works, it is still truly a gene 'for uncapping' provided that, other things being equal, bees possessing the gene end up by uncapping, and bees not possessing the gene do not uncap.

Secondly it illustrates the fact that genes 'cooperate' in their effects on the behaviour of the communal survival machine. The throwing-out gene is useless unless it is accompanied by the uncapping gene and vice versa. Yet the genetic experiments show equally clearly that the two genes are in principle quite separable in their journey through the generations. As far as their useful work is concerned you can think of them as a single cooperating unit, but as replicating genes they are two free and independent agents.

For purposes of argument it will be necessary to speculate about genes 'for' doing all sorts of improbable things. If I speak, for example, of a hypothetical gene 'for saving companions from drowning', and you find such a concept incredible, remember the story of the hygienic bees. Recall that we are not talking about the gene as the sole antecedent cause of all the complex muscular contractions, sensory integrations, and even conscious decisions that are involved in saving somebody from drowning. We are saying nothing about the question of whether learning, experience, or environmental influences enter into the development of the behaviour. All you have to concede is that it is possible for a single gene, other things being equal and lots of other essential genes and environmental factors being present, to make a body more likely to save somebody from drowning than its allele would. The difference between the two genes may turn out at bottom to be a slight difference in some simple quantitative variable. The details of the embryonic developmental process, interesting as they may be, are irrelevant to evolutionary considerations. Konrad Lorenz has put this point well.

The genes are master programmers, and they are programming for their lives. They are judged according to the success of their programs in coping with all the hazards that life throws at their survival machines, and the judge is the ruthless judge of the court of survival. We shall come later to ways in which gene survival can be fostered by what appears to be altruistic behaviour. But the obvious first priorities of a survival machine, and of the brain that takes the decisions for it, are individual survival and reproduction. All the genes in the 'colony' would agree about these priorities. Animals therefore go to elaborate lengths to find and catch food; to avoid being caught and eaten themselves; to avoid disease and accident; to protect themselves from unfavourable climatic conditions; to find members of the opposite sex and persuade them to mate; and to confer on their children advantages similar to those they enjoy themselves. I shall not give examples—if you want one just look carefully at the next wild animal that you see. But I do want to mention one particular kind of behaviour because we shall need to refer to it again when we come to speak of altruism and selfishness. This is the behaviour that can be broadly labelled *communication*.*

A survival machine may be said to have communicated with another one when it influences its behaviour or the state of its nervous system. This is not a definition I should like to have to defend for very long, but it is good enough for present purposes. By influence I mean direct causal influence. Examples of communication are numerous: song in birds, frogs, and crickets; tail-wagging and hackle-raising in dogs; 'grinning' in chimpanzees; human gestures and language. A great number of survival-machine actions promote their genes' welfare indirectly by influencing the behaviour of other survival machines. Animals go to great lengths to make this communication effective. The songs of birds enchant and mystify successive generations of men. I have already referred to the even more elaborate and mysterious song of the humpback whale, with its prodigious range, its frequencies spanning the whole of human hearing from subsonic rumblings to ultrasonic squeaks. Mole-crickets amplify their song to stentorian loudness by singing down in a burrow which they carefully dig in the shape of a double exponential horn, or megaphone. Bees dance in the dark to give other bees accurate information about the direction and distance of food, a feat of communication rivalled only by human language itself.

The traditional story of ethologists is that communication signals evolve for the mutual benefit of both sender and recipient. For instance, baby chicks influence their mother's behaviour by giving high piercing cheeps when they are lost or cold. This usually has the immediate effect of summoning the mother, who leads the chick back to the main clutch. This behaviour could be said to have evolved for mutual benefit, in the sense that natural selection has favoured babies that cheep when they are lost, and also mothers that respond appropriately to the cheeping.

If we wish to (it is not really necessary), we can regard signals such as the cheep call as having a meaning, or as carrying information: in this case 'I am lost.' The alarm call given by small birds, which I mentioned in Chapter 1, could be said to convey the information 'There is a hawk.' Animals who receive this information and act on it are benefited. Therefore the information can be said to be true. But do animals ever communicate false information; do they ever tell lies?

The notion of an animal telling a lie is open to misunderstanding, so I must try to forestall this. I remember attending a lecture given by Beatrice and Allen Gardner about their famous 'talking' chimpanzee Washoe (she uses American Sign Language, and her achievement is of great potential interest to students of language). There were some philosophers in the audience, and in the discussion after the lecture they were much exercised by the question of whether Washoe could tell a lie. I suspected that the Gardners thought there were more interesting things to talk about, and I agreed with them. In this book I am using words like 'deceive' and 'lie' in a much more straight-forward sense than those philosophers. They were interested in conscious intention to deceive. I am talking simply about having an effect functionally equivalent to deception. If a bird used the 'There is a hawk' signal when there was no hawk, thereby frightening his colleagues away, leaving him to eat all their food, we might say he had told a lie. We would not mean he had deliberately intended consciously to deceive. All that is implied is that the liar gained food at the other birds' expense, and the reason the other birds flew away was that they reacted to the liar's cry in a way appropriate to the presence of a hawk.

Many edible insects, like the butterflies of the previous chapter, derive protection by mimicking the external appearance of other distasteful or stinging insects. We ourselves are often fooled into thinking that yellow and black striped hover-flies are wasps. Some bee-mimicking flies are even more perfect in their deception. Predators too tell lies. Angler fish wait patiently on the bottom of the sea, blending in with the background. The only conspicuous part is a wriggling worm-like piece of flesh on the end of a long 'fishing rod', projecting from the top of the head. When a small prey fish comes near, the angler will dance its worm-like bait in front of the little fish, and lure it down to the region of the angler's own concealed mouth. Suddenly it opens its jaws, and the little fish is sucked in and eaten. The angler is telling a lie, exploiting the little fish's tendency to approach wriggling worm-like objects. He is saying 'Here is a worm', and any little fish who 'believes' the lie is quickly eaten.

Some survival machines exploit the sexual desires of others. Bee orchids induce bees to copulate with their flowers, because of their strong resemblance to female bees. What the orchid has to gain from this deception is pollination, for a bee who is fooled by two orchids will incidentally carry pollen from one to the other. Fireflies (which are really beetles) attract their mates by flashing lights at them. Each species has its own particular dot-dash flashing pattern, which prevents confusion between species, and consequent harmful hybridization. Just as sailors look out for the flash patterns of particular lighthouses, so fireflies seek the coded flash patterns of their own species. Females of the genus *Photuris* have 'discovered' that they can lure males of the genus *Photinus* if they imitate the flashing code of a *Photinus* female. This they do, and when a *Photinus* male is fooled by the lie into approaching, he is summarily eaten by the *Photuris* female. Sirens and Lorelei spring to mind as analogies, but Cornishmen will prefer to think of the wreckers of the old days, who used lanterns to lure ships on to the rocks, and then plundered the cargoes that spilled out of the wrecks.

Whenever a system of communication evolves, there is always the danger that some will exploit the system for their own ends. Brought up as we have been on the 'good of the species' view of evolution, we naturally think first of liars and deceivers as belonging to different species: predators, prey, parasites, and so on. However, we must expect lies and deceit, and selfish exploitation of communication to arise whenever the interests of the genes of different individuals diverge. This will include individuals of the same species. As we shall see, we must even expect that children will deceive their parents, that husbands will cheat on wives, and that brother will lie to brother.

Even the belief that animal communication signals originally evolve to foster mutual benefit, and then afterwards become exploited by malevolent parties, is too simple. It may well be that all animal communication contains an element of deception right from the start, because all animal interactions involve at least some conflict of interest. The next chapter introduces a powerful way of thinking about conflicts of interest from an evolutionary point of view.

AGGRESSION
Stability and the selfish machine

This chapter is mostly about the much-misunderstood topic of aggression. We shall continue to treat the individual as a selfish machine, programmed to do whatever is best for its genes as a whole. This is the language of convenience. At the end of the chapter we return to the language of single genes.

To a survival machine, another survival machine (which is not its own child or another close relative) is part of its environment, like a rock or a river or a lump of food. It is something that gets in the way, or something that can be exploited. It differs from a rock or a river in one important respect: it is inclined to hit back. This is because it too is a machine that holds its immortal genes in trust for the future, and it too will stop at nothing to preserve them. Natural selection favours genes that control their survival machines in such a way that they make the best use of their environment. This includes making the best use of other survival machines, both of the same and of different species.

In some cases survival machines seem to impinge rather little on each others' lives. For instance moles and blackbirds do not eat each other, mate with each other, or compete with each other for living space. Even so, we must not treat them as completely insulated. They may compete for something, perhaps earthworms.

This does not mean you will ever see a mole and a blackbird engaged in a tug of war over a worm; indeed a blackbird may never set eyes on a mole in its life. But if you wiped out the population of moles, the effect on blackbirds might be dramatic, although I could not hazard a guess as to what the details might be, nor by what tortuously indirect routes the influence might travel.

Survival machines of different species influence each other in a variety of ways. They may be predators or prey, parasites or hosts, competitors for some scarce resource. They may be exploited in special ways, as for instance when bees are used as pollen carriers by flowers.

Survival machines of the same species tend to impinge on each others' lives more directly. This is for many reasons. One is that half the population of one's own species may be potential mates, and potentially hard-working and exploitable parents to one's children. Another reason is that members of the same species, being very similar to each other, being machines for preserving genes in the same kind of place, with the same kind of way of life, are particularly direct competitors for all the resources necessary for life. To a blackbird, a mole may be a competitor, but it is not nearly so important a competitor as another blackbird. Moles and blackbirds may compete for worms, but blackbirds and blackbirds compete with each other for worms *and* for everything else. If they are members of the same sex, they may also compete for mating partners. For reasons that we shall see, it is usually the males who compete with each other for females. This means that a male might benefit his own genes if he does something detrimental to another male with whom he is competing.

The logical policy for a survival machine might therefore seem to be to murder its rivals, and then, preferably, to eat them. Although murder and cannibalism do occur in nature, they are not as common as a naïve interpretation of the selfish gene theory might predict. Indeed Konrad Lorenz, in *On Aggression*, stresses the restrained and gentlemanly nature of animal fighting. For him the notable thing about animal fights is that they are formal tournaments, played according to rules like those of boxing or fencing. Animals fight with gloved fists and blunted foils. Threat and bluff take the place of deadly earnest. Gestures of surrender are recognized by victors, who then refrain from dealing the killing blow or bite that our naïve theory might predict.

This interpretation of animal aggression as being restrained and formal can be disputed. In particular, it is certainly wrong to condemn poor old *Homo sapiens* as the only species to kill his own kind, the only inheritor of the mark of Cain, and similar melodramatic charges. Whether a naturalist stresses the violence or the restraint of animal aggression depends partly on the kinds of animals he is used to watching, and partly on his evolutionary preconceptions—Lorenz is, after all, a 'good of the species' man. Even if it has been exaggerated, the gloved fist view of animal fights seems to have at least some truth. Superficially this looks like a form of altruism. The selfish gene theory must face up to the difficult task of explaining it. Why is it that animals do not go all out to kill rival members of their species at every possible opportunity?

The general answer to this is that there are costs as well as benefits resulting from outright pugnacity, and not only the obvious costs in time and energy. For instance, suppose that B and C are both my rivals, and I happen to meet B. It might seem sensible for me as a selfish individual to try to kill him. But wait. C is also my rival, and C is also B's rival. By killing B, I am potentially doing a good turn to C by removing one of his rivals. I might have done better to let B live, because he might then have competed or fought with C, thereby benefiting me indirectly. The moral of this simple hypothetical example is that there is no obvious merit in indiscriminately trying to kill rivals. In a large and complex system of rivalries, removing one rival from the scene does not necessarily do any good: other rivals may be more likely to benefit from his death than oneself. This is the kind of hard lesson that has been learned by pest-control officers. You have a serious agricultural pest, you discover a good way to exterminate it and you gleefully do so, only to find that another pest benefits from the extermination even more than human agriculture does, and you end up worse off than you were before.

On the other hand, it might seem a good plan to kill, or at least fight with, certain particular rivals in a discriminating way. If B is an elephant seal in possession of a large harem full of females, and if I, another elephant seal, can acquire his harem by killing him, I might be well advised to attempt to do so. But there are costs and risks even in selectivity pugnacity. It is to B's advantage to fight back, to defend his valuable property. If I start a fight, I am just as likely to end up dead as he is. Perhaps even more so. He holds a valuable resource, that is why I want to fight him. But why does he hold it? Perhaps he won it in combat. He has probably beaten off other challengers before me. He is probably a good fighter. Even if I win the fight and gain the harem, I may be so badly mauled in the process that I cannot enjoy the benefits. Also, fighting uses up time and energy. These might be better conserved for the time being. If I concentrate on feeding and on keeping out of trouble for a time, I shall grow bigger and stronger. I'll fight him for the harem in the end, but I may have a better chance of winning eventually if I wait, rather than rush in now.

This subjective soliloquy is just a way of pointing out that the decision whether or not to fight should ideally be preceded by a complex, if unconscious, 'cost–benefit' calculation. The potential benefits are not all stacked up on the side of fighting, although undoubtedly some of them are. Similarly, during a fight, each tactical decision over whether to escalate the fight or cool it has costs and benefits which could, in principle, be analysed. This has long been realized by ethologists in a vague sort of way, but it has taken J. Maynard Smith, not normally regarded as an ethologist, to express the idea forcefully and clearly. In collaboration with G. R. Price and G. A. Parker, he uses the branch of mathematics known as Game Theory. Their elegant ideas can be expressed in words without mathematical symbols, albeit at some cost in rigour.

The essential concept Maynard Smith introduces is that of the *evolutionarily stable strategy*, an idea that he traces back to W. D. Hamilton and R. H. MacArthur. A 'strategy' is a pre-programmed behavioural policy. An example of a strategy is: 'Attack opponent; if he flees pursue him; if he retaliates run away.' It is important to realize that we are not thinking of the strategy as being consciously worked out by the individual. Remember that we are picturing the animal as a robot survival machine with a pre-programmed computer controlling the muscles. To write the strategy out as a set of simple instructions in English is just a convenient way for us to think about it. By some unspecified mechanism, the animal behaves as if he were following these instructions.

An evolutionarily stable strategy or ESS is defined as a strategy which, if most members of a population adopt it, cannot be bettered by an alternative strategy.* It is a subtle and important idea. Another way of putting it is to say that the best strategy for an individual depends on what the majority of the population are doing. Since the rest of the population consists of individuals, each one trying to maximize his *own* success, the only strategy that persists will be one which, once evolved, cannot be bettered by any deviant individual. Following a major environmental change there may be a brief period of evolutionary instability, perhaps even oscillation in the population. But once an ESS is achieved it will stay: selection will penalize deviation from it.

To apply this idea to aggression, consider one of Maynard Smith's simplest hypothetical cases. Suppose that there are only two sorts of fighting strategy in a population of a particular species, named *hawk* and *dove*. (The names refer to conventional human usage and have no connection with the habits of the birds from whom the names are derived: doves are in fact rather aggressive birds.) Any individual of our hypothetical population is classifed as a hawk or a dove. Hawks always fight as hard and as unrestrainedly as they can, retreating only when seriously injured. Doves merely threaten in a dignified conventional way, never hurting anybody. If a hawk fights a dove the dove quickly runs away, and so does not get hurt. If a hawk fights a hawk they go on until one of them is seriously injured or dead. If a dove meets a dove nobody gets hurt; they go on posturing at each other for a long time until one of them tires or decides not to bother any more, and therefore backs down. For the time being, we assume that there is no way in which an individual can tell, in advance, whether a particular rival is a hawk or a dove. He only discovers this by fighting him, and he has no memory of past fights with particular individuals to guide him.

Now as a purely arbitrary convention we allot contestants 'points'. Say 50 points for a win, 0 for losing, –100 for being seriously injured, and –10 for wasting time over a long contest. These points can be thought of as being directly convertible into the currency of gene survival. An individual who scores high points, who has a high average 'pay-off', is an individual who leaves many genes behind him in the gene pool. Within broad limits the actual numerical values do not matter for the analysis, but they help us to think about the problem.

The important thing is that we are *not* interested in whether hawks will tend to beat doves when they fight them. We already know the answer to that: hawks will always win. We want to know whether either hawk or dove is an evolutionarily stable strategy. If one of them is an ESS and the other is not, we must expect that the one which is the ESS will evolve. It is theoretically possible for there to be two ESSs. This would be true if, whatever the majority strategy of the population happened to be, whether hawk or dove, the best strategy for any given individual was to follow suit. In this case the population would tend to stick at whichever one of its two stable states it happened to reach first. However, as we shall now see, neither of these two strategies, hawk or dove, would in fact be evolutionarily stable on its own, and we should therefore not expect either of them to evolve. To show this we must calculate average pay-offs.

Suppose we have a population consisting entirely of doves. Whenever they fight, nobody gets hurt. The contests consist of prolonged ritual tournaments, staring matches perhaps, which end only when one rival backs down. The winner then scores 50 points for gaining the resource in dispute, but he pays a penalty of −10 for wasting time over a long staring match, so scores 40 in all. The loser also is penalized −10 points for wasting time. On average, any one individual dove can expect to win half his contests and lose half. Therefore his average pay-off per contest is the average of +40 and −10, which is +15. Therefore, every individual dove in a population of doves seems to be doing quite nicely.

But now suppose a mutant hawk arises in the population. Since he is the only hawk around, every fight he has is against a dove. Hawks always beat doves, so he scores +50 every fight, and this is his average pay-off. He enjoys an enormous advantage over the doves, whose net pay-off is only +15. Hawk genes will rapidly spread through the population as a result. But now each hawk can no longer count on every rival he meets being a dove. To take an extreme example, if the hawk gene spread so successfully that the entire population came to consist of hawks, all fights would now be hawk fights. Things are now very different. When hawk meets hawk, one of them is seriously injured, scoring –100, while the winner scores +50. Each hawk in a population of hawks can expect to win half his fights and lose half his fights. His average expected pay-off per fight is therefore halfway between +50 and –100, which is –25. Now consider a single dove in a population of hawks. To be sure, he loses all his fights, but on the other hand he never gets hurt. His average pay-off is 0 in a population of hawks, whereas the average pay-off for a hawk in a population of hawks is –25. Dove genes will therefore tend to spread through the population.

The way I have told the story it looks as if there will be a continuous oscillation in the population. Hawk genes will sweep to ascendancy; then, as a consequence of the hawk majority, dove genes will gain an advantage and increase in numbers until once again hawk genes start to prosper, and so on. However, it need not be an oscillation like this. There is a stable ratio of hawks to doves. For the particular arbitrary points system we are using, the stable ratio, if you work it out, turns out to be $\frac{5}{12}$ doves to $\frac{7}{12}$ hawks. When this stable ratio is reached, the average pay-off for hawks is exactly equal to the average pay-off for doves. Therefore selection does not favour either one of them over the other. If the number of hawks in the population started to drift upwards so that the ratio was no longer $\frac{7}{12}$, doves would start to gain an extra advantage, and the ratio would swing back to the stable state. Just as we shall find the stable sex ratio to be 50:50, so the stable hawk to dove ratio in this hypothetical example is 7:5. In either case, if there are oscillations about the stable point, they need not be very large ones.

Superficially, this sounds a little like group selection, but it is really nothing of the kind. It sounds like group selection because it enables us to think of a population as having a stable equilibrium to which it tends to return when disturbed. But the ESS is a much more subtle concept than group selection. It has nothing to do with some groups being more successful than others. This can be nicely illustrated using the arbitrary points system of our hypothetical example. The average pay-off to an individual in a stable population consisting of $\frac{7}{12}$ hawks and $\frac{5}{12}$ doves turns out to be $6\frac{1}{4}$. This is true whether the individual is a hawk or a dove. Now $6\frac{1}{4}$ is much less than the average pay-off for a dove in a population of doves (15). If *only* everybody would agree to be a dove, every single individual would benefit. By simple group selection, any group in which all individuals mutually agree to be doves would be far more successful than a rival group sitting at the ESS ratio. (As a matter of fact, a conspiracy of nothing but doves is not quite the most successful possible group. In a group consisting of $\frac{1}{6}$ hawks and $\frac{5}{6}$ doves, the average pay-off per contest is $16\frac{2}{3}$. This is the most successful possible conspiracy, but for present purposes we can ignore it. A simpler all-dove conspiracy, with its average pay-off for each individual of 15, is far better for every single individual than the ESS would be.) Group selection theory would therefore predict a tendency to evolve towards an all-dove conspiracy, since a *group* that contained a $\frac{7}{12}$ proportion of hawks would be less successful. But the trouble with conspiracies, even those that are to everybody's advantage in the long run, is that they are open to abuse. It is true that everybody does better in an all-dove group than he would in an ESS group. But unfortunately, in conspiracies of doves, a single hawk does so extremely well that nothing could stop the evolution of hawks. The conspiracy is therefore bound to be broken by treachery from within. An ESS is stable, not because it is particularly good for the individuals participating in it, but simply because it is immune to treachery from within.

It is possible for humans to enter into pacts or conspiracies that are to every individual's advantage, even if these are not stable in the ESS sense. But this is only possible because every individual uses his *conscious* foresight, and is able to see that it is in his own long-term interests to obey the rules of the pact. Even in human pacts there is a constant danger that individuals will stand to gain so much in the *short term* by breaking the pact that the temptation to do so will be overwhelming. Perhaps the best example of this is price-fixing. It is in the long-term interests of all individual garage owners to standardize the price of petrol at some artificially high value. Price rings, based on conscious estimation of long-term best interests, can survive for quite long periods. Every so often, however, an individual gives in to the temptation to make a quick killing by cutting his prices. Immediately, his neighbours follow suit, and a wave of price cutting spreads over the country. Unfortunately for the rest of us, the conscious foresight of the garage owners then reasserts itself, and they enter into a new price-fixing pact. So, even in man, a species with the gift of conscious foresight, pacts or conspiracies based on long-term best interests teeter constantly on the brink of collapse due to treachery from within. In wild animals, controlled by the struggling genes, it is even more difficult to see ways in which group benefit or conspiracy strategies could possibly evolve. We must expect to find evolutionarily stable strategies everywhere.

In our hypothetical example we made the simple assumption that any one individual was either a hawk or a dove. We ended up with an evolutionarily stable ratio of hawks to doves. In practice, what this means is that a stable ratio of hawk genes to dove genes would be achieved in the gene pool. The genetic technical term for this state is stable polymorphism. As far as the maths are concerned, an exactly equivalent ESS can be achieved without polymorphism as follows. If *every individual* is capable of behaving either like a hawk or like a dove in each particular contest, an ESS can be achieved in which all individuals have the same *probability* of behaving like a hawk, namely $\frac{7}{12}$ in our particular example. In practice this would mean that each individual enters each contest having made a random decision whether to behave on this occasion like a hawk or like a dove; random, but with a 7:5 bias in favour of hawk. It is very important that the decisions, although biased towards hawk, should be random in the sense that a rival has no way of guessing how his opponent is going to behave in any particular contest. It is no good, for instance, playing hawk seven fights in a row, then dove five fights in a row and so on. If any individual adopted such a simple sequence, his rivals would quickly catch on and take advantage. The way to take advantage of a simple sequence strategist is to play hawk against him only when you know he is going to play dove.

The hawk and dove story is, of course, naïvely simple. It is a 'model', something that does not really happen in nature, but which helps us to understand things that do happen in nature. Models can be very simple, like this one, and still be useful for understanding a point, or getting an idea. Simple models can be elaborated and gradually made more complex. If all goes well, as they get more complex they come to resemble the real world more. One way in which we can begin to develop the hawk and dove model is to introduce some more strategies. Hawk and dove are not the only possibilities. A more complex strategy which Maynard Smith and Price introduced is called *Retaliator*.

A retaliator plays like a dove at the beginning of every fight. That is, he does not mount an all-out savage attack like a hawk, but has a conventional threatening match. If his opponent attacks him, however, he retaliates. In other words, a retaliator behaves like a hawk when he is attacked by a hawk, and like a dove when he meets a dove. When he meets another retaliator he plays like a dove. A retaliator is a *conditional strategist*. His behaviour depends on the behaviour of his opponent.

Another conditional strategist is called *Bully*. A bully goes around behaving like a hawk until somebody hits back. Then he immediately runs away. Yet another conditional strategist is *Prober–retaliator*. A prober–retaliator is basically like a retaliator, but he occasionally tries a brief experimental escalation of the contest. He persists in this hawk-like behaviour if his opponent does not fight back. If, on the other hand, his opponent does fight back he reverts to conventional threatening like a dove. If he is attacked, he retaliates just like an ordinary retaliator.

If all the five strategies I have mentioned are turned loose upon one another in a computer simulation, only one of them, retaliator, emerges as evolutionarily stable.* Prober–retaliator is nearly stable. Dove is not stable, because a population of doves would be invaded by hawks and bullies. Hawk is not stable, because a population of hawks would be invaded by doves and bullies. Bully is not stable, because a population of bullies would be invaded by hawks. In a population of retaliators, no other strategy would invade, since there is no other strategy that does better than retaliator itself. However, dove does equally well in a population of retaliators. This means that, other things being equal, the numbers of doves could slowly drift upwards. Now if the numbers of doves drifted up to any significant extent, prober–retaliators (and, incidentally, hawks and bullies) would start to have an advantage, since they do better against doves than retaliators do. Prober–retaliator itself, unlike hawk and bully, is almost an ESS, in the sense that, in a population of prober–retaliators, only one other strategy, retaliator, does better, and then only slightly. We might expect, therefore, that a mixture of retaliators and prober–retaliators would tend to predominate, with perhaps even a gentle oscillation between the two, in association with an oscillation in the size of a small dove minority. Once again, we don't have to think in terms of a polymorphism in which every individual always plays one strategy or another. Each individual could play a complex mixture between retaliator, prober–retaliator, and dove.

This theoretical conclusion is not far from what actually happens in most wild animals. We have in a sense explained the 'gloved fist' aspect of animal aggression. Of course the details depend on the exact numbers of 'points' awarded for winning, being injured, wasting time, and so on. In elephant seals the prize for winning may be near-monopoly rights over a large harem of females. The pay-off for winning must therefore be rated as very high. Small wonder that fights are vicious and the probability of serious injury is also high. The cost of wasting time should presumably be regarded as small in comparison with the cost of being injured and the benefit of winning. For a small bird in a cold climate, on the other hand, the cost of wasting time may be paramount. A great tit when feeding nestlings needs to catch an average of one prey per thirty seconds. Every second of daylight is precious. Even the comparatively short time wasted in a hawk/hawk fight should perhaps be regarded as more serious than the risk of injury to such a bird. Unfortunately, we know too little at present to assign realistic numbers to the costs and benefits of various outcomes in nature.* We must be careful not to draw conclusions that result simply from our own arbitrary choice of numbers. The general conclusions which are important are that ESSs will tend to evolve, that an ESS is not the same as the optimum that could be achieved by a group conspiracy, and that common sense can be misleading.

Another kind of war game that Maynard Smith has considered is the 'war of attrition'. This can be thought of as arising in a species that never engages in dangerous combat, perhaps a well-armoured species in which injury is very unlikely. All disputes in this species are settled by conventional posturing. A contest always ends in one rival or the other backing down. To win, all you have to do is stand your ground and glare at the opponent until he finally turns tail. Obviously no animal can afford to spend infinite time threatening; there are important things to be done elsewhere. The resource he is competing for may be valuable, but it is not infinitely valuable. It is only worth so much time and, as at an auction sale, each individual is prepared to spend only so much on it. Time is the currency of this two-bidder auction.

Suppose all such individuals worked out in advance exactly how much time they thought a particular kind of resource, say a female, was worth. A mutant individual who was prepared to go on just a little bit longer would always win. So the strategy of maintaining a fixed bidding limit is unstable. Even if the value of the resource can be very finely estimated, and all individuals bid exactly the right value, the strategy is unstable. Any two individuals bidding according to this maximum strategy would give up at exactly the same instant, and neither would get the resource! It would then pay an individual to give up right at the start rather than waste any time in contests at all. The important difference between the war of attrition and a real auction sale is, after all, that in the war of attrition *both* contestants pay the price but only one of them gets the goods. In a population of maximum bidders, therefore, a strategy of giving up at the beginning would be successful and would spread through the population. As a consequence of this some benefit would start to accrue to individuals who did not give up immediately, but waited for a few seconds before giving up. This strategy would pay when played against the immediate retreaters who now predominate in the population. Selection would then favour a progressive extension of the giving-up time until it once more approached the maximum allowed by the true economic worth of the resource under dispute.

Once again, by using words, we have talked ourselves into picturing an oscillation in a population. Once again, mathematical analysis shows that this is not correct. There is an evolutionarily stable strategy, which can be expressed as a mathematical formula, but in words what it amounts to is this. Each individual goes on for an *unpredictable* time. Unpredictable on any particular occasion, that is, but averaging the true value of the resource. For example, suppose the resource is really worth five minutes of display. At the ESS, any particular individual may go on for more than five minutes or he may go on for less than five minutes, or he may even go on for exactly five minutes. The important thing is that his opponent has no way of knowing how long he is prepared to persist on this particular occasion.

Obviously, it is vitally important in the war of attrition that individuals should give no inkling of when they are going to give up. Anybody who betrayed, by the merest flicker of a whisker, that he was beginning to think of throwing in the sponge, would be at an instant disadvantage. If, say, whisker-flickering happened to be a reliable sign that retreat would follow within one minute, there would be a very simple winning strategy: 'If your opponent's whiskers flicker, wait one more minute, regardless of what your own previous plans for giving up might have been. If your opponent's whiskers have not yet flickered, and you are within one minute of the time when you intend to give up anyway, give up immediately and don't waste any more time. Never flicker your own whiskers.' So natural selection would quickly penalize whisker-flickering and any analogous betrayals of future behaviour. The poker face would evolve.

Why the poker face rather than out-and-out lies? Once again, because lying is not stable. Suppose it happened to be the case that the majority of individuals raised their hackles only when they were truly intending to go on for a very long time in the war of attrition. The obvious counterploy would evolve: individuals would give up immediately when an opponent raised his hackles. But now, liars might start to evolve. Individuals who really had no intention of going on for a long time would raise their hackles on every occasion, and reap the benefits of easy and quick victory. So liar genes would spread. When liars became the majority, selection would now favour individuals who called their bluff. Therefore liars would decrease in numbers again. In the war of attrition, telling lies is no more evolutionarily stable than telling the truth. The poker face is evolutionarily stable. Surrender, when it finally comes, will be sudden and unpredictable.

So far we have considered only what Maynard Smith calls 'symmetric' contests. This means we have assumed that the contestants are identical in all respects except their fighting strategy. Hawks and doves are assumed to be equally strong, to be equally well endowed with weapons and with armour, and to have an equal amount to gain from winning. This is a convenient assumption to make for a model, but it is not very realistic. Parker and Maynard Smith went on to consider asymmetric contests. For example, if individuals vary in size and fighting ability, and each individual is capable of gauging a rival's size in comparison to his own, does this affect the ESS that emerges? It most certainly does.

There seem to be three main sorts of asymmetry. The first we have just met: individuals may differ in their size or fighting equipment. Secondly, individuals may differ in how much they have to gain from winning. For instance an old male, who has not long to live anyway, might have less to lose if he is injured than a young male with the bulk of his reproductive life ahead of him.

Thirdly, it is a strange consequence of the theory that a purely arbitrary, apparently irrelevant, asymmetry can give rise to an ESS, since it can be used to settle contests quickly. For instance it will usually be the case that one contestant happens to arrive at the location of the contest earlier than the other. Call them 'resident' and 'intruder' respectively. For the sake of argument, I am assuming that there is no general advantage attached to being a resident or an intruder. As we shall see, there are practical reasons why this assumption may not be true, but that is not the point. The point is that even if there were no general reason to suppose that residents have an advantage over intruders, an ESS depending on the asymmetry itself would be likely to evolve. A simple analogy is to humans who settle a dispute quickly and without fuss by tossing a coin.

The conditional strategy 'If you are the resident, attack; if you are the intruder, retreat' could be an ESS. Since the asymmetry is assumed to be arbitrary, the opposite strategy 'If resident, retreat; if intruder, attack' could also be stable. Which of the two ESSs is adopted in a particular population would depend on which one happens to reach a majority first. Once a majority of individuals is playing one of these two conditional strategies, deviants from it are penalized. Hence, by definition, it is an ESS.

For instance, suppose all individuals are playing 'resident wins, intruder runs away'. This means they will win half their fights and lose half their fights. They will never be injured and they will never waste time, since all disputes are instantly settled by arbitrary convention. Now consider a new mutant rebel. Suppose he plays a pure hawk strategy, always attacking and never retreating. He will win when his opponent is an intruder. When his opponent is a resident he will run a grave risk of injury. On average he will have a lower pay-off than individuals playing according to the arbitrary rules of the ESS. A rebel who tries the reverse convention 'if resident run away, if intruder attack', will do even worse. Not only will he frequently be injured, he will also seldom win a contest. Suppose, though, that by some chance events individuals playing this reverse convention managed to become the majority. In this case their strategy would then become the stable norm, and deviation from it would be penalized. Conceivably, if we watched a population for many generations we would see a series of occasional flips from one stable state to the other.

However, in real life, truly arbitrary asymmetries probably do not exist. For instance, residents probably tend to have a practical advantage over intruders. They have better knowledge of local terrain. An intruder is perhaps more likely to be out of breath because he moved into the battle area, whereas the resident was there all the time. There is a more abstract reason why, of the two stable states, the 'resident wins, intruder retreats' one is the more probable in nature. This is that the reverse strategy, 'intruder wins, resident retreats' has an inherent tendency to self-destruction—it is what Maynard Smith would call a paradoxical strategy. In any population sitting at this paradoxical ESS, individuals would always be striving never to be caught as residents: they would always be trying to be the intruder in any encounter. They could only achieve this by ceaseless, and otherwise pointless, moving around! Quite apart from the costs in time and energy that would be incurred, this evolutionary trend would, of itself, tend to lead to the category 'resident' ceasing to exist. In a population sitting at the other stable state, 'resident wins, intruder retreats', natural selection would favour individuals who strove to be residents. For each individual, this would mean holding on to a particular piece of ground, leaving it as little as possible, and appearing to 'defend' it. As is now well known, such behaviour is commonly observed in nature, and goes by the name of 'territorial defence'.

The neatest demonstration I know of this form of behavioural asymmetry was provided by the great ethologist Niko Tinbergen, in an experiment of characteristically ingenious simplicity.* He had a fish-tank containing two male sticklebacks. The males had each built nests, at opposite ends of the tank, and each 'defended' the territory around his own nest. Tinbergen placed each of the two males in a large glass test tube, and he held the two tubes next to each other and watched the males trying to fight each other through the glass. Now comes the interesting result. When he moved the two tubes into the vicinity of male *A*'s nest, male *A* assumed an attacking posture, and male *B* attempted to retreat. But when he moved the two tubes into male *B*'s territory, the tables were turned. By simply moving the two tubes from one end of the tank to the other, Tinbergen was able to dictate which male attacked and which retreated. Both males were evidently playing the simple conditional strategy: 'if resident, attack; if intruder, retreat'.

Biologists often ask what the biological 'advantages' of territorial behaviour are. Numerous suggestions have been made, some of which will be mentioned later. But we can now see that the very question may be superfluous. Territorial 'defence' may simply be an ESS which arises because of the asymmetry in time of arrival that usually characterizes the relationship between two individuals and a patch of ground.

Presumably the most important kind of non-arbitrary asymmetry is in size and general fighting ability. Large size is not necessarily always the most important quality needed to win fights, but it is probably one of them. If the larger of two fighters always wins, and if each individual knows for certain whether he is larger or smaller than his opponent, only one strategy makes any sense: 'If your opponent is larger than you, run away. Pick fights with people smaller than you are.' Things are a bit more complicated if the importance of size is less certain. If large size confers only a slight advantage, the strategy I have just mentioned is still stable. But if the risk of injury is serious there may also be a second, 'paradoxical strategy'. This is: 'Pick fights with people larger than you are and run away from people smaller than you are'! It is obvious why this is called paradoxical. It seems completely counter to common sense. The reason it can be stable is this. In a population consisting entirely of paradoxical strategists, nobody ever gets hurt. This is because in every contest one of the participants, the larger, always runs away. A mutant of average size who plays the 'sensible' strategy of picking on smaller opponents is involved in a seriously escalated fight with half the people he meets. This is because, if he meets somebody smaller than him, he attacks; the smaller individual fights back fiercely, because he is playing paradoxical; although the sensible strategist is more likely to win than the paradoxical one, he still runs a substantial risk of losing and of being seriously injured. Since the majority of the population are paradoxical, a sensible strategist is more likely to be injured than any single paradoxical strategist.

Even though a paradoxical strategy can be stable, it is probably only of academic interest. Paradoxical fighters will only have a higher average pay-off if they very heavily out-number sensible ones. It is hard to imagine how this state of affairs could ever arise in the first place. Even if it did, the ratio of sensibles to paradoxicals in the population only has to drift a little way towards the sensible side before reaching the 'zone of attraction' of the other ESS, the sensible one. The zone of attraction is the set of population ratios at which, in this case, sensible strategists have the advantage: once a population reaches this zone, it will be sucked inevitably towards the sensible stable point. It would be exciting to find an example of a paradoxical ESS in nature, but I doubt if we can really hope to do so. (I spoke too soon. After I had written this last sentence, Professor Maynard Smith called my attention to the following description of the behaviour of the Mexican social spider, *Oecobius civitas*, by J. W. Burgess: 'If a spider is disturbed and driven out of its retreat, it darts across the rock and, in the absence of a vacant crevice to hide in, may seek refuge in the hiding place of another spider of the same species. If the other spider is in residence when the intruder enters, it does not attack but darts out and seeks a new refuge of its own. Thus once the first spider is disturbed the process of sequential displacement from web to web may continue for several seconds, often causing a majority of the spiders in the aggregation to shift from their home refuge to an alien one' ('Social Spiders', *Scientific American*, March 1976). This is paradoxical in the sense of page 103.)*

What if individuals retain some memory of the outcome of past fights? This depends on whether the memory is specific or general. Crickets have a general memory of what happened in past fights. A cricket that has recently won a large number of fights becomes more hawkish. A cricket that has recently had a losing streak becomes more dovish. This was neatly shown by R. D. Alexander. He used a model cricket to beat up real crickets. After this treatment the real crickets became more likely to lose fights against other real crickets. Each cricket can be thought of as constantly updating his own estimate of his fighting ability, relative to that of an average individual in his population. If animals such as crickets, who work with a general memory of past fights, are kept together in a closed group for a time, a kind of dominance hierarchy is likely to develop.* An observer can rank the individuals in order. Individuals lower in the order tend to give in to individuals higher in the order. There is no need to suppose that the individuals recognize each other. All that happens is that individuals who are accustomed to winning become even more likely to win, while individuals who are accustomed to losing become steadily more likely to lose. Even if the individuals started by winning or losing entirely at random, they would tend to sort themselves out into a rank order. This incidentally has the effect that the number of serious fights in the group gradually dies down.

I have to use the phrase 'kind of dominance hierarchy', because many people reserve the term dominance hierarchy for cases in which individual recognition is involved. In these cases, memory of past fights is specific rather than general. Crickets do not recognize each other as individuals, but hens and monkeys do. If you are a monkey, a monkey who has beaten you in the past is likely to beat you in the future. The best strategy for an individual is to be relatively dovish towards an individual who has previously beaten him. If a batch of hens who have never met before are introduced to each other, there is usually a great deal of fighting. After a time the fighting dies down. Not for the same reason as in the crickets, though. In the case of the hens it is because each individual 'learns her place' relative to each other individual. This is incidentally good for the group as a whole. As an indicator of this it has been noticed that in established groups of hens, where fierce fighting is rare, egg production is higher than in groups of hens whose membership is continually being changed, and in which fights are consequently more frequent. Biologists often speak of the biological advantage or 'function' of dominance hierarchies as being to reduce overt aggression in the group. However, this is the wrong way to put it. A dominance hierarchy per se cannot be said to have a 'function' in the evolutionary sense, since it is a property of a group, not of an individual. The individual behaviour patterns that manifest themselves in the form of dominance hierarchies when viewed at the group level may be said to have functions. It is, however, even better to abandon the word 'function' altogether, and to think about the matter in terms of ESSs in asymmetric contests where there is individual recognition and memory.

We have been thinking of contests between members of the same species. What about inter-specific contests? As we saw earlier, members of different species are less direct competitors than members of the same species. For this reason we should expect fewer disputes between them over resources, and our expectation is borne out. For instance, robins defend territories against other robins, but not against great tits. One can draw a map of the territories of different individual robins in a wood and one can superimpose a map of the territories of individual great tits. The territories of the two species overlap in an entirely indiscriminate way. They might as well be on different planets.

But there are other ways in which the interests of individuals from different species conflict very sharply. For instance a lion wants to eat an antelope's body, but the antelope has very different plans for its body. This is not normally regarded as competition for a resource, but logically it is hard to see why not. The resource in question is meat. The lion genes 'want' the meat as food for their survival machine. The antelope genes want the meat as working muscle and organs for their survival machine. These two uses for the meat are mutually incompatible; therefore there is conflict of interest.

Members of one's own species are made of meat too. Why is cannibalism relatively rare? As we saw in the case of black-headed gulls, adults do sometimes eat the young of their own species. Yet adult carnivores are never to be seen actively pursuing other adults of their own species with a view to eating them. Why not? We are still so used to thinking in terms of the 'good of the species' view of evolution that we often forget to ask perfectly reasonable questions like: Why don't lions hunt other lions? Another good question of a type which is seldom asked is: Why do antelopes run away from lions instead of hitting back?

The reason lions do not hunt lions is that it would not be an ESS for them to do so. A cannibal strategy would be unstable for the same reason as the hawk strategy in the earlier example. There is too much danger of retaliation. This is less likely to be true in contests between members of different species, which is why so many prey animals run away instead of retaliating. It probably stems originally from the fact that in an interaction between two animals of different species there is a built-in asymmetry which is greater than that between members of the same species. Whenever there is strong asymmetry in a contest, ESSs are likely to be conditional strategies dependent on the asymmetry. Strategies analogous to 'if smaller, run away; if larger, attack' are very likely to evolve in contests between members of different species because there are so many available asymmetries. Lions and antelopes have reached a kind of stability by evolutionary divergence, which has accentuated the original asymmetry of the contest in an ever-increasing fashion. They have become highly proficient in the arts of, respectively, chasing, and running away. A mutant antelope that adopted a 'stand and fight' strategy against lions would be less successful than rival antelopes disappearing over the horizon.

I have a hunch that we may come to look back on the invention of the ESS concept as one of the most important advances in evolutionary theory since Darwin.* It is applicable wherever we find conflict of interest, and that means almost everywhere. Students of animal behaviour have got into the habit of talking about something called 'social organization'. Too often the social organization of a species is treated as an entity in its own right, with its own biological 'advantage'. An example I have already given is that of the 'dominance hierarchy'. I believe it is possible to discern hidden group selectionist assumptions lying behind a large number of the statements that biologists make about social organization. Maynard Smith's concept of the ESS will enable us, for the first time, to see clearly how a collection of independent selfish entities can come to resemble a single organized whole. I think this will be true not only of social organizations within species, but also of 'ecosystems' and 'communities' consisting of many species. In the long term, I expect the ESS concept to revolutionize the science of ecology.

We can also apply it to a matter that was deferred from Chapter 3, arising from the analogy of oarsmen in a boat (representing genes in a body) needing a good team spirit. Genes are selected, not as 'good' in isolation, but as good at working against the background of the other genes in the gene pool. A good gene must be compatible with, and complementary to, the other genes with whom it has to share a long succession of bodies. A gene for plant-grinding teeth is a good gene in the gene pool of a herbivorous species, but a bad gene in the gene pool of a carnivorous species.

It is possible to imagine a compatible combination of genes as being selected together *as a unit*. In the case of the butterfly mimicry example of Chapter 3, this seems to be exactly what happened. But the power of the ESS concept is that it can now enable us to see how the same kind of result could be achieved by selection purely at the level of the independent gene. The genes do not have to be linked on the same chromosome.

The rowing analogy is really not up to explaining this idea. The nearest we can come to it is this. Suppose it is important in a really successful crew that the rowers should coordinate their activities by means of speech. Suppose further that, in the pool of oarsmen at the coach's disposal, some speak only English and some speak only German. The English are not consistently better or worse rowers than the Germans. But because of the importance of communication, a mixed crew will tend to win fewer races than either a pure English crew or a pure German crew.

The coach does not realize this. All he does is shuffle his men around, giving credit points to individuals in winning boats, marking down individuals in losing boats. Now if the pool available to him just happens to be dominated by Englishmen it follows that any German who gets into a boat is likely to cause it to lose, because communications break down. Conversely, if the pool happened to be dominated by Germans, an Englishman would tend to cause any boat in which he found himself to lose. What will emerge as the overall best crew will be one of the two stable states—pure English or pure German, but not mixed. Superficially it looks as though the coach is selecting whole language groups *as units*. This is not what he is doing. He is selecting individual oarsmen for their apparent ability to win races. It so happens that the tendency for an individual to win races depends on which other individuals are present in the pool of candidates. Minority candidates are automatically penalized, not because they are bad rowers, but simply because they are minority candidates. Similarly, the fact that genes are selected for mutual compatibility does not necessarily mean we *have* to think of groups of genes as being selected as units, as they were in the case of the butterflies. Selection at the low level of the single gene can give the impression of selection at some higher level.

In this example, selection favours simple conformity. More interestingly, genes may be selected because they complement each other. In terms of the analogy, suppose an ideally balanced crew would consist of four right-handers and four left-handers. Once again assume that the coach, unaware of this fact, selects blindly on 'merit'. Now if the pool of candidates happens to be dominated by right-handers, any individual left-hander will tend to be at an advantage: he is likely to cause any boat in which he finds himself to win, and he will therefore appear to be a good oarsman. Conversely, in a pool dominated by left-handers, a right-hander would have an advantage. This is similar to the case of a hawk doing well in a population of doves, and a dove doing well in a population of hawks. The difference is that there we were talking about interactions between individual bodies—selfish machines—whereas here we are talking, by analogy, about interactions between genes within bodies.

The coach's blind selection of 'good' oarsmen will lead in the end to an ideal crew consisting of four left-handers and four right-handers. It will look as though he selected them all together as a complete, balanced unit. I find it more parsimonious to think of him as selecting at a lower level, the level of the independent candidates. The evolutionarily stable state ('strategy' is misleading in this context) of four left-handers and four right-handers will emerge simply as a consequence of low-level selection on the basis of apparent merit.

The gene pool is the long-term environment of the gene. 'Good' genes are blindly selected as those that survive in the gene pool. This is not a theory; it is not even an observed fact: it is a tautology. The interesting question is: What makes a gene good? As a first approximation I said that what makes a gene good is the ability to build efficient survival machines—bodies. We must now amend that statement. The gene pool will become an *evolutionarily stable set* of genes, defined as a gene pool that cannot be invaded by any new gene. Most new genes that arise, either by mutation or reassortment or immigration, are quickly penalized by natural selection: the evolutionarily stable set is restored. Occasionally a new gene does succeed in invading the set: it succeeds in spreading through the gene pool. There is a transitional period of instability, terminating in a new evolutionarily stable set—a little bit of evolution has occurred. By analogy with the aggression strategies, a population might have more than one alternative stable point, and it might occasionally flip from one to another. Progressive evolution may be not so much a steady upward climb as a series of discrete steps from stable plateau to stable plateau.* It may look as though the population as a whole is behaving like a single self-regulating unit. But this illusion is produced by selection going on at the level of the single gene. Genes are selected on 'merit'. But merit is judged on the basis of performance against the background of the evolutionarily stable set which is the current gene pool.

By focussing on aggressive interactions between whole individuals, Maynard Smith was able to make things very clear. It is easy to think of stable ratios of hawk bodies and dove bodies, because bodies are large things which we can see. But such interactions between genes sitting in *different* bodies are only the tip of the iceberg. The vast majority of significant interactions between genes in the evolutionarily stable set—the gene pool—go on *within* individual bodies. These interactions are difficult to see, for they take place within cells, notably the cells of developing embryos. Well-integrated bodies exist because they are the product of an evolutionarily stable set of selfish genes.

But I must return to the level of interactions between whole animals which is the main subject of this book. For understanding aggression it was convenient to treat individual animals as independent selfish machines. This model breaks down when the individuals concerned are close relatives—brothers and sisters, cousins, parents and children. This is because relatives share a substantial proportion of their genes. Each selfish gene therefore has its loyalties divided between different bodies. This is explained in the next chapter.

GENESMANSHIP

What is the selfish gene? It is not just one single physical bit of DNA. Just as in the primeval soup, it is *all replicas* of a particular bit of DNA, distributed throughout the world. If we allow ourselves the licence of talking about genes as if they had conscious aims, always reassuring ourselves that we could translate our sloppy language back into respectable terms if we wanted to, we can ask the question: What is a single selfish gene trying to do? It is trying to get more numerous in the gene pool. Basically it does this by helping to program the bodies in which it finds itself to survive and to reproduce. But now we are emphasizing that 'it' is a distributed agency, existing in many different individuals at once. The key point of this chapter is that a gene might be able to assist *replicas* of itself that are sitting in other bodies. If so, this would appear as individual altruism but it would be brought about by gene selfishness.

Consider the gene for being an albino in man. In fact several genes exist that can give rise to albinism, but I am talking about just one of them. It is recessive; that is, it has to be present in double dose in order for the person to be an albino. This is true of about 1 in 20,000 of us. But it is also present, in single dose, in about 1 in 70 of us, and these individuals are not albinos. Since it is distributed in many individuals, a gene such as the albino gene could, in theory, assist its own survival in the gene pool by pro-gramming its bodies to behave altruistically towards other albino bodies, since these are known to contain the same gene. The albino gene should be quite happy if some of the bodies that it inhabits die, provided that in doing so they help other bodies containing the same gene to survive. If the albino gene could make one of its bodies save the lives of ten albino bodies, then even the death of the altruist is amply compensated by the increased numbers of albino genes in the gene pool.

Should we then expect albinos to be especially nice to each other? Actually the answer is probably no. In order to see why not, we must temporarily abandon our metaphor of the gene as a conscious agent, because in this context it becomes posi-tively misleading. We must translate back into respectable, if more longwinded terms. Albino genes do not really 'want' to survive or to help other albino genes. But if the albino gene just happened to cause its bodies to behave altruistically towards other albinos, then automatically, willy-nilly, it would tend to become more numerous in the gene pool as a result. But, in order for this to happen, the gene would have to have two independent effects on bodies. Not only must it confer its usual effect of a very pale complexion. It must also confer a tendency to be selectively altruistic towards individuals with a very pale complexion. Such a double-effect gene could, if it existed, be very successful in the population.

Now it is true that genes do have multiple effects, as I empha-sized in Chapter 3. It is theoretically possible that a gene could arise which conferred an externally visible 'label', say a pale skin, or a green beard, or anything conspicuous, and also a tendency to be specially nice to bearers of that conspicuous label. It is possible, but not particularly likely. Green beardedness is just as likely to be linked to a tendency to develop ingrowing toenails or any other trait, and a fondness for green beards is just as likely to go together with an inability to smell freesias. It is not very probable that one and the same gene would produce both the right label and the right sort of altruism. Nevertheless, what may be called the Green Beard Altruism Effect is a theoretical possibility.

An arbitrary label like a green beard is just one way in which a gene might 'recognize' copies of itself in other individuals. Are there any other ways? A particularly direct possible way is the following. The possessor of an altruistic gene might be recognized simply by the fact that he does altruistic acts. A gene could prosper in the gene pool if it 'said' the equivalent of: 'Body, if A is drowning as a result of trying to save someone else from drowning, jump in and rescue A.' The reason such a gene could do well is that there is a greater than average chance that A contains the same life-saving altruistic gene. The fact that A is seen to be trying to rescue somebody else is a label, equivalent to a green beard. It is less arbitrary than a green beard, but it still seems rather implausible. Are there any plausible ways in which genes might 'recognize' their copies in other individuals?

The answer is yes. It is easy to show that *close relatives*—kin— have a greater than average chance of sharing genes. It has long been clear that this must be why altruism by parents towards their young is so common. What R. A. Fisher, J. B. S. Haldane, and especially W. D. Hamilton realized was that the same applies to other close relations—brothers and sisters, nephews and nieces, close cousins. If an individual dies in order to save ten close relatives, one copy of the kin-altruism gene may be lost, but a larger number of copies of the same gene is saved.

'A larger number' is a bit vague. So is 'close relatives'. We can do better than that, as Hamilton showed. His two papers of 1964 are among the most important contributions to social ethology ever written, and I have never been able to understand why they have been so neglected by ethologists (his name does not even appear in the index of two major text-books of ethology, both published in 1970).* Fortunately there are recent signs of a revival of interest in his ideas. Hamilton's papers are rather mathematical, but it is easy to grasp the basic principles intuitively, without rigorous mathematics, though at the cost of some over-simplification. The thing we want to calculate is the probability, or odds, that two individuals, say two sisters, share a particular gene.

For simplicity I shall assume that we are talking about genes that are rare in the gene pool as a whole.* Most people share 'the gene for not being an albino', whether they are related to each other or not. The reason this gene is so common is that in nature albinos are less likely to survive than non-albinos because, for example, the sun dazzles them and makes them relatively unlikely to see an approaching predator. We are not concerned with explaining the prevalence in the gene pool of such obviously 'good' genes as the gene for not being an albino. We are interested in explaining the success of genes specifically as a result of their altruism. We can therefore assume that, at least in the early stages of this process of evolution, these genes are rare. Now the important point is that even a gene that is rare in the population as a whole is common within a family. I contain a number of genes that are rare in the population as a whole, and you also contain genes that are rare in the population as a whole. The chance that we both contain the same rare genes is very small indeed. But the chances are good that my sister contains a particular rare gene that I contain, and the chances are equally good that your sister contains a rare gene in common with you. The odds are in this case exactly 50 per cent, and it is easy to explain why.

Suppose you contain one copy of the gene G. You must have received it either from your father or from your mother (for convenience we can neglect various infrequent possibilities—

that G is a new mutation, that both your parents had it, or that either of your parents had two copies of it). Suppose it was your father who gave you the gene. Then every one of his ordinary body cells contained one copy of G. Now you will remember that when a man makes a sperm he doles out half his genes to it. There is therefore a 50 per cent chance that the sperm that begot your sister received the gene G. If, on the other hand, you received G from your mother, exactly parallel reasoning shows that half of her eggs must have contained G; once again, the chances are 50 per cent that your sister contains G. This means that if you had 100 brothers and sisters, approximately 50 of them would contain any particular rare gene that you contain. It also means that if you have 100 rare genes, approximately 50 of them are in the body of any one of your brothers or sisters.

You can do the same kind of calculation for any degree of kinship you like. An important relationship is that between parent and child. If you have one copy of gene H, the chance that any particular one of your children has it is 50 per cent, because half your sex cells contain H, and any particular child was made from one of those sex cells. If you have one copy of gene J, the chance that your father also had J is 50 per cent, because you received half your genes from him, and half from your mother. For convenience we use an index of *relatedness*, which expresses the chance of a gene being shared between two relatives. The relatedness between two brothers is $\frac{1}{2}$, since half the genes possessed by one brother will be found in the other. This is an average figure: by the luck of the meiotic draw, it is possible for particular pairs of brothers to share more or fewer genes than this. The relatedness between parent and child is always exactly $\frac{1}{2}$.

It is rather tedious going through the calculations from first principles every time, so here is a rough and ready rule for working out the relatedness between any two individuals A and B.

You may find it useful in making your will, or in interpreting apparent resemblances in your own family. It works for all simple cases, but breaks down where incestuous mating occurs, and in certain insects, as we shall see.

First identify all the *common ancestors* of A and B. For instance, the common ancestors of a pair of first cousins are their shared grandfather and grandmother. Once you have found a common ancestor, it is of course logically true that all his ancestors are common to A and B as well. However, we ignore all but the most recent common ancestors. In this sense, first cousins have only two common ancestors. If B is a lineal descendant of A, for instance his great grandson, then A himself is the 'common ancestor' we are looking for.

Having located the common ancestor(s) of A and B, count the *generation distance* as follows. Starting at A, climb up the family tree until you hit a common ancestor, and then climb down again to B. The total number of steps up the tree and then down again is the generation distance. For instance, if A is B's uncle, the generation distance is 3. The common ancestor is A's father (say) and B's grandfather. Starting at A you have to climb up one generation in order to hit the common ancestor. Then to get down to B you have to descend two generations on the other side. Therefore the generation distance is $1 + 2 = 3$.

Having found the generation distance between A and B via a particular common ancestor, calculate that part of their relatedness for which that ancestor is responsible. To do this, multiply $\frac{1}{2}$ by itself once for each step of the generation distance. If the generation distance is 3, this means calculate $\frac{1}{2} \times \frac{1}{2} \times \frac{1}{2}$ or $(\frac{1}{2})^3$. If the generation distance via a particular ancestor is equal to g steps, the portion of relatedness due to that ancestor is $(\frac{1}{2})^g$.

But this is only part of the relatedness between A and B. If they have more than one common ancestor we have to add on the equivalent figure for each ancestor. It is usually the case that the generation distance is the same for all common ancestors of a pair of individuals. Therefore, having worked out the relatedness between A and B due to any one of the ancestors, all you have to do in practice is to multiply by the number of ancestors. First cousins, for instance, have two common ancestors, and the generation distance via each one is 4. Therefore their relatedness is $2 \times (\frac{1}{2})^4 = \frac{1}{8}$. If A is B's great-grandchild, the generation distance is 3 and the number of common 'ancestors' is 1 (B himself), so the relatedness is $1 \times (\frac{1}{2})^3 = \frac{1}{8}$. Genetically speaking, your first cousin is equivalent to a great-grandchild. Similarly, you are just as likely to 'take after' your uncle (relatedness $= 2 \times (\frac{1}{2})^3 = \frac{1}{4}$) as after your grandfather (relatedness $= 1 \times (\frac{1}{2})^2 = \frac{1}{4}$).

For relationships as distant as third cousin ($2 \times (\frac{1}{2})^8 = \frac{1}{128}$, we are getting down near the baseline probability that a particular gene possessed by A will be shared by any random individual taken from the population. A third cousin is not far from being equivalent to any old Tom, Dick, or Harry as far as an altruistic gene is concerned. A second cousin (relatedness $= \frac{1}{32}$) is only a little bit special; a first cousin somewhat more so ($\frac{1}{8}$). Full brothers and sisters, and parents and children are very special ($\frac{1}{2}$), and identical twins (relatedness $= 1$) just as special as oneself. Uncles and aunts, nephews and nieces, grandparents and grandchildren, and half brothers and half sisters, are intermediate with a relatedness of $\frac{1}{4}$.

Now we are in a position to talk about genes for kin-altruism much more precisely. A gene for suicidally saving five cousins would not become more numerous in the population, but a gene for saving five brothers or ten first cousins would. The minimum requirement for a suicidal altruistic gene to be successful is that it should save more than two siblings (or children or parents), or more than four half-siblings (or uncles, aunts, nephews, nieces, grandparents, grandchildren), or more than eight first cousins, etc. Such a gene, on average, tends to live on in the bodies of enough individuals saved by the altruist to compensate for the death of the altruist itself.

If an individual could be sure that a particular person was his identical twin, he should be exactly as concerned for his twin's welfare as for his own. Any gene for twin altruism is bound to be carried by both twins; therefore if one dies heroically to save the other the gene lives on. Nine-banded armadillos are born in a litter of identical quadruplets. As far as I know, no feats of heroic self-sacrifice have been reported for young armadillos, but it has been pointed out that some strong altruism is definitely to be expected, and it would be well worth somebody's while going out to South America to have a look.*

We can now see that parental care is just a special case of kin altruism. Genetically speaking, an adult should devote just as much care and attention to its orphaned baby brother as it does to one of its own children. Its relatedness to both infants is exactly the same, $\frac{1}{2}$. In gene selection terms, a gene for big sister altruistic behaviour should have just as good a chance of spreading through the population as a gene for parental altruism. In practice, this is an over-simplification for various reasons which we shall come to later, and brotherly or sisterly care is nothing like so common in nature as parental care. But the point I am making here is that there is nothing special *genetically* speaking about the parent/child relationship as against the brother/sister relationship. The fact that parents actually hand on genes to children, but sisters do not hand on genes to each other is irrelevant, since the sisters both receive identical replicas of the same genes from the same parents.

Some people use the term *kin selection* to distinguish this kind of natural selection from group selection (the differential survival of groups) and individual selection (the differential survival of individuals). Kin selection accounts for within-family altruism; the closer the relationship, the stronger the selection. There is nothing wrong with this term, but unfortunately it may have to be abandoned because of recent gross misuses of it, which are likely to muddle and confuse biologists for years to come. E. O. Wilson, in his otherwise admirable *Sociobiology: The New Synthesis*, defines kin selection as a special case of group selection. He has a diagram which clearly shows that he thinks of it as intermediate between 'individual selection' and 'group selection' in the conventional sense—the sense that I used in Chapter 1. Now group selection—even by Wilson's own definition—means the differential survival of *groups* of individuals. There is, to be sure, a sense in which a family is a special kind of group. But the whole point of Hamilton's argument is that the distinction between family and non-family is not hard and fast, but a matter of mathematical probability. It is no part of Hamilton's theory that animals should behave altruistically towards all 'members of the family', and selfishly to everybody else. There are no definite lines to be drawn between family and non-family. We do not have to decide whether, say, second cousins should count as inside the family group or outside it: we simply expect that second cousins should be $\frac{1}{16}$ as likely to receive altruism as offspring or siblings. Kin selection is emphatically *not* a special case of group selection.* It is a special consequence of gene selection.

There is an even more serious shortcoming in Wilson's definition of kin selection. He deliberately excludes offspring: they don't count as kin!* Now of course he knows perfectly well that offspring are kin to their parents, but he prefers not to invoke the theory of kin selection in order to explain altruistic care by parents of their own offspring. He is, of course, entitled to define a word however he likes, but this is a most confusing definition, and I hope that Wilson will change it in future editions of his justly influential book. Genetically speaking, parental care and brother/sister altruism evolve for exactly the same reason: in both cases there is a good chance that the altruistic gene is present in the body of the beneficiary.

I ask the general reader's indulgence for this little diatribe, and return hastily to the main story. So far, I have over-simplified somewhat, and it is now time to introduce some qualifications. I have talked in elemental terms of suicidal genes for saving the lives of particular numbers of kin of exactly known relatedness. Obviously, in real life, animals cannot be expected to count exactly how many relatives they are saving, nor to perform Hamilton's calculations in their heads even if they had some way of knowing exactly who their brothers and cousins were. In real life, certain suicide and absolute 'saving' of life must be replaced by *statistical risks* of death, one's own and other people's. Even a third cousin may be worth saving, if the risk to yourself is very small. Then again, both you and the relative you are thinking of saving are going to die one day in any case. Every individual has an 'expectation of life' which an actuary could calculate with a certain probability of error. To save the life of a relative who is soon going to die of old age has less of an impact on the gene pool of the future than to save the life of an equally close relative who has the bulk of his life ahead of him.

Our neat symmetrical calculations of relatedness have to be modified by messy actuarial weightings. Grandparents and grandchildren have, genetically speaking, equal reason to behave altruistically to each other, since they share $\frac{1}{4}$ of each other's genes. But if the grandchildren have the greater expectation of life, genes for grandparent to grandchild altruism have a higher selective advantage than genes for grandchild to grandparent altruism. It is quite possible for the net benefit of assisting a young distant relative to exceed the net benefit of assisting an old close relative. (Incidentally, it is not, of course, necessarily the case that grandparents have a shorter expectation of life than grandchildren. In species with a high infant-mortality rate, the reverse may be true.)

To extend the actuarial analogy, individuals can be thought of as life-insurance underwriters. An individual can be expected to invest or risk a certain proportion of his own assets in the life of another individual. He takes into account his relatedness to the other individual, and also whether the individual is a 'good risk' in terms of his life expectancy compared with the insurer's own. Strictly we should say 'reproduction expectancy' rather than 'life expectancy', or to be even more strict, 'general capacity to benefit own genes in the future expectancy'. Then in order for altruistic behaviour to evolve, the net risk to the altruist must be less than the net benefit to the recipient multiplied by the relatedness. Risks and benefits have to be calculated in the complex actuarial way I have outlined.

But what a complicated calculation to expect a poor survival machine to do, especially in a hurry!* Even the great mathematical biologist J. B. S. Haldane (in a paper of 1955 in which he anticipated Hamilton by postulating the spread of a gene for saving close relatives from drowning) remarked: '...on the two occasions when I have pulled possibly drowning people out of the water (at an infinitesimal risk to myself) I had no time to make such calculations.' Fortunately, however, as Haldane well knew, it is not necessary to assume that survival machines do the sums consciously in their heads. Just as we may use a slide rule without appreciating that we are, in effect, using logarithms, so an animal may be pre-programmed in such a way that it behaves *as if* it had made a complicated calculation.

This is not so difficult to imagine as it appears. When a man throws a ball high in the air and catches it again, he behaves as if he had solved a set of differential equations in predicting the trajectory of the ball. He may neither know nor care what a differential equation is, but this does not affect his skill with the ball. At some subconscious level, something functionally equivalent to the mathematical calculations is going on. Similarly, when a man takes a difficult decision, after weighing up all the pros and cons, and all the consequences of the decision that he can imagine, he is doing the functional equivalent of a large 'weighted sum' calculation, such as a computer might perform.

If we were to program a computer to simulate a model survival machine making decisions about whether to behave altruistically, we should probably proceed roughly as follows. We should make a list of all the alternative things the animal might do. Then for each of these alternative behaviour patterns we program a weighted sum calculation. All the various benefits will have a plus sign; all the risks will have a minus sign; both benefits and risks will be *weighted* by being multiplied by the appropriate index of relatedness before being added up. For simplicity we can, to begin with, ignore other weightings, such as those for age and health. Since an individual's 'relatedness' with himself is 1 (i.e. he has 100 per cent of his own genes—obviously), risks and benefits to himself will not be devalued at all, but will be given their full weight in the calculation. The whole sum for any one of the alternative behaviour patterns will look like this: Net benefit of behaviour pattern = Benefit to self − Risk to self $+\frac{1}{2}$ Benefit to brother $-\frac{1}{2}$ Risk to brother $+\frac{1}{2}$ Benefit to other brother $-\frac{1}{2}$ Risk to other brother $+\frac{1}{8}$ Benefit to first cousin $-\frac{1}{8}$ Risk to first cousin $+\frac{1}{2}$ Benefit to child $-\frac{1}{2}$ Risk to child + etc.

The result of the sum will be a number called the net benefit score of that behaviour pattern. Next, the model animal computes the equivalent sum for each alternative behaviour pattern in his repertoire. Finally he chooses to perform the behaviour pattern which emerges with the largest net benefit. Even if all the scores come out negative, he should still choose the action with the highest one, the least of evils. Remember that any positive action involves consumption of energy and time, both of which could have been spent doing other things. If doing nothing emerges as the 'behaviour' with the highest net benefit score, the model animal will do nothing.

Here is a very over-simplified example, this time expressed in the form of a subjective soliloquy rather than a computer simulation. I am an animal who has found a clump of eight mushrooms. After taking account of their nutritional value, and subtracting something for the slight risk that they might be poisonous, I estimate that they are worth +6 units each (the units are arbitrary pay-offs as in the previous chapter). The mushrooms are so big I could eat only three of them. Should I inform anybody else about my find, by giving a 'food call'? Who is within earshot? Brother B (his relatedness to me is $\frac{1}{2}$), cousin C (relatedness to me $= \frac{1}{8}$), and D (no particular relation: his relatedness to me is some small number which can be treated as zero for practical purposes). The net benefit score to me if I keep quiet about my find will be +6 for each of the three mushrooms I eat, that is + 18 in all. My net benefit score if I give the food call needs a bit of figuring. The eight mushrooms will be shared equally between the four of us. The pay-off to me from the two that I eat myself will be the full +6 units each, that is +12 in all. But I shall also get some pay-off when my brother and cousin eat their two mushrooms each, because of our shared genes. The actual score comes to $(1 \times 12) + (\frac{1}{2} \times 12) + (\frac{1}{8} \times 12) + (0 \times 12) = +19\frac{1}{2}$. The corresponding net benefit for the selfish behaviour was +18: it is a close-run thing, but the verdict is clear. I should give the food call; altruism on my part would in this case pay my selfish genes.

I have made the simplifying assumption that the individual animal works out what is best for his genes. What really happens is that the gene pool becomes filled with genes that influence bodies in such a way that they behave as if they had made such calculations.

In any case the calculation is only a very preliminary first approximation to what it ideally should be. It neglects many things, including the ages of the individuals concerned. Also, if I have just had a good meal, so that I can only find room for one mushroom, the net benefit of giving the food call will be greater than it would be if I was famished. There is no end to the progressive refinements of the calculation that could be achieved in the best of all possible worlds. But real life is not lived in the best of all possible worlds. We cannot expect real animals to take every last detail into account in coming to an optimum decision. We shall have to discover, by observation and experiment in the wild, how closely real animals actually come to achieving an ideal cost–benefit analysis.

Just to reassure ourselves that we have not become too carried away with subjective examples, let us briefly return to gene language. Living bodies are machines programmed by genes that have survived. The genes that have survived have done so in conditions that tended *on average* to characterize the environment of the species in the past. Therefore 'estimates' of costs and benefits are based on past 'experience', just as they are in human decision-making. However, experience in this case has the special meaning of gene experience or, more precisely, conditions of past gene survival. (Since genes also endow survival machines with the capacity to learn, some cost–benefit estimates could be said to be taken on the basis of individual experience as well.) So long as conditions do not change too drastically, the estimates will be good estimates, and survival machines will tend to make the right decisions on average. If conditions change radically, survival machines will tend to make erroneous decisions, and their genes will pay the penalty. Just so; human decisions based on outdated information tend to be wrong.

Estimates of relatedness are also subject to error and uncertainty. In our over-simplified calculations so far, we have talked as if survival machines *know* who is related to them, and how closely. In real life such certain knowledge is occasionally possible, but more usually the relatedness can only be estimated as an average number. For example, suppose that A and B could equally well be either half brothers or full brothers. Their relatedness is either $\frac{1}{4}$ or $\frac{1}{2}$, but since we do not know whether they are half or full brothers, the effectively usable figure is the average, $\frac{3}{8}$. If it is certain that they have the same mother, but the odds that they have the same father are only 1 in 10, then it is 90 per cent certain that they are half brothers, and 10 per cent certain that they are full brothers, and the effective relatedness is $\frac{1}{10} \times \frac{1}{2} + \frac{9}{10} \times \frac{1}{4} = 0.275$.

But when we say something like 'it' is 90 per cent certain, what 'it' are we referring to? Do we mean a human naturalist after a long field study is 90 per cent certain, or do we mean the animals are 90 per cent certain? With a bit of luck these two may amount to nearly the same thing. To see this, we have to think how animals might actually go about estimating who their close relations are.*

We know who our relations are because we are told, because we give them names, because we have formal marriages, and because we have written records and good memories. Many social anthropologists are preoccupied with 'kinship' in the societies which they study. They do not mean real genetic kinship, but subjective and cultural ideas of kinship. Human customs and tribal rituals commonly give great emphasis to kinship; ancestor worship is widespread, family obligations and loyalties dominate

much of life. Blood-feuds and inter-clan warfare are easily inter-
pretable in terms of Hamilton's genetic theory. Incest taboos
testify to the great kinship-consciousness of man, although the
genetical advantage of an incest taboo is nothing to do with
altruism; it is presumably concerned with the injurious effects of
recessive genes which appear with inbreeding. (For some reason
many anthropologists do not like this explanation.)*

How could wild animals 'know' who their kin are, or in other
words, what behavioural rules could they follow which would
have the indirect effect of making them seem to know about kinship? The rule 'be nice to your relations' begs the question of how
relations are to be recognized in practice. Animals have to be
given by their genes a simple rule for action, a rule that does not
involve all-wise cognition of the ultimate purpose of the action,
but a rule that works nevertheless, at least in average conditions.
We humans are familiar with rules, and so powerful are they that
if we are small minded we obey a rule itself, even when we can see
perfectly well that it is not doing us, or anybody else, any good.
For instance, some orthodox Jews and Muslims would starve rather
than break their rule against eating pork. What simple practical
rules could animals obey which, under normal conditions, would
have the indirect effect of benefiting their close relations?

If animals had a tendency to behave altruistically towards individuals who physically resembled them, they might indirectly be
doing their kin a bit of good. Much would depend on details of
the species concerned. Such a rule would, in any case, only lead
to 'right' decisions in a statistical sense. If conditions changed, for
example if a species started living in much larger groups, it could
lead to wrong decisions. Conceivably, racial prejudice could be
interpreted as an irrational generalization of a kin-selected tendency to identify with individuals physically resembling oneself,
and to be nasty to individuals different in appearance.

In a species whose members do not move around much, or whose members move around in small groups, the chances may be good that any random individual you come across is fairly close kin to you. In this case the rule 'Be nice to any member of the species whom you meet' could have positive survival value, in the sense that a gene predisposing its possessors to obey the rule might become more numerous in the gene pool. This may be why altruistic behaviour is so frequently reported in troops of monkeys and schools of whales. Whales and dolphins drown if they are not allowed to breathe air. Baby whales and injured individuals who cannot swim to the surface have been seen to be rescued and held up by companions in the school. It is not known whether whales have ways of knowing who their close relatives are, but it is possible that it does not matter. It may be that the overall probability that a random member of the school is a relation is so high that the altruism is worth the cost. Incidentally, there is at least one well-authenticated story of a drowning human swimmer being rescued by a wild dolphin. This could be regarded as a misfiring of the rule for saving drowning members of the school. The rule's 'definition' of a member of the school who is drowning might be something like: 'A long thing thrashing about and choking near the surface.'

Adult male baboons have been reported to risk their lives defending the rest of the troop against predators such as leopards. It is quite probable that any adult male has, on average, a fairly large number of genes tied up in other members of the troop. A gene that 'says', in effect, 'Body, if you happen to be an adult male, defend the troop against leopards', could become more numerous in the gene pool. Before leaving this often-quoted example, it is only fair to add that at least one respected authority has reported very different facts. According to her, adult males are the first over the horizon when a leopard appears.

Baby chicks feed in family clutches, all following their mother. They have two main calls. In addition to the loud piercing cheep which I have already mentioned, they give short melodious twitters when feeding. The cheeps, which have the effect of summoning the mother's aid, are ignored by the other chicks. The twitters, however, are attractive to chicks. This means that when one chick finds food, its twitters attract other chicks to the food as well: in the terms of the earlier hypothetical example, the twitters are 'food calls'. As in that case, the apparent altruism of the chicks can easily be explained by kin selection. Since, in nature, the chicks would be all full brothers and sisters, a gene for giving the food twitter would spread, provided the cost to the twitterer is less than half the net benefit to the other chicks. As the benefit is shared out between the whole clutch, which normally numbers more than two, it is not difficult to imagine this condition being realized. Of course the rule misfires in domestic or farm situations when a hen is made to sit on eggs not her own, even turkey or duck eggs. But neither the hen nor her chicks can be expected to realize this. Their behaviour has been shaped under the conditions that normally prevail in nature, and in nature strangers are not normally found in your nest.

Mistakes of this sort may, however, occasionally happen in nature. In species that live in herds or troops, an orphaned youngster may be adopted by a strange female, most probably one who has lost her own child. Monkey-watchers sometimes use the word 'aunt' for an adopting female. In most cases there is no evidence that she really is an aunt, or indeed any kind of relative: if monkey-watchers were as gene-conscious as they might be, they would not use an important word like 'aunt' so uncritically. In most cases we should probably regard adoption, however touching it may seem, as a misfiring of a built-in rule. This is because the generous female is doing her own genes no good by caring for the orphan. She is wasting time and energy which she could be investing in the lives of her own kin, particularly future children of her own. It is presumably a mistake that happens too seldom for natural selection to have 'bothered' to change the rule by making the maternal instinct more selective. In many cases, by the way, such adoptions do not occur, and an orphan is left to die.

There is one example of a mistake which is so extreme that you may prefer to regard it not as a mistake at all, but as evidence against the selfish gene theory. This is the case of bereaved monkey mothers who have been seen to steal a baby from another female, and look after it. I see this as a double mistake, since the adopter not only wastes her own time; she also releases a rival female from the burden of child-rearing, and frees her to have another child more quickly. It seems to me a critical example which deserves some thorough research. We need to know how often it happens; what the average relatedness between adopter and child is likely to be; and what the attitude of the real mother of the child is—it is, after all, to her advantage that her child *should* be adopted; do mothers deliberately try to deceive naïve young females into adopting their children? (It has also been suggested that adopters and baby-snatchers might benefit by gaining valuable practice in the art of child-rearing.)

An example of a deliberately engineered misfiring of the maternal instinct is provided by cuckoos, and other 'brood-parasites'—birds that lay their eggs in somebody else's nest. Cuckoos exploit the rule built into bird parents: 'Be nice to any small bird sitting in the nest that you built.' Cuckoos apart, this rule will normally have the desired effect of restricting altruism to immediate kin, because it happens to be a fact that nests are so isolated from each other that the contents of your own nest are almost bound to be your own chicks. Adult herring gulls do not recognize their own eggs, and will happily sit on other gull eggs, and even crude wooden dummies if these are substituted by a human experimenter. In nature, egg recognition is not important for gulls, because eggs do not roll far enough to reach the vicinity of a neighbour's nest, some yards away. Gulls do, however, recognize their own chicks: chicks, unlike eggs, wander, and can easily end up near the nest of a neighbouring adult, often with fatal results, as we saw in Chapter 1.

Guillemots, on the other hand, do recognize their own eggs by means of the speckling pattern, and actively discriminate in favour of them when incubating. This is presumably because they nest on flat rocks, where there is a danger of eggs rolling around and getting muddled up. Now, it might be said, why do they bother to discriminate and sit only on their own eggs? Surely if everybody saw to it that she sat on somebody's egg, it would not matter whether each particular mother was sitting on her own or somebody else's. This is the argument of a group selectionist. Just consider what would happen if such a group baby-sitting circle did develop. The average clutch size of the guillemot is one. This means that if the mutual baby-sitting circle is to work successfully, every adult would have to sit on an average of one egg. Now suppose somebody cheated, and refused to sit on an egg. Instead of wasting time sitting, she could spend her time laying more eggs. And the beauty of the scheme is that the other, more altruistic, adults would look after them for her. They would go on faithfully obeying the rule 'If you see a stray egg near your nest, haul it in and sit on it.' So the gene for cheating the system would spread through the population, and the nice friendly baby-sitting circle would break down.

'Well', it might be said, 'what if the honest birds retaliated by refusing to be blackmailed, and resolutely decided to sit on one egg and only one egg? That should foil the cheaters, because they would see their own eggs lying out on the rocks with nobody incubating them. That should soon bring them into line.' Alas, it would not. Since we are postulating that the sitters are not discriminating one egg from another, if the honest birds put into practice this scheme for resisting cheating, the eggs that ended up being neglected would be just as likely to be their own eggs as those of the cheaters. The cheaters would still have the advantage, because they would lay more eggs and have more surviving children. The only way an honest guillemot could beat the cheaters would be to discriminate actively in favour of her own eggs. That is, to cease being altruistic and look after her own interests.

To use the language of Maynard Smith, the altruistic adoption 'strategy' is not an evolutionarily stable strategy. It is unstable in the sense that it can be bettered by a rival selfish strategy of laying more than one's fair share of eggs, and then refusing to sit on them. This latter selfish strategy is in its turn unstable, because the altruistic strategy which it exploits is unstable, and will disappear. The only evolutionarily stable strategy for a guillemot is to recognize its own egg, and sit exclusively on its own egg, and this is exactly what happens.

The song-bird species that are parasitized by cuckoos have fought back, not in this case by learning the appearance of their own eggs, but by discriminating instinctively in favour of eggs with the species-typical markings. Since they are not in danger of being parasitized by members of their own species, this is effective.* But the cuckoos have retaliated in their turn by making their eggs more and more like those of the host species in colour, size, and markings. This is an example of a lie, and it often works. The result of this evolutionary arms race has been a remarkable perfection of mimicry on the part of the cuckoo eggs. We may suppose that a proportion of cuckoo eggs and chicks are 'found out', and those that are not found out are the ones who live to lay the next generation of cuckoo eggs. So genes for more effective deception spread through the cuckoo gene pool. Similarly, those host birds with eyes sharp enough to detect any slight imperfection in the cuckoo eggs' mimicry are the ones that contribute most to their own gene pool. Thus sharp and sceptical eyes are passed on to their next generation. This is a good example of how natural selection can sharpen up active discrimination, in this case discrimination against another species whose members are doing their best to foil the discriminators.

Now let us return to the comparison between an animal's 'estimate' of its kinship with other members of its group, and the corresponding estimate of an expert field naturalist. Brian Bertram has spent many years studying the biology of lions in the Serengeti National Park. On the basis of his knowledge of their reproductive habits, he has estimated the average relatedness between individuals in a typical lion pride. The facts that he uses to make his estimates are things like this. A typical pride consists of seven adult females who are its more permanent members, and two adult males who are itinerant. About half the adult females give birth as a batch at the same time, and rear their cubs together so that it is difficult to tell which cub belongs to whom. The typical litter size is three cubs. The fathering of litters is shared equally between the adult males in the pride. Young females remain in the pride and replace old females who die or leave. Young males are driven out when adolescent. When they grow up, they wander around from pride to pride in small related gangs or pairs, and are unlikely to return to their original family.

Using these and other assumptions, you can see that it would be possible to compute an average figure for the relatedness of two individuals from a typical lion pride. Bertram arrives at a figure of 0.22 for a pair of randomly chosen males, and 0.15 for a pair of females. That is to say, males within a pride are on average slightly less close than half brothers, and females slightly closer than first cousins.

Now, of course, any particular pair of individuals might be full brothers, but Bertram had no way of knowing this, and it is a fair bet that the lions did not know it either. On the other hand, the average figures that Bertram estimated are available to the lions themselves in a certain sense. If these figures really are typical for an average lion pride, then any gene that predisposed males to behave towards other males as if they were nearly half brothers would have positive survival value. Any gene that went too far and made males behave in a friendly way more appropriate to full brothers would on average be penalized, as would a gene for not being friendly enough, say treating other males like second cousins. If the facts of lion life are as Bertram says, and, just as important, if they have been like that for a large number of generations, then we may expect that natural selection will have favoured a degree of altruism appropriate to the average degree of relatedness in a typical pride. This is what I meant when I said that the kinship estimates of animal and of good naturalist might end up rather the same.*

So we conclude that the 'true' relatedness may be less important in the evolution of altruism than the best *estimate* of relatedness that animals can get. This fact is probably a key to understanding why parental care is so much more common and more devoted than brother/sister altruism in nature, and also why animals may value themselves more highly even than several brothers. Briefly, what I am saying is that, in addition to the index of relatedness, we should consider something like an index of 'certainty'. Although the parent/child relationship is no closer genetically than the brother/sister relationship, its certainty is greater. It is normally possible to be much more certain who your children are than who your brothers are. And you can be more certain still who you yourself are!

We considered cheaters among guillemots, and we shall have more to say about liars and cheaters and exploiters in following chapters. In a world where other individuals are constantly on the alert for opportunities to exploit kin-selected altruism, and use it for their own ends, a survival machine has to consider who it can trust, who it can be really sure of. *If* B is really my baby brother, then I should care for him up to half as much as I care for myself, and fully as much as I care for my own child. But can I be as sure of him as I can of my own child? How do I know he is my baby brother?

If C is my identical twin, then I should care for him twice as much as I care for any of my children; indeed I should value his life no less than my own.* But can I be sure of him? He looks like me to be sure, but it could be that we just happen to share the genes for facial features. No, I will not give up my life for him, because although it is *possible* that he bears 100 per cent of my genes, I absolutely *know* that I contain 100 per cent of my genes, so I am worth more to me than he is. I am the only individual that any one of my selfish genes can be sure of. And although ideally a gene for individual selfishness could be displaced by a rival gene for altruistically saving at least one identical twin, two children or brothers, or at least four grandchildren, etc., the gene for individual selfishness has the enormous advantage of *certainty* of individual identity. The rival kin-altruistic gene runs the risk of making mistakes of identity, either genuinely accidental or deliberately engineered by cheats and parasites. We therefore must expect individual selfishness in nature, to an extent greater than would be predicted by considerations of genetic relatedness alone.

In many species a mother can be more sure of her young than a father can. The mother lays the visible, tangible egg, or bears the child. She has a good chance of knowing for certain the bearers of her own genes. The poor father is much more vulnerable to deception. It is therefore to be expected that fathers will put less effort than mothers into caring for young. We shall see that there are other reasons to expect the same thing, in the chapter on the Battle of the Sexes (Chapter 9). Similarly, maternal grandmothers can be more sure of their grandchildren than paternal grandmothers can, and might be expected to show more altruism than paternal grandmothers. This is because they can be sure of their daughter's children, but their son may have been cuckolded. Maternal grandfathers are just as sure of their grandchildren as paternal grandmothers are, since both can reckon on one generation of certainty and one generation of uncertainty. Similarly, uncles on the mother's side should be more interested in the welfare of nephews and nieces than uncles on the father's side, and in general should be just as altruistic as aunts are. Indeed in a society with a high degree of marital infidelity, maternal uncles should be more altruistic than 'fathers' since they have more grounds for confidence in their relatedness to the child. They know that the child's mother is at least their half-sister. The 'legal' father knows nothing. I do not know of any evidence bearing on these predictions, but I offer them in the hope that others may, or may start looking for evidence. In particcular, perhaps social anthropologists might have interesting things to say.*

Returning to the fact that parental altruism is more common than fraternal altruism, it does seem reasonable to explain this in terms of the 'identification problem'. But this does not explain the fundamental asymmetry in the parent/child relationship itself. Parents care more for their children than children do for their parents, although the genetic relationship is symmetrical, and certainty of relatedness is just as great both ways. One reason is that parents are in a better practical position to help their young, being older and more competent at the business of living. Even if a baby wanted to feed its parents, it is not well equipped to do so in practice.

There is another asymmetry in the parent/child relationship which does not apply to the brother/sister one. Children are always younger than their parents. This often, though not always means they have a longer expectation of life. As I emphasized above, expectation of life is an important variable which, in the best of all possible worlds, should enter into an animal's 'calculation' when it is 'deciding' whether to behave altruistically or not. In a species in which children have a longer average life expectancy than parents, any gene for child altruism would be labouring under a disadvantage. It would be engineering altruistic self-sacrifice for the benefit of individuals who are nearer to dying of old age than the altruist itself. A gene for parent altruism, on the other hand, would have a corresponding advantage as far as the life-expectancy terms in the equation were concerned.

One sometimes hears it said that kin selection is all very well as a theory, but there are few examples of its working in practice. This criticism can only be made by someone who does not understand what kin selection means. The truth is that all examples of child-protection and parental care, and all associated bodily organs, milk-secreting glands, kangaroo pouches, and so on, are examples of the working in nature of the kin selection principle. The critics are of course familiar with the widespread existence of parental care, but they fail to understand that parental care is no less an example of kin selection than brother/sister altruism. When they say they want examples, they mean that they want examples other than parental care, and it is true that such examples are less common. I have suggested reasons why this might be so. I could have gone out of my way to quote examples of brother/sister altruism—there are in fact quite a few. But I don't want to do this, because it would reinforce the erroneous idea (favoured, as we have seen, by Wilson) that kin selection is specifically about relationships *other than* the parent/child relationship.

The reason this error has grown up is largely historical. The evolutionary advantage of parental care is so obvious that we did not have to wait for Hamilton to point it out. It has been understood ever since Darwin. When Hamilton demonstrated the genetic equivalence of other relationships, and their evolutionary significance, he naturally had to lay stress on these other relationships. In particular, he drew examples from the social insects such as ants and bees, in which the sister/sister relationship is particularly important, as we shall see in a later chapter. I have even heard people say that they thought Hamilton's theory applied *only* to the social insects!

If anybody does not want to admit that parental care is an example of kin selection in action, then the onus is on him to formulate a general theory of natural selection that predicts parental altruism, but that does *not* predict altruism between collateral kin. I think he will fail.

FAMILY PLANNING

It is easy to see why some people have wanted to separate parental care from the other kinds of kin-selected altruism. Parental care looks like an integral part of reproduction whereas, for example, altruism toward a nephew is not. I think there really is an important distinction hidden here, but that people have mistaken what the distinction is. They have put reproduction and parental care on one side, and other sorts of altruism on the other. But I wish to make a distinction between *bringing new individuals into the world*, on the one hand, and *caring for existing individuals* on the other. I shall call these two activities respectively child-bearing and child-caring. An individual survival machine has to make two quite different sorts of decisions, caring decisions and bearing decisions. I use the word decision to mean unconscious strategic move. The caring decisions are of this form: 'There is a child; its degree of relatedness to me is so and so; its chances of dying if I do not feed it are such and such; shall I feed it?' Bearing decisions, on the other hand, are like this: 'Shall I take whatever steps are necessary in order to bring a new individual into the world; shall I reproduce?' To some extent, caring and bearing are bound to compete with each other for an individual's time and other resources: the individual may have to make a choice: 'Shall I care for this child or shall I bear a new one?'

Depending on the ecological details of the species, various mixes of caring and bearing strategies can be evolutionarily stable. The one thing that cannot be evolutionarily stable is a *pure* caring strategy. If all individuals devoted themselves to caring for existing children to such an extent that they never brought any new ones into the world, the population would quickly become invaded by mutant individuals who specialized in bearing. Caring can only be evolutionarily stable as part of a mixed strategy—at least some bearing has to go on.

The species with which we are most familiar—mammals and birds—tend to be great carers. A decision to bear a new child is usually followed by a decision to care for it. It is because bearing and caring so often go together in practice that people have muddled the two things up. But from the point of view of the selfish genes there is, as we have seen, no distinction in principle between caring for a baby brother and caring for a baby son. Both infants are equally closely related to you. If you have to choose between feeding one or the other, there is no genetic reason why you should choose your own son. But on the other hand you cannot, by definition, bear a baby brother. You can only care for him once somebody else has brought him into the world. In the last chapter we looked at how individual survival machines ideally should decide whether to behave altruistically towards other individuals who already exist. In this chapter we look at how they should decide whether to bring new individuals into the world.

It is over this matter that the controversy about 'group selection', which I mentioned in Chapter 1, has chiefly raged. This is because Wynne-Edwards, who has been mainly responsible for promulgating the idea of group selection, did so in the context of a theory of 'population regulation'.* He suggested that individual animals deliberately and altruistically reduce their birth rates for the good of the group as a whole.

This is a very attractive hypothesis, because it fits so well with what individual humans ought to do. Mankind is having too many children. Population size depends upon four things: births, deaths, immigrations, and emigrations. Taking the world population as a whole, immigrations and emigrations do not occur, and we are left with births and deaths. So long as the average number of children per couple is larger than two surviving to reproduce, the numbers of babies born will tend to increase over the years at an ever-accelerating rate. In each generation the population, instead of going up by a fixed amount, increases by something more like a fixed proportion of the size that it has already reached. Since this size is itself getting bigger, the size of the increment gets bigger. If this kind of growth was allowed to go on unchecked, a population would reach astronomical proportions surprisingly quickly.

Incidentally, a thing that is sometimes not realized even by people who worry about population problems is that population growth depends on *when* people have children, as well as on how many they have. Since populations tend to increase by a certain proportion *per generation*, it follows that if you space the generations out more, the population will grow at a slower rate per year. Banners that read 'Stop at Two' could equally well be changed to 'Start at Thirty'! But in any case, accelerating population growth spells serious trouble.

We have probably all seen examples of the startling calculations that can be used to bring this home. For instance, the present population of Latin America is around 300 million, and already many of them are under-nourished. But if the population continued to increase at the present rate, it would take less than 500 years to reach the point where the people, packed in a standing position, formed a solid human carpet over the whole area of the continent. This is so, even if we assume them to be very skinny—a not unrealistic assumption. In 1,000 years from now they would be standing on each other's shoulders more than a million deep. By 2,000 years, the mountain of people, travelling outwards at the speed of light, would have reached the edge of the known universe.

It will not have escaped you that this is a hypothetical calculation! It will not really happen like that for some very good practical reasons. The names of some of these reasons are famine, plague, and war; *or*, if we are lucky, birth control. It is no use appealing to advances in agricultural science—'green revolutions' and the like. Increases in food production may temporarily alleviate the problem, but it is mathematically certain that they cannot be a long-term solution; indeed, like the medical advances that have precipitated the crisis, they may well make the problem worse, by speeding up the rate of the population expansion. It is a simple logical truth that, short of mass emigration into space, with rockets taking off at the rate of several million per second, uncontrolled birth rates are bound to lead to horribly increased death rates. It is hard to believe that this simple truth is not understood by those leaders who forbid their followers to use effective contraceptive methods. They express a preference for 'natural' methods of population limitation, and a natural method is exactly what they are going to get. It is called starvation.

But of course the unease that such long-term calculations arouse is based on concern for the future welfare of our species as a whole. Humans (some of them) have the conscious foresight to see ahead to the disastrous consequences of over-population. It is the basic assumption of this book that survival machines in general are guided by selfish genes, who most certainly cannot be expected to see into the future, nor to have the welfare of the whole species at heart. This is where Wynne-Edwards parts company with orthodox evolutionary theorists. He thinks there is a way in which genuine altruistic birth control can evolve.

A point that is not emphasized in the writings of Wynne-Edwards, or in Ardrey's popularization of his views, is that there is a large body of agreed facts that are not in dispute. It is an obvious fact that wild animal populations do not grow at the astronomical rates of which they are theoretically capable. Sometimes wild animal populations remain rather stable, with birth rates and death rates roughly keeping pace with each other. In many cases, lemmings being a famous example, the population fluctuates widely, with violent explosions alternating with crashes and near extinction. Occasionally the result is outright extinction, at least of the population in a local area. Sometimes, as in the case of the Canadian lynx—where estimates are obtained from the numbers of pelts sold by the Hudson's Bay Company in successive years—the population seems to oscillate rhythmically. The one thing animal populations do not do is go on increasing indefinitely.

Wild animals almost never die of old age: starvation, disease, or predators catch up with them long before they become really senile. Until recently this was true of man too. Most animals die in childhood, many never get beyond the egg stage. Starvation and other causes of death are the ultimate reasons why populations cannot increase indefinitely. But as we have seen for our own species, there is no necessary reason why it ever has to come to that. If only animals would regulate their *birth rates*, starvation need never happen. It is Wynne-Edwards's thesis that that is exactly what they do. But even here there is less disagreement than you might think from reading his book. Adherents of the selfish gene theory would readily agree that animals *do* regulate their birth rates. Any given species tends to have a rather fixed clutch size or litter size: no animal has an infinite number of children. The disagreement comes not over *whether* birth rates are regulated. The disagreement is over *why* they are regulated: by what process of natural selection has family planning evolved? In a nutshell, the disagreement is over whether animal birth control is altruistic, practised for the good of the group as a whole; or selfish, practised for the good of the individual doing the reproducing. I will deal with the two theories in order.

Wynne-Edwards supposed that individuals have fewer children than they are capable of, for the benefit of the group as a whole. He recognized that normal natural selection cannot possibly give rise to the evolution of such altruism: the natural selection of lower-than-average reproductive rates is, on the face of it, a contradiction in terms. He therefore invoked group selection, as we saw in Chapter 1. According to him, groups whose individual members restrain their own birth rates are less likely to go extinct than rival groups whose individual members reproduce so fast that they endanger the food supply. Therefore the world becomes populated by groups of restrained breeders. The individual restraint that Wynne-Edwards is suggesting amounts in a general sense to birth control, but he is more specific than this, and indeed comes up with a grand conception in which the whole of social life is seen as a mechanism of population regulation. For instance, two major features of social life in many species of animals are *territoriality* and *dominance hierarchies*, already mentioned in Chapter 5.

Many animals devote a great deal of time and energy to apparently defending an area of ground which naturalists call a territory. The phenomenon is very widespread in the animal kingdom, not only in birds, mammals, and fish, but in insects and even sea anemones. The territory may be a large area of woodland which is the principal foraging ground of a breeding pair, as in the case of robins. Or, in herring gulls for instance, it may be a small area containing no food, but with a nest at its centre. Wynne-Edwards believes that animals who fight over territory are fighting over a *token* prize, rather than an actual prize like a bit of food. In many cases females refuse to mate with males who do not possess a territory. Indeed it often happens that a female whose mate is defeated and his territory conquered promptly attaches herself to the victor. Even in apparently faithful monogamous species, the female may be wedded to a male's territory rather than to him personally.

If the population gets too big, some individuals will not get territories, and therefore will not breed. Winning a territory is therefore, to Wynne-Edwards, like winning a ticket or licence to breed. Since there is a finite number of territories available, it is as if a finite number of breeding licences is issued. Individuals may fight over who gets these licences, but the total number of babies that the population can have as a whole is limited by the number of territories available. In some cases, for instance in red grouse, individuals do, at first sight, seem to show restraint, because those who cannot win territories not only do not breed; they also appear to give up the struggle to win a territory. It is as though they all accepted the rules of the game: that if, by the end of the competition season, you have not secured one of the official tickets to breed, you voluntarily refrain from breeding and leave the lucky ones unmolested during the breeding season, so that they can get on with propagating the species.

Wynne-Edwards interprets dominance hierarchies in a similar way. In many groups of animals, especially in captivity, but also in some cases in the wild, individuals learn each other's identity, and they learn whom they can beat in a fight, and who usually beats them. As we saw in Chapter 5, they tend to submit without a struggle to individuals who they 'know' are likely to beat them anyway. As a result a naturalist is able to describe a dominance hierarchy or 'peck order' (so called because it was first described for hens)—a rank-ordering of society in which everybody knows his place, and does not get ideas above his station. Of course sometimes real earnest fights do take place, and sometimes individuals can win promotion over their former immediate bosses. But as we saw in Chapter 5, the overall effect of the automatic submission by lower-ranking individuals is that few prolonged fights actually take place, and serious injuries seldom occur.

Many people think of this as a 'good thing' in some vaguely group selectionist way. Wynne-Edwards has an altogether more daring interpretation. High-ranking individuals are more likely to breed than low-ranking individuals, either because they are preferred by females or because they physically prevent low-ranking males from getting near females. Wynne-Edwards sees high social rank as another ticket of entitlement to reproduce. Instead of fighting directly over females themselves, individuals fight over social status, and then accept that if they do not end up high on the social scale they are not entitled to breed. They restrain themselves where females are directly concerned, though they may try even now and then to win higher status, and therefore could be said to compete *indirectly* over females. But, as in the case of territorial behaviour, the result of this 'voluntary acceptance' of the rule that only high-status males should breed is, according to Wynne-Edwards, that populations do not grow too fast. Instead of actually having too many children, and then finding out the hard way that it was a mistake, populations use formal contests over status and territory as a means of limiting their size slightly below the level at which starvation itself actually takes its toll.

Perhaps the most startling of Wynne-Edwards's ideas is that of *epideictic* behaviour, a word that he coined himself. Many animals spend a great deal of time in large flocks, herds, or shoals. Various more or less common-sense reasons why such aggregating behaviour should have been favoured by natural selection have been suggested, and I will talk about some of them in Chapter 10. Wynne-Edwards's idea is quite different. He proposes that when huge flocks of starlings mass at evening, or crowds of midges dance over a gate-post, they are performing a census of their population. Since he is supposing that individuals restrain their birth rates in the interests of the group as a whole, and have fewer babies when the population density is high, it is reasonable that they should have some way of measuring the population density. Just so; a thermostat needs a thermometer as an integral part of its mechanism. For Wynne-Edwards, epideictic behaviour is deliberate massing in crowds to facilitate population estimation. He is not suggesting conscious population estimation, but an automatic nervous or hormonal mechanism linking the individuals' sensory perception of the density of their population with their reproductive systems.

I have tried to do justice to Wynne-Edwards's theory, even if rather briefly. If I have succeeded, you should now be feeling persuaded that it is, on the face of it, rather plausible. But the earlier chapters of this book should have prepared you to be sceptical to the point of saying that, plausible as it may sound, the evidence for Wynne-Edwards's theory had better be good, or else.... And unfortunately the evidence is not good. It consists of a large number of examples which could be interpreted in his way, but which could equally well be interpreted on more orthodox 'selfish gene' lines.

Although he would never have used that name, the chief architect of the selfish gene theory of family planning was the great ecologist David Lack. He worked especially on clutch size in wild birds, but his theories and conclusions have the merit of being generally applicable. Each bird species tends to have a typical clutch size. For instance, gannets and guillemots incubate one egg at a time, swifts three, great tits half a dozen or more. There is variation in this: some swifts lay only two at a time, great tits may lay twelve. It is reasonable to suppose that the number of eggs a female lays and incubates is at least partly under genetic control, like any other characteristic. That is say there may be a gene for laying two eggs, a rival allele for laying three, another allele for laying four, and so on, although in practice it is unlikely to be quite as simple as this. Now the selfish gene theory requires us to ask which of these genes will become more numerous in the gene pool. Superficially it might seem that the gene for laying four eggs is bound to have an advantage over the genes for laying three or two. A moment's reflection shows that this simple 'more means better' argument cannot be true, however. It leads to the expectation that five eggs should be better than four, ten better still, 100 even better, and infinity best of all. In other words it leads logically to an absurdity. Obviously there are *costs* as well as benefits in laying a large number of eggs. Increased bearing is bound to be paid for in less efficient caring. Lack's essential point is that for any given species, in any given environmental situation, there must be an optimal clutch size. Where he differs from Wynne-Edwards is in his answer to the question: 'Optimal from whose point of view?'. Wynne-Edwards would say the important optimum, to which all individuals should aspire, is the optimum for the group as a whole. Lack would say each selfish individual chooses the clutch size that maximizes the number of children she rears. If three is the optimum clutch size for swifts, what this means, for Lack, is that any individual who tries to rear four will probably end up with fewer children than rival, more cautious individuals who only try to rear three. The obvious reason for this would be that the food is so thinly spread between the four babies that few of them survive to adulthood. This would be true both of the original allocation of yolk to the four eggs and of the food given to the babies after hatching. According to Lack, therefore, individuals regulate their clutch size for reasons that are anything but altruistic. They are not practising birth control in order to avoid over-exploiting the group's resources. They are practising birth control in order to maximize the number of surviving children they actually have, an aim which is the very opposite of that which we normally associate with birth control.

Rearing baby birds is a costly business. The mother has to invest a large quantity of food and energy in manufacturing eggs. Possibly with her mate's help, she invests a large effort in building a nest to hold her eggs and protect them. Parents spend weeks patiently sitting on the eggs. Then, when the babies hatch out, the parents work themselves nearly to death fetching food for them, more or less non-stop without resting. As we have already seen, a parent great tit brings an average of one item of food to the nest every 30 seconds of daylight. Mammals such as ourselves do it in a slightly different way, but the basic idea of reproduction being a costly affair, especially for the mother, is no less true. It is obvious that if a parent tries to spread her limited resources of food and effort among too many children, she will end up rearing fewer than if she had set out with more modest ambitions. She has to strike a balance between bearing and caring. The total amount of food and other resources which an individual female, or a mated pair, can muster is the limiting factor determining the number of children they can rear. Natural selection, according to the Lack theory, adjusts initial clutch size (litter size, etc.) so as to take maximum advantage of these limited resources.

Individuals who have too many children are penalized, not because the whole population goes extinct, but simply because fewer of their children survive. Genes for having too many children are just not passed on to the next generation in large numbers, because few of the children bearing these genes reach adulthood. What has happened in modern civilized man is that family sizes are no longer limited by the finite resources that the individual parents can provide. If a husband and wife have more children than they can feed, the state, which means the rest of the population, simply steps in and keeps the surplus children alive and healthy. There is, in fact, nothing to stop a couple with no material resources at all having and rearing precisely as many children as the woman can physically bear. But the welfare state is a very unnatural thing. In nature, parents who have more children than they can support do not have many grandchildren, and their genes are not passed on to future generations. There is no *need* for altruistic restraint in the birth rate, because there is no welfare state in nature. Any gene for over-indulgence is promptly punished: the children containing that gene starve. Since we humans do not want to return to the old selfish ways where we let the children of too-large families starve to death, we have abolished the family as a unit of economic self-sufficiency, and substituted the state. But the privilege of guaranteed support for children should not be abused.

Contraception is sometimes attacked as 'unnatural'. So it is, very unnatural. The trouble is, so is the welfare state. I think that most of us believe the welfare state is highly desirable. But you cannot have an unnatural welfare state, unless you also have unnatural birth control; otherwise, the end result will be misery even greater than that which obtains in nature. The welfare state is perhaps the greatest altruistic system the animal kingdom has ever known. But any altruistic system is inherently unstable, because it is open to abuse by selfish individuals, ready to exploit it. Individual humans who have more children than they are capable of rearing are probably too ignorant in most cases to be accused of conscious malevolent exploitation. Powerful institutions and leaders who deliberately encourage them to do so seem to me less free from suspicion.

Returning to wild animals, the Lack clutch size argument can be generalized to all the other examples Wynne-Edwards uses: territorial behaviour, dominance hierarchies, and so on. Take, for instance, the red grouse that he and his colleagues have worked on. These birds eat heather, and they parcel out the moors in territories containing apparently more food than the territory owners actually need. Early in the season they fight over territories, but after a while the losers seem to accept that they have failed, and do not fight any more. They become outcasts who never get territories, and by the end of the season they have mostly starved to death. Only territory owners breed. That non-territory owners are physically capable of breeding is shown by the fact that if a territory owner is shot his place is promptly filled by one of the former outcasts, who then breeds. Wynne-Edwards's interpretation of this extreme territorial behaviour is, as we have seen, that the outcasts 'accept' that they have failed to gain a ticket or licence to breed; they do not try to breed.

On the face of it, this seems an awkward example for the selfish gene theory to explain. Why don't the outcasts try, try, and try again to oust a territory holder, until they drop from exhaustion? They would seem to have nothing to lose. But wait, perhaps they do have something to lose. We have already seen that if a territory-holder should happen to die, an outcast has a chance of taking his place, and therefore of breeding. If the odds of an outcast's succeeding to a territory in this way are greater than the odds of his gaining one by fighting, then it may pay him, as a selfish individual, to wait in the hope that somebody will die, rather than squander what little energy he has in futile fighting. For Wynne-Edwards, the role of the outcasts in the welfare of the group is to wait in the wings as understudies, ready to step into the shoes of

any territory holder who dies on the main stage of group reproduction. We can now see that this may also be their best strategy
purely as selfish individuals. As we saw in Chapter 4, we can
regard animals as gamblers. The best strategy for a gambler may
sometimes be a wait-and-hope strategy, rather than a bull-at-a-
gate strategy.

Similarly, the many other examples where animals appear to
'accept' non-reproductive status passively can be explained quite
easily by the selfish gene theory. The general form of the explanation is always the same: the individual's best bet is to restrain
himself for the moment, in the hope of better chances in the
future. A seal who leaves the harem-holders unmolested is not
doing it for the good of the group. He is biding his time, waiting
for a more propitious moment. Even if the moment never comes
and he ends up without descendants, the gamble *might* have paid
off, though, with hindsight we can see that for him it did not.
And when lemmings flood in their millions away from the centre
of a population explosion, they are not doing it in order to reduce
the density of the area they leave behind! They are seeking, every
selfish one of them, a less crowded place in which to live. The fact
that any particular one may fail to find it, and dies, is something
we can see with hindsight. It does not alter the likelihood that to
stay behind would have been an even worse gamble.

It is a well-documented fact that over-crowding sometimes
reduces birth rates. This is sometimes taken to be evidence for
Wynne-Edwards's theory. It is nothing of the kind. It is compatible with his theory, and it is also just as compatible with the
selfish gene theory. For example, in one experiment mice were
put in an outdoor enclosure with plenty of food, and allowed to
breed freely. The population grew up to a point, then levelled off.
The reason for the levelling-off turned out to be that the females
became less fertile as a consequence of over-crowding: they had

fewer babies. This kind of effect has often been reported. Its immediate cause is often called 'stress', although giving it a name like that does not of itself help to explain it. In any case, whatever its immediate cause may be, we still have to ask about its ultimate, or evolutionary explanation. Why does natural selection favour females who reduce their birth rate when their population is over-crowded?

Wynne-Edwards's answer is clear. Group selection favours groups in which the females measure the population and adjust their birth rates so that food supplies are not over-exploited. In the condition of the experiment, it so happened that food was never going to be scarce, but the mice could not be expected to realize that. They are programmed for life in the wild, and it is likely that in natural conditions over-crowding is a reliable indicator of future famine.

What does the selfish gene theory say? Almost exactly the same thing, but with one crucial difference. You will remember that, according to Lack, animals will tend to have the optimum number of children from their own selfish point of view. If they *bear* too few or too many, they will end up *rearing* fewer than they would have if they had hit on just the right number. Now, 'just the right number' is likely to be a smaller number in a year when the population is over-crowded than in a year when the population is sparse. We have already agreed that over-crowding is likely to foreshadow famine. Obviously, if a female is presented with reliable evidence that a famine is to be expected, it is in her own selfish interests to reduce her own birth rate. Rivals who do not respond to the warning signs in this way will end up rearing fewer babies, even if they actually bear more. We therefore end up with almost exactly the same conclusion as Wynne-Edwards, but we get there by an entirely different type of evolutionary reasoning.

The selfish gene theory has no trouble even with 'epideictic displays'. You will remember that Wynne-Edwards hypothesized that animals deliberately display together in large crowds in order to make it easy for all the individuals to conduct a census, and regulate their birth rates accordingly. There is no direct evidence that any aggregations are in fact epideictic, but just suppose some such evidence were found. Would the selfish gene theory be embarrassed? Not a bit.

Starlings roost together in huge numbers. Suppose it were shown, not only that over-crowding in winter reduced fertility in the following spring, but that this was directly due to the birds' listening to each other's calls. It might be demonstrated experimentally that individuals exposed to a tape-recording of a dense and very loud starling roost laid fewer eggs than individuals exposed to a recording of a quieter, less dense, roost. By definition, this would indicate that the calls of starlings constituted an epideictic display. The selfish gene theory would explain it in much the same way as it handled the case of the mice.

Again, we start from the assumption that genes for having a larger family than you can support are automatically penalized, and become less numerous in the gene pool. The task of an efficient egg-layer is one of predicting what is going to be the optimum clutch size for her, as a selfish individual, in the coming breeding season. You will remember from Chapter 4 the special sense in which we are using the word prediction. Now how can a female bird predict her optimum clutch size? What variables should influence her prediction? It may be that many species make a fixed prediction, which does not change from year to year. Thus on average the optimum clutch size for a gannet is one. It is possible that in particular bumper years for fish the true optimum for an individual might temporarily rise to two eggs. If there is no way for gannets to know in advance whether a particular year is going to be a bumper one, we cannot expect individual females to take the risk of wasting their resources on two eggs, when this would damage their reproductive success in an average year.

But there may be other species, perhaps starlings, in which it is in principle possible to predict in winter whether the following spring is going to yield a good crop of some particular food resource. Country people have numerous old sayings suggesting that such clues as the abundance of holly berries may be good predictors of the weather in the coming spring. Whether any particular old wives' tale is accurate or not, it remains logically possible that there are such clues, and that a good prophet could in theory adjust her clutch size from year to year to her own advantage. Holly berries may be reliable predictors or they may not but, as in the case of the mice, it does seem quite likely that population density would be a good predictor. A female starling can in principle know that, when she comes to feed her babies in the coming spring, she will be competing for food with rivals of the same species. If she can somehow estimate the local density of her own species in winter, this could provide her with a powerful means of predicting how difficult it is going to be to get food for babies next spring. If she found the winter population to be particularly high, her prudent policy, from her own selfish point of view, might well be to lay relatively few eggs: her estimate of her own optimum clutch size would have been reduced.

Now the moment it becomes true that individuals are reducing their clutch size on the basis of their estimate of population density, it will immediately be to the advantage of each selfish individual to pretend to rivals that the population is large, whether it really is or not. If starlings are estimating population size by the volume of noise in a winter roost, it would pay each individual to shout as loudly as possible, in order to sound more like two starlings than one. This idea of animals pretending to be several animals at once has been suggested in another context by J. R. Krebs, and is named the *Beau Geste Effect* after the novel in which a similar tactic was used by a unit of the French Foreign Legion. The idea in our case is to try to induce neighbouring starlings to reduce *their* clutch size to a level lower than the true optimum. If you are a starling who succeeds in doing this, it is to your selfish advantage, since you are reducing the numbers of individuals who do not bear your genes. I therefore conclude that Wynne-Edwards's idea of epideictic displays may actually be a good idea: he may have been right all along, but for the wrong reasons. More generally, the Lack type of hypothesis is powerful enough to account, in selfish gene terms, for all evidence that might seem to support the group selection theory, should any such evidence turn up.

Our conclusion from this chapter is that individual parents practise family planning, but in the sense that they optimize their birth rates rather than restrict them for public good. They try to maximize the number of surviving children that they have, and this means having neither too many babies nor too few. Genes that make an individual have too many babies tend not to persist in the gene pool, because children containing such genes tend not to survive to adulthood.

So much, then, for quantitative considerations of family size. We now come on to conflicts of interest within families. Will it always pay a mother to treat all her children equally, or might she have favourites? Should the family function as a single cooperating whole, or are we to expect selfishness and deception even within the family? Will all members of a family be working towards the same optimum, or will they 'disagree' about what the optimum is? These are the questions we try to answer in the next chapter. The related question of whether there may be conflict of interest between mates, we postpone until Chapter 9.

BATTLE OF THE
GENERATIONS

Let us begin by tackling the first of the questions posed at the end of the last chapter. Should a mother have favourites, or should she be equally altruistic towards all her children? At the risk of being boring, I must yet again throw in my customary warning. The word 'favourite' carries no subjective connotations, and the word 'should' no moral ones. I am treating a mother as a machine programmed to do everything in its power to propagate copies of the genes which ride inside it. Since you and I are humans who know what it is like to have conscious purposes, it is convenient for me to use the language of purpose as a metaphor in explaining the behaviour of survival machines.

In practice, what would it mean to say a mother had a favourite child? It would mean she would invest her resources unequally among her children. The resources that a mother has available to invest consist of a variety of things. Food is the obvious one, together with the effort expended in gathering food, since this in itself costs the mother something. Risk undergone in protecting young from predators is another resource which the mother can 'spend' or refuse to spend. Energy and time devoted to nest or home maintenance, protection from the elements, and, in some species, time spent in teaching children are valuable resources which a parent can allocate to children, equally or unequally as she 'chooses'.

It is difficult to think of a common currency in which to measure all these resources that a parent can invest. Just as human societies use money as a universally convertible currency which can be translated into food or land or labouring time, so we require a currency in which to measure resources that an individual survival machine may invest in another individual's life, in particular a child's life. A measure of energy such as the calorie is tempting, and some ecologists have devoted themselves to the accounting of energy costs in nature. This is inadequate though, because it is only loosely convertible into the currency that really matters, the 'gold-standard' of evolution, gene survival. R. L. Trivers, in 1972, neatly solved the problem with his concept of *parental investment* (although, reading between the close-packed lines, one feels that Sir Ronald Fisher, the greatest biologist of the twentieth century, meant much the same thing in 1930 by his 'parental expenditure').*

Parental investment (P.I.) is defined as 'any investment by the parent in an individual offspring that increases the offspring's chance of surviving (and hence reproductive success) at the cost of the parent's ability to invest in other offspring'. The beauty of Trivers's parental investment is that it is measured in units very close to the units that really matter. When a child uses up some of its mother's milk, the amount of milk consumed is measured not in pints, not in calories, but in units of detriment to other children of the same mother. For instance, if a mother has two babies, X and Y, and X drinks one pint of milk, a major part of the P.I. that this pint represents is measured in units of increased probability that Y will die because he did not drink that pint. P.I. is measured in units of decrease in life expectancy of other children, born or yet to be born.

Parental investment is not quite an ideal measure, because it overemphasizes the importance of parentage, as against other genetic relationships. Ideally we should use a generalized *altruism investment* measure. Individual A may be said to invest in individual B, when A increases B's chance of surviving, at the cost of A's ability to invest in other individuals including herself, all costs being weighted by the appropriate relatedness. Thus a parent's investment in any one child should ideally be measured in terms of detriment to life expectancy not only of other children, but also of nephews, nieces, herself, etc. In many respects, however, this is just a quibble, and Trivers's measure is well worth using in practice.

Now any particular adult individual has, in her whole lifetime, a certain total quantity of P.I. available to invest in children (and other relatives and in herself, but for simplicity we consider only children). This represents the sum of all the food she can gather or manufacture in a lifetime of work, all the risks she is prepared to take, and all the energy and effort that she is able to put into the welfare of children. How should a young female, setting out on her adult life, invest her life's resources? What would be a wise investment policy for her to follow? We have already seen from the Lack theory that she should not spread her investment too thinly among too many children. That way she will lose too many genes: she won't have enough grandchildren. On the other hand, she must not devote all her investment to too few children— spoilt brats. She may virtually guarantee herself *some* grandchildren, but rivals who invest in the optimum number of children will end up with more grandchildren. So much for even-handed investment policies. Our present interest is in whether it could ever pay a mother to invest unequally among her children, i.e. in whether she should have favourites.

The answer is that there is no genetic reason for a mother to have favourites. Her relatedness to all her children is the same, $\frac{1}{2}$.

Her optimal strategy is to invest *equally* in the largest number of children that she can rear to the age when they have children of their own. But, as we have already seen, some individuals are better life insurance risks than others. An under-sized runt bears just as many of his mother's genes as his more thriving litter mates. But his life expectation is less. Another way to put this is that he *needs* more than his fair share of parental investment, just to end up equal to his brothers. Depending on the circumstances, it may pay a mother to refuse to feed a runt, and allocate all of his share of her parental investment to his brothers and sisters. Indeed it may pay her to feed him to his brothers and sisters, or to eat him herself, and use him to make milk. Mother pigs do sometimes devour their young, but I do not know whether they pick especially on runts.

Runts constitute a particular example. We can make some more general predictions about how a mother's tendency to invest in a child might be affected by his age. If she has a straight choice between saving the life of one child or saving the life of another, and if the one she does not save is bound to die, she should prefer the older one. This is because she stands to lose a higher proportion of her life's parental investment if he dies than if his little brother dies. Perhaps a better way to put this is that if she saves the little brother she will still have to invest some costly resources in him just to get him up to the age of the big brother.

On the other hand, if the choice is not such a stark life or death choice, her best bet might be to prefer the younger one. For instance, suppose her dilemma is whether to give a particular morsel of food to a little child or a big one. The big one is likely to be more capable of finding his own food unaided. Therefore if she stopped feeding him he would not necessarily die. On the other hand, the little one who is too young to find food for himself would be more likely to die if his mother gave the food to his big brother. Now, even though the mother would prefer the little brother to die rather than the big brother, she may still give the food to the little one, because the big one is unlikely to die anyway. This is why mammal mothers wean their children, rather than going on feeding them indefinitely throughout their lives. There comes a time in the life of a child when it pays the mother to divert investment from him into future children. When this moment comes, she will want to wean him. A mother who had some way of knowing that she had had her last child might be expected to continue to invest all her resources in him for the rest of her life, and perhaps suckle him well into adulthood. Nevertheless, she should 'weigh up' whether it would not pay her more to invest in grandchildren or nephews and nieces, since although these are half as closely related to her as her own children, their capacity to benefit from her investment may be more than double that of one of her own children.

This seems a good moment to mention the puzzling phenomenon known as the menopause, the rather abrupt termination of a human female's reproductive fertility in middle age. This may not have occurred too commonly in our wild ancestors, since not many women would have lived that long anyway. But still, the difference between the abrupt change of life in women and the gradual fading out of fertility in men suggests that there is something genetically 'deliberate' about the menopause—that it is an 'adaptation'. It is rather difficult to explain. At first sight we might expect that a woman should go on having children until she dropped, even if advancing years made it progressively less likely that any individual child would survive. Surely it would seem always worth trying? But we must remember that she is also related to her grandchildren, though half as closely.

For various reasons, perhaps connected with the Medawar theory of ageing (page 51), women in the natural state became gradually less efficient at bringing up children as they got older. Therefore the life expectancy of a child of an old mother was less than that of a child of a young mother. This means that, if a woman had a child and a grandchild born on the same day, the grandchild could expect to live longer than the child. When a woman reached the age where the average chance of each child reaching adulthood was just less than half the chance of each grandchild of the same age reaching adulthood, any gene for investing in grandchildren in preference to children would tend to prosper. Such a gene is carried by only one in four grandchildren, whereas the rival gene is carried by one in two children, but the greater expectation of life of the grandchildren outweighs this, and the 'grandchild altruism' gene prevails in the gene pool. A woman could not invest fully in her grandchildren if she went on having children of her own. Therefore genes for becoming reproductively infertile in middle age became more numerous, since they were carried in the bodies of grandchildren whose survival was assisted by grandmotherly altruism.

This is a possible explanation of the evolution of the menopause in females. The reason why the fertility of males tails off gradually rather than abruptly is probably that males do not invest so much as females in each individual child anyway. Provided he can sire children by young women, it will always pay even a very old man to invest in children rather than in grandchildren.

So far, in this chapter and in the last, we have seen everything from the parent's point of view, largely the mother's. We have asked whether parents can be expected to have favourites, and in general what is the best investment policy for a parent. But perhaps each child can influence how much his parents invest in him as against his brothers and sisters. Even if parents do not 'want' to show favouritism among their children, could it be that children grab favoured treatment for themselves? Would it pay them to do so? More strictly, would genes for selfish grabbing among children become more numerous in the gene pool than rival genes for accepting no more than one's fair share? This matter has been brilliantly analysed by Trivers, in a paper of 1974 called 'Parent-Offspring Conflict'.

A mother is equally related to all her children, born and to be born. On genetic grounds alone she should have no favourites, as we have seen. If she does show favouritism it should be based on differences in expectation of life, depending on age and other things. The mother, like any individual, is twice as closely 'related' to herself as she is to any of her children. Other things being equal, this means that she should invest most of her resources selfishly in herself, but other things are not equal. She can do her genes more good by investing a fair proportion of her resources in her children. This is because these are younger and more help-less than she is, and they can therefore benefit more from each unit of investment than she can herself. Genes for investing in more helpless individuals in preference to oneself can prevail in the gene pool, even though the beneficiaries may share only a proportion of one's genes. This is why animals show parental altruism, and indeed why they show any kind of kin-selected altruism.

Now look at it from the point of view of a particular child. He is just as closely related to each of his brothers and sisters as his mother is to them. The relatedness is $\frac{1}{2}$ in all cases. Therefore he 'wants' his mother to invest some of her resources in his brothers and sisters. Genetically speaking, he is just as altruistically dis-posed to them as his mother is. But again, he is twice as closely related to himself as he is to any brother or sister, and this will dispose him to want his mother to invest in him more than in any particular brother or sister, other things being equal. In this case other things might indeed be equal. If you and your brother are the same age, and both are in a position to benefit equally from a pint of mother's milk, you 'should' try to grab more than your fair share, and he should try to grab more than his fair share. Have you ever heard a litter of piglets squealing to be first on the scene when the mother sow lies down to feed them? Or little boys fighting over the last slice of cake? Selfish greed seems to characterize much of child behaviour.

But there is more to it than this. If I am competing with my brother for a morsel of food, and if he is much younger than me so that he could benefit from the food more than I could, it might pay my genes to let him have it. An elder brother may have exactly the same grounds for altruism as a parent: in both cases, as we have seen, the relatedness is $\frac{1}{2}$, and in both cases the younger individual can make better use of the resource than the elder. If I possess a gene for giving up food, there is a 50 per cent chance that my baby brother contains the same gene. Although the gene has double the chance of being in my own body—100 per cent, it *is* in my body—my need of the food may be less than half as urgent. In general, a child 'should' grab more than his share of parental investment, but only up to a point. Up to what point? Up to the point where the resulting net cost to his brothers and sisters, born and potentially to be born, is just double the benefit of the grabbing to himself.

Consider the question of when weaning should take place. A mother wants to stop suckling her present child so that she can prepare for the next one. The present child, on the other hand, does not want to be weaned yet, because milk is a convenient, trouble-free source of food, and he does not want to have to go out and work for his living. To be more exact, he does want eventually to go out and work for his living, but only when he can do his genes more good by leaving his mother free to rear his little brothers and sisters, than by staying behind himself. The older a child is, the less relative benefit does he derive from each pint of milk. This is because he is bigger, and a pint of milk is therefore a smaller proportion of his requirement, and also he is becoming more capable of fending for himself if he is forced to. Therefore when an old child drinks a pint that could have been invested in a younger child, he is taking relatively more parental invest-ment for himself than when a young child drinks a pint. As a child grows older, there will come a moment when it would pay his mother to stop feeding him, and invest in a new child instead. Somewhat later there will come a time when the old child too would benefit his genes most by weaning himself. This is the moment when a pint of milk can do more good to the copies of his genes that *may be* present in his brothers and sisters than it can to the genes that *are* present in himself.

The disagreement between mother and child is not an abso-lute one, but a quantitative one, in this case a disagreement over timing. The mother wants to go on suckling her present child up to the moment when investment in him reaches his 'fair' share, taking into account his expectation of life and how much she has already invested in him. Up to this point there is no disagree-ment. Similarly, both mother and child agree in not wanting him to go on sucking after the point when the cost to future children is more than double the benefit to him. But there is disagreement between mother and child during the intermediate period, the period when the child is getting more than his share as the mother sees it, but when the cost to other children is still less than double the benefit to him.

Weaning time is just one example of a matter of dispute between mother and child. It could also be regarded as a dispute between one individual and all his future unborn brothers and sisters, with the mother taking the part of her future unborn children. More directly there may be competition between contemporary rivals for her investment, between litter mates or nest mates. Here, once again, the mother will normally be anxious to see fair play.

Many baby birds are fed in the nest by their parents. They all gape and scream, and the parent drops a worm or other morsel in the open mouth of one of them. The loudness with which each baby screams is, ideally, proportional to how hungry he is. Therefore, if the parent always gives the food to the loudest screamer, they should all tend to get their fair share, since when one has had enough he will not scream so loudly. At least that is what would happen in the best of all possible worlds, if individuals did not cheat. But in the light of our selfish gene concept we must expect that individuals *will* cheat, *will* tell lies about how hungry they are. This will escalate, apparently rather pointlessly because it might seem that if they are all lying by screaming too loudly, this level of loudness will become the norm, and will cease, in effect, to be a lie. However, it cannot de-escalate, because any individual who takes the first step in decreasing the loudness of his scream will be penalized by being fed less, and is more likely to starve. Baby bird screams do not become infinitely loud, because of other considerations. For example, loud screams tend to attract predators, and they use up energy.

Sometimes, as we have seen, one member of a litter is a runt, much smaller than the rest. He is unable to fight for food as strongly as the rest, and runts often die. We have considered the conditions under which it would actually pay a mother to let a runt die. We might suppose intuitively that the runt himself should go on struggling to the last, but the theory does not necessarily predict this. As soon as a runt becomes so small and weak that his expectation of life is reduced to the point where benefit to him due to parental investment is less than half the benefit that the same investment could potentially confer on the other babies, the runt should die gracefully and willingly. He can benefit his genes most by doing so. That is to say, a gene that gives the instruction 'Body, if you are very much smaller than your litter-mates, give up the struggle and die' could be successful in the gene pool, because it has a 50 per cent chance of being in the body of each brother and sister saved, and its chances of surviving in the body of the runt are very small anyway. There should be a point of no return in the career of a runt. Before he reaches this point he should go on struggling. As soon as he reaches it he should give up and preferably let himself be eaten by his litter-mates or his parents.

I did not mention it when we were discussing Lack's theory of clutch size, but the following is a reasonable strategy for a parent who is undecided as to what is her optimum clutch size for the current year. She might lay one more egg than she actually 'thinks' is likely to be the true optimum. Then, if the year's food crop should turn out to be a better one than expected, she will rear the extra child. If not, she can cut her losses. By being careful always to feed the young in the same order, say in order of size, she sees to it that one, perhaps a runt, quickly dies, and not too much food is wasted on him, beyond the initial investment of egg yolk or equivalent. From the mother's point of view, this may be the explanation of the runt phenomenon. He represents the hedging of the mother's bets. This has been observed in many birds.

Using our metaphor of the individual animal as a survival machine behaving as if it had the 'purpose' of preserving its genes, we can talk about a conflict between parents and young, a battle of the generations. The battle is a subtle one, and no holds are barred on either side. A child will lose no opportunity of cheating. It will pretend to be hungrier than it is, perhaps younger than it is, more in danger than it really is. It is too small and weak to bully its parents physically, but it uses every psychological weapon at its disposal: lying, cheating, deceiving, exploiting, right up to the point where it starts to penalize its relatives more than its genetic relatedness to them should allow. Parents, on the other hand, must be alert to cheating and deceiving, and must try not to be fooled by it. This might seem an easy task. If the parent knows that its child is likely to lie about how hungry it is, it might employ the tactic of feeding it a fixed amount and no more, even though the child goes on screaming. One trouble with this is that the child may not have been lying, and if it dies as a result of not being fed the parent would have lost some of its precious genes. Wild birds can die after being starved for only a few hours.

A. Zahavi has suggested a particularly diabolical form of child blackmail: the child screams in such a way as to attract predators deliberately to the nest. The child is 'saying' 'Fox, fox, come and get me.' The only way the parent can stop it screaming is to feed it. So the child gains more than its fair share of food, but at a cost of some risk to itself. The principle of this ruthless tactic is the same as that of the hijacker threatening to blow up an aeroplane, with himself on board, unless he is given a ransom. I am sceptical about whether it could ever be favoured in evolution, not because it is too ruthless, but because I doubt if it could ever pay the blackmailing baby. He has too much to lose if a predator really came. This is clear for an only child, which is the case Zahavi himself considers. No matter how much his mother may already have invested in him, he should still value his own life more than his mother values it, since she has only half of his genes. Moreover, the tactic would not pay even if the blackmailer was one of a clutch of vulnerable babies, all in the nest together, since the blackmailer has a 50 per cent genetic 'stake' in each of his endangered brothers and sisters, as well as a 100 per cent stake in himself. I suppose the theory might conceivably work if the predominant predator had the habit of only taking the largest nestling from a nest. Then it might pay a smaller one to use the threat of summoning a predator, since it would not be greatly endangering itself. This is analogous to holding a pistol to your brother's head rather than threatening to blow yourself up.

More plausibly, the blackmail tactic might pay a baby cuckoo. As is well known, cuckoo females lay one egg in each of several 'foster' nests, and then leave the unwitting foster parents, of a quite different species, to rear the cuckoo young. Therefore a baby cuckoo has no genetic stake in his foster brothers and sisters. (Some species of baby cuckoo will not have any foster brothers and sisters, for a sinister reason which we shall come to. For the moment I assume we are dealing with one of those species in which foster brothers and sisters co-exist alongside the baby cuckoo.) If a baby cuckoo screamed loudly enough to attract predators, it would have a lot to lose—its life—but the foster mother would have even more to lose, perhaps four of her young. It could therefore pay her to feed it more than its share, and the advantage of this to the cuckoo might outweigh the risk.

This is one of those occasions when it would be wise to translate back into respectable gene language, just to reassure ourselves that we have not become too carried away with subjective metaphors. What does it really mean to set up the hypothesis that baby cuckoos 'blackmail' their foster parents by screaming 'Predator, predator, come and get me and all my little brothers and sisters'? In gene terms it means the following.

Cuckoo genes for screaming loudly became more numerous in the cuckoo gene pool because the loud screams increased the probability that the foster parents would feed the baby cuckoos. The reason the foster parents responded to the screams in this way was that genes for responding to the screams had spread through the gene pool of the foster species. The reason these genes spread was that individual foster parents who did not feed the cuckoos extra food reared fewer of their own children—fewer than rival parents who did feed their cuckoos extra. This was because predators were attracted to the nest by the cuckoo cries. Although cuckoo genes for not screaming were less likely to end up in the bellies of predators than screaming genes, the non-screaming cuckoos paid the greater penalty of not being fed extra rations. Therefore the screaming genes spread through the cuckoo gene pool.

A similar chain of genetic reasoning, following the more subjective argument given above, would show that although such a blackmailing gene could conceivably spread through a cuckoo gene pool, it is unlikely to spread through the gene pool of an ordinary species, at least not for the specific reason that it attracted predators. Of course, in an ordinary species there could be other reasons for screaming genes to spread, as we have already seen, and these would *incidentally* have the effect of occasionally attracting predators. But here the selective influence of predation would be, if anything, in the direction of making the cries quieter. In the hypothetical case of the cuckoos, the net influence of predators, paradoxical as it sounds at first, could be to make the cries louder.

There is no evidence, one way or the other, on whether cuckoos, and other birds of similar 'brood-parasitic' habit, actually employ the blackmail tactic. But they certainly do not lack ruthlessness. For instance, there are honeyguides who, like cuckoos, lay their eggs in the nests of other species. The baby honeyguide is equipped with a sharp, hooked beak. As soon as he hatches out, while he is still blind, naked, and otherwise helpless, he scythes and slashes his foster brothers and sisters to death: dead brothers do not compete for food! The familiar British cuckoo achieves the same result in a slightly different way. It has a short incubation-time, and so the baby cuckoo manages to hatch out before its foster brothers and sisters. As soon as it hatches, blindly and mechanically, but with devastating effectiveness, it throws the other eggs out of the nest. It gets underneath an egg, fitting it into a hollow in its back. Then it slowly backs up the side of the nest, balancing the egg between its wing-stubs, and topples the egg out on to the ground. It does the same with all the other eggs, until it has the nest, and therefore the attention of its foster parents, entirely to itself.

One of the most remarkable facts I have learned in the past year was reported from Spain by F. Alvarez, L. Arias de Reyna, and H. Segura. They were investigating the ability of potential foster parents—potential victims of cuckoos—to detect intruders, cuckoo eggs or chicks. In the course of their experiments they had occasion to introduce into magpie nests the eggs and chicks of cuckoos, and, for comparison, eggs and chicks of other species such as swallows. On one occasion they introduced a baby swallow into a magpie's nest. The next day they noticed one of the magpie eggs lying on the ground under the nest. It had not broken, so they picked it up, replaced it, and watched. What they saw is utterly remarkable. The baby swallow, behaving exactly as if it was a baby cuckoo, threw the egg out. They replaced the egg again, and exactly the same thing happened. The baby swallow used the cuckoo method of balancing the egg on its back between its wing-stubs, and walking backwards up the side of the nest until the egg toppled out.

Perhaps wisely, Alvarez and his colleagues made no attempt to explain their astonishing observation. How could such behaviour evolve in the swallow gene pool? It must correspond to something in the normal life of a swallow. Baby swallows are not accustomed to finding themselves in magpie nests. They are never normally found in any nest except their own. Could the behaviour represent an evolved anti-cuckoo adaptation? Has the natural selection been favouring a policy of counter-attack in the swallow gene pool, genes for hitting the cuckoo with his own weapons? It seems to be a fact that swallows' nests are not normally parasitized by cuckoos. Perhaps this is why. According to this theory, the magpie eggs of the experiment would be incidentally getting the same treatment, perhaps because, like cuckoo eggs, they are bigger than swallow eggs. But if baby swallows can tell the difference between a large egg and a normal swallow egg, surely the mother should be able to as well. In this case why is it not the mother who ejects the cuckoo egg, since it would be so much easier for her to do so than the baby? The same objection applies to the theory that the baby swallow's behaviour normally functions to remove addled eggs or other debris from the nest. Once again, this task could be—and is—performed better by the parent. The fact that the difficult and skilled egg-rejecting operation was seen to be performed by a weak and helpless baby swallow, whereas an adult swallow could surely do it much more easily, compels me to the conclusion that, from the parent's point of view, the baby is up to no good.

It seems to me just conceivable that the true explanation has nothing to do with cuckoos at all. The blood may chill at the thought, but could this be what baby swallows do to each other? Since the firstborn is going to compete with his yet unhatched brothers and sisters for parental investment, it could be to his advantage to begin his life by throwing out one of the other eggs.

The Lack theory of clutch size considered the optimum from the parent's point of view. If I am a mother swallow, the optimum clutch size from my point of view is, say, five. But if I am a baby swallow, the optimum clutch size as I see it may well be a smaller number, provided I am one of them! The parent has a certain amount of parental investment, which she 'wishes' to distribute even-handedly among five young. But each baby wants more than his allotted one-fifth share. Unlike a cuckoo, he does not want all of it, because he is related to the other babies. But he does want more than one-fifth. He can acquire a $\frac{1}{4}$ share simply by tipping out one egg; a $\frac{1}{3}$ share by tipping out another. Translating into gene language, a gene for fratricide could conceivably spread through the gene pool, because it has 100 per cent chance of being in the body of the fratricidal individual, and only a 50 per cent chance of being in the body of his victim.

The chief objection to this theory is that it is very difficult to believe that nobody would have seen this diabolical behaviour if it really occurred. I have no convincing explanation for this. There are different races of swallow in different parts of the world. It is known that the Spanish race differs from, for example, the British one, in certain respects. The Spanish race has not been subjected to the same degree of intensive observation as the British one, and I suppose it is just conceivable that fratricide occurs but has been overlooked.

My reason for suggesting such an improbable idea as the fratricide hypothesis here is that I want to make a general point. This is that the ruthless behaviour of a baby cuckoo is only an extreme case of what must go on in any family. Full brothers are more closely related to each other than a baby cuckoo is to its foster brothers, but the difference is only a matter of degree. Even if we cannot believe that outright fratricide could evolve, there must be numerous lesser examples of selfishness where the cost to the child, in the form of losses to his brothers and sisters, is outweighed, more than two to one, by the benefit to himself. In such cases, as in the example of weaning time, there is a real conflict of interest between parent and child.

Who is most likely to win this battle of the generations? R. D. Alexander has written an interesting paper in which he suggests that there is a general answer to this question. According to him the parent will always win.* Now if this is the case, you have been wasting your time reading this chapter. If Alexander is right, much that is of interest follows. For instance, altruistic behaviour could evolve, not because of benefit to the genes of the individual himself, but solely because of benefit to his parents' genes. Parental manipulation, to use Alexander's term, becomes an alternative evolutionary cause of altruistic behaviour, independent of straightforward kin selection. It is therefore important that we examine Alexander's reasoning, and convince ourselves that we understand why he is wrong. This should really be done mathematically, but we are avoiding explicit use of mathematics in this book, and it is possible to give an intuitive idea of what is wrong with Alexander's thesis.

His fundamental genetic point is contained in the following abridged quotation. 'Suppose that a juvenile...cause(s) an uneven distribution of parental benefits in its own favor, thereby reducing the mother's own overall reproduction. A gene which in this fashion improves an individual's fitness when it is a juvenile cannot fail to lower its fitness more when it is an adult, for such mutant genes will be present in an increased proportion of the mutant individual's offspring.' The fact that Alexander is considering a newly mutated gene is not fundamental to the argument. It is better to think of a rare gene inherited from one of the parents. 'Fitness' has the special technical meaning of reproductive success. What Alexander is basically saying is this. A gene that made a child grab more than his fair share when he was a child, at the expense of his parent's total reproductive output, might indeed increase his chances of surviving. But he would pay the penalty when he came to be a parent himself, because his own children would tend to inherit the same selfish gene, and this would reduce his overall reproductive success. He would be hoist with his own petard. Therefore the gene cannot succeed, and parents must always win the conflict.

Our suspicions should be immediately aroused by this argument, because it rests on the assumption of a genetic asymmetry which is not really there. Alexander is using the words 'parent' and 'offspring' as though there was a fundamental genetic difference between them. As we have seen, although there are *practical* differences between parent and child, for instance parents are older than children, and children come out of parents' bodies, there is really no fundamental *genetic* asymmetry. The relatedness is 50 per cent, whichever way round you look at it. To illustrate what I mean, I am going to repeat Alexander's words, but with 'parent', 'juvenile', and other appropriate words reversed. 'Suppose that a *parent* has a gene that tends to cause an *even* distribution of parental benefits. A gene which in this fashion improves an individual's fitness when it is a *parent* could not fail to have lowered its fitness more when it was a *juvenile*.' We therefore reach the opposite conclusion to Alexander, namely that in any parent–offspring conflict, the child must win!

Obviously something is wrong here. Both arguments have been put too simply. The purpose of my reverse quotation is not to prove the opposite point to Alexander, but simply to show that you cannot argue in that kind of artificially asymmetrical way. Both Alexander's argument, and my reversal of it, erred through looking at things from the point of view of an *individual*—in Alexander's case, the parent, in my case, the child. I believe this kind of error is all too easy to make when we use the technical term 'fitness'. This is why I have avoided using the word in this book. There is really only one entity whose point of view matters in evolution, and that entity is the selfish gene. Genes in juvenile bodies will be selected for their ability to outsmart parental bodies; genes in parental bodies will be selected for their ability to outsmart the young. There is no paradox in the fact that the very same genes successively occupy a juvenile body and a parental body. Genes are selected for their ability to make the best use of the levers of power at their disposal: they will exploit their practical opportunities. When a gene is sitting in a juvenile body its practical opportunities will be different from when it is sitting in a parental body. Therefore its optimum policy will be different in the two stages in its body's life history. There is no reason to suppose, as Alexander does, that the later optimum policy should necessarily overrule the earlier.

There is another way of putting the argument against Alexander. He is tacitly assuming a false asymmetry between the parent/ child relationship, on the one hand, and the brother/sister relationship on the other. You will remember that, according to Trivers, the cost to a selfish child of grabbing more than his share, the reason why he only grabs up to a point, is the danger of loss of his brothers and sisters who each bear half his genes. But brothers and sisters are only a special case of relatives with a 50 per cent relatedness. The selfish child's own future children are no more and no less 'valuable' to him than his brothers and sisters. Therefore the total net cost of grabbing more than your fair share of resources should really be measured, not only in lost brothers and sisters, but also in lost future offspring due to their selfishness among themselves. Alexander's point about the disadvantage of juvenile selfishness spreading to your own children, thereby reducing your own long-term reproductive output, is well taken, but it simply means we must add this in to the cost side of the equation. An individual child will still do well to be selfish so long as the net benefit to him is at least half the net cost to close relatives. But 'close relatives' should be read as including not just brothers and sisters, but future children of one's own as well. An individual should reckon his own welfare as twice as valuable as that of his brothers, which is the basic assumption Trivers makes. But he should also value himself twice as highly as one of his own future children. Alexander's conclusion that there is a built-in advantage on the parent's side in the conflict of interests is not correct.

In addition to his fundamental genetic point, Alexander also has more practical arguments, stemming from undeniable asymmetries in the parent/child relationship. The parent is the active partner, the one who actually does the work to get the food, etc., and is therefore in a position to call the tune. If the parent decides to withdraw its labour, there is not much that the child can do about it, since it is smaller, and cannot hit back. Therefore the parent is in a position to impose its will, regardless of what the child may want. This argument is not obviously wrong, since in this case the asymmetry that it postulates is a real one. Parents really are bigger, stronger, and more worldly-wise than children. They seem to hold all the good cards. But the young have a few aces up their sleeves too. For example, it is important for a parent to know how hungry each of its children is, so that it can most efficiently dole out the food. It could of course ration the food exactly equally between them all, but in the best of all possible worlds this would be less efficient than a system of giving a little bit more to those that could genuinely use it best. A system whereby each child told the parent how hungry he was would be ideal for the parent, and, as we have seen, such a system seems to have evolved. But the young are in a strong position to lie, because they *know* exactly how hungry they are, while the parent can only *guess* whether they are telling the truth or not. It is almost impossible for a parent to detect a small lie, although it might see through a big one.

Then again, it is of advantage to parent to know when a baby is happy, and it is a good thing for a baby to be able to tell its parents when it is happy. Signals like purring and smiling may have been selected because they enable parents to learn which of their actions are most beneficial to their children. The sight of her child smiling, or the sound of her kitten purring, is rewarding to a mother, in the same sense as food in the stomach is rewarding to a rat in a maze. But once it becomes true that a sweet smile or a loud purr are rewarding, the child is in a position to use the smile or the purr in order to manipulate the parent, and gain more than its fair share of parental investment.

There is, then, no general answer to the question of who is more likely to win the battle of the generations. What will finally emerge is a compromise between the ideal situation desired by the child and that desired by the parent. It is a battle comparable to that between cuckoo and foster parent, not such a fierce battle to be sure, for the enemies do have some genetic interests in common—they are only enemies up to a point, or during certain sensitive times. However, many of the tactics used by cuckoos, tactics of deception and exploitation, may be employed by a parent's own young, although the parent's own young will stop short of the total selfishness that is to be expected of a cuckoo.

This chapter, and the next in which we discuss conflict between mates, could seem horribly cynical, and might even be distressing to human parents, devoted as they are to their children, and to each other. Once again I must emphasize that I am not talking about conscious motives. Nobody is suggesting that children deliberately and consciously deceive their parents because of the selfish genes within them. And I must repeat that when I say something like 'A child should lose no opportunity of cheating... lying, deceiving, exploiting...', I am using the word 'should' in a special way. I am not advocating this kind of behaviour as moral or desirable. I am simply saying that natural selection will tend to favour children who do act in this way, and that therefore when we look at wild populations we may expect to see cheating and selfishness within families. The phrase 'the child should cheat' means that genes that tend to make children cheat have an advantage in the gene pool. If there is a human moral to be drawn, it is that we must *teach* our children altruism, for we cannot expect it to be part of their biological nature.

BATTLE OF THE SEXES

If there is conflict of interest between parents and children, who share 50 per cent of each other's genes, how much more severe must be the conflict between mates, who are not related to each other?* All that they have in common is a 50 per cent genetic shareholding in the same children. Since father and mother are both interested in the welfare of different halves of the same children, there may be some advantage for both of them in cooperating with each other in rearing those children. If one parent can get away with investing less than his or her fair share of costly resources in each child, however, he will be better off, since he will have more to spend on other children by other sexual partners, and so propagate more of his genes. Each partner can therefore be thought of as trying to exploit the other, trying to force the other one to invest more. Ideally, what an individual would 'like' (I don't mean physically enjoy, although he might) would be to copulate with as many members of the opposite sex as possible, leaving the partner in each case to bring up the children. As we shall see, this state of affairs is achieved by the males of a number of species, but in other species the males are obliged to share an equal part of the burden of bringing up children. This view of sexual partnership, as a relationship of mutual mistrust and mutual exploitation, has been stressed especially by Trivers. It is

a comparatively new one to ethologists. We had usually thought
of sexual behaviour, copulation, and the courtship that precedes
it as essentially a cooperative venture undertaken for mutual
benefit, or even for the good of the species!

Let us go right back to first principles, and inquire into the fundamental nature of maleness and femaleness. In Chapter 3 we
discussed sexuality without stressing its basic asymmetry. We
simply accepted that some animals are called male, and others
female, without asking what these words really meant. But what
is the essence of maleness? What, at bottom, defines a female?
We as mammals see the sexes defined by whole syndromes of
characteristics—possession of a penis, bearing of the young,
suckling by means of special milk glands, certain chromosomal
features, and so on. These criteria for judging the sex of an individual are all very well for mammals but, for animals and plants
generally, they are no more reliable than is the tendency to wear
trousers as a criterion for judging human sex. In frogs, for
instance, neither sex has a penis. Perhaps, then, the words male
and female have no general meaning. They are, after all, only
words, and if we do not find them helpful for describing frogs,
we are quite at liberty to abandon them. We could arbitrarily
divide frogs into Sex 1 and Sex 2 if we wished. However, there is
one fundamental feature of the sexes which can be used to label
males as males, and females as females, throughout animals and
plants. This is that the sex cells or 'gametes' of males are much
smaller and more numerous than the gametes of females. This is
true whether we are dealing with animals or plants. One group
of individuals has large sex cells, and it is convenient to use the
word female for them. The other group, which it is convenient
to call male, has small sex cells. The difference is especially pronounced in reptiles and in birds, where a single egg cell is big
enough and nutritious enough to feed a developing baby for

several weeks. Even in humans, where the egg is microscopic, it is still many times larger than the sperm. As we shall see, it is possible to interpret all the other differences between the sexes as stemming from this one basic difference.

In certain primitive organisms, for instance some fungi, maleness and femaleness do not occur, although sexual reproduction of a kind does. In the system known as isogamy the individuals are not distinguishable into two sexes. Anybody can mate with anybody else. There are not two different sorts of gametes—sperms and eggs—but all sex cells are the same, called isogametes. New individuals are formed by the fusion of two isogametes, each produced by meiotic division. If we have three isogametes, A, B, and C, A could fuse with B or C, B could fuse with A or C. The same is never true of normal sexual systems. If A is a sperm and it can fuse with B or C, then B and C must be eggs and B cannot fuse with C.

When two isogametes fuse, both contribute equal numbers of genes to the new individual, and they also contribute equal amounts of food reserves. Sperms and eggs too contribute equal numbers of genes, but eggs contribute far more in the way of food reserves: indeed, sperms make no contribution at all and are simply concerned with transporting their genes as fast as possible to an egg. At the moment of conception, therefore, the father has invested less than his fair share (i.e. 50 per cent) of resources in the offspring. Since each sperm is so tiny, a male can afford to make many millions of them every day. This means he is potentially able to beget a very large number of children in a very short period of time, using different females. This is only possible because each new embryo is endowed with adequate food by the mother in each case. This therefore places a limit on the number of children a female can have, but the number of children a male can have is virtually unlimited. Female exploitation begins here.*

Parker and others showed how this asymmetry might have evolved from an originally isogamous state of affairs. In the days when all sex cells were interchangeable and of roughly the same size, there would have been some that just happened to be slightly bigger than others. In some respects a big isogamete would have an advantage over an average-sized one, because it would get its embryo off to a good start by giving it a large initial food supply. There might therefore have been an evolutionary trend towards larger gametes. But there was a catch. The evolution of isogametes that were larger than was strictly necessary would have opened the door to selfish exploitation. Individuals who produced *smaller* than average gametes could cash in, provided they could ensure that their small gametes fused with extra-big ones. This could be achieved by making the small ones more mobile, and able to seek out large ones actively. The advantage to an individual of producing small, rapidly moving gametes would be that he could afford to make a larger number of gametes, and therefore could potentially have more children. Natural selection favoured the production of sex cells that were small and that actively sought out big ones to fuse with. So we can think of two divergent sexual 'strategies' evolving. There was the large-investment or 'honest' strategy. This automatically opened the way for a small-investment exploitative strategy. Once the divergence between the two strategies had started, it would have continued in runaway fashion. Medium-sized intermediates would have been penalized, because they did not enjoy the advantages of either of the two more extreme strategies. The exploiters would have evolved smaller and smaller size, and faster mobility. The honest ones would have evolved larger and larger size, to compensate for the ever-smaller investment contributed by the exploiters, and they became immobile because they would always be actively chased by the exploiters anyway. Each honest one would 'prefer' to fuse with another honest one. But the selection pressure to lock out exploiters would have been weaker than the pressure on exploiters to duck under the barrier: the exploiters had more to lose, and they therefore won the evolutionary battle. The honest ones became eggs, and the exploiters became sperms.

Males, then, seem to be pretty worthless fellows, and on simple 'good of the species' grounds, we might expect that males would become less numerous than females. Since one male can theoretically produce enough sperms to service a harem of 100 females we might suppose that females should outnumber males in animal populations by 100 to 1. Other ways of putting this are that the male is more 'expendable', and the female more 'valuable' to the species. Of course, looked at from the point of view of the species as a whole, this is perfectly true. To take an extreme example, in one study of elephant seals, 4 per cent of the males accounted for 88 per cent of all the copulations observed. In this case, and in many others, there is a large surplus of bachelor males who probably never get a chance to copulate in their whole lives. But these extra males live otherwise normal lives, and they eat up the population's food resources no less hungrily than other adults. From a 'good of the species' point of view this is horribly wasteful; the extra males might be regarded as social parasites. This is just one more example of the difficulties that the group selection theory gets into. The selfish gene theory, on the other hand, has no trouble in explaining the fact that the numbers of males and females tend to be equal, even when the males who actually reproduce may be a small fraction of the total number. The explanation was first offered by R. A. Fisher.

The problem of how many males and how many females are born is a special case of a problem in parental strategy. Just as we discussed the optimal family size for an individual parent trying to maximize her gene survival, we can also discuss the optimal sex ratio. Is it better to entrust your precious genes to sons or to daughters? Suppose a mother invested all her resources in sons, and therefore had none left to invest in daughters: would she on average contribute more to the gene pool of the future than a rival mother who invested in daughters? Do genes for preferring sons become more or less numerous than genes for preferring daughters? What Fisher showed is that under normal circumstances the stable sex ratio is 50:50. In order to see why, we must first know a little bit about the mechanics of sex determination.

In mammals, sex is determined genetically as follows. All eggs are capable of developing into either a male or a female. It is the sperms that carry the sex-determining chromosomes. Half the sperms produced by a man are female-producing, or X-sperms, and half are male-producing, or Y-sperms. The two sorts of sperms look alike. They differ with respect to one chromosome only. A gene for making a father have nothing but daughters could achieve its object by making him manufacture nothing but X-sperms. A gene for making a mother have nothing but daughters could work by making her secrete a selective spermicide, or by making her abort male embryos. What we seek is something equivalent to an evolutionarily stable strategy (ESS), although here, even more than in the chapter on aggression, strategy is just a figure of speech. An individual cannot literally choose the sex of his children. But genes for tending to have children of one sex or the other are possible. If we suppose that such genes, favouring unequal sex ratios, exist, are any of them likely to become more numerous in the gene pool than their rival alleles, which favour an equal sex ratio?

Suppose that in the elephant seals mentioned above, a mutant gene arose that tended to make parents have mostly daughters. Since there is no shortage of males in the population, the daughters would have no trouble finding mates, and the daughter-manufacturing gene could spread. The sex ratio in the population might then start to shift towards a surplus of females. From the point of view of the good of the species, this would be all right, because just a few males are quite capable of providing all the sperms needed for even a huge surplus of females, as we have seen. Superficially, therefore, we might expect the daughter-producing gene to go on spreading until the sex ratio was so unbalanced that the few remaining males, working flat out, could just manage. But now, think what an enormous genetic advantage is enjoyed by those few parents who have sons. Anyone who invests in a son has a very good chance of being the grandparent of hundreds of seals. Those who are producing nothing but daughters are assured of a safe few grandchildren, but this is nothing compared to the glorious genetic possibilities that open up before anyone specializing in sons. Therefore genes for producing sons will tend to become more numerous, and the pendulum will swing back.

For simplicity I have talked in terms of a pendulum swing. In practice the pendulum would never have been allowed to swing that far in the direction of female domination, because the pressure to have sons would have started to push it back as soon as the sex ratio became unequal. The strategy of producing equal numbers of sons and daughters is an evolutionarily stable strategy, in the sense that any gene for departing from it makes a net loss.

I have told the story in terms of numbers of sons versus numbers of daughters. This is to make it simple, but strictly it should be worked out in terms of parental investment, meaning all the food and other resources that a parent has to offer, measured in the way discussed in the previous chapter. Parents should *invest* equally in sons and daughters. This usually means they should have numerically as many sons as they have daughters. But there could be unequal sex ratios that were evolutionarily stable, provided correspondingly unequal amounts of resources were invested in sons and daughters. In the case of the elephant seals, a policy of having three times as many daughters as sons, but of making each son a supermale by investing three times as much food and other resources in him, could be stable. By investing more food in a son and making him big and strong, a parent might increase his chances of winning the supreme prize of a harem. But this is a special case. Normally the amount invested in each son will roughly equal the amount invested in each daughter, and the sex ratio, in terms of numbers, is usually one to one.

In its long journey down the generations therefore, an average gene will spend approximately half its time sitting in male bodies, and the other half sitting in female bodies. Some gene effects show themselves only in bodies of one sex. These are called sex-limited gene effects. A gene controlling penis-length expresses this effect only in male bodies, but it is carried about in female bodies too and may have some quite different effect on female bodies. There is no reason why a man should not inherit a tendency to develop a long penis from his mother.

In whichever of the two sorts of body it finds itself, we can expect a gene to make the best use of the opportunities offered by that sort of body. These opportunities may well differ according to whether the body is male or female. As a convenient approximation, we can once again assume that each individual body is a selfish machine, trying to do the best for all its genes. The best policy for such a selfish machine will often be one thing if it is male, and quite a different thing if it is female. For brevity, we shall again use the convention of thinking of the individual as though it had a conscious purpose. As before, we shall hold in the back of our mind that this is just a figure of speech. A body is really a machine blindly programmed by its selfish genes.

Consider again the mated pair with which we began the chapter. Both partners, as selfish machines, 'want' sons and daughters in equal numbers. To this extent they agree. Where they disagree is in who is going to bear the brunt of the cost of rearing each one of those children. Each individual wants as many surviving children as possible. The less he or she is obliged to invest in any one of those children, the more children he or she can have. The obvious way to achieve this desirable state of affairs is to induce your sexual partner to invest more than his or her fair share of resources in each child, leaving you free to have other children with other partners. This would be a desirable strategy for either sex, but it is more difficult for the female to achieve. Since she starts by investing more than the male, in the form of her large, food-rich egg, a mother is already at the moment of conception 'committed' to each child more deeply than the father is. She stands to lose more if the child dies than the father does. More to the point, she would have to invest more than the father *in the future* in order to bring a new substitute child up to the same level of development. If she tried the tactic of leaving the father holding the baby, while she went off with another male, the father might, at relatively small cost to himself, retaliate by abandoning the baby too. Therefore, at least in the early stages of child development, if any abandoning is going to be done, it is likely to be the father who abandons the mother rather than the other way around. Similarly, females can be expected to invest more in children than males, not only at the outset, but throughout development. So, in mammals for example, it is the female who incubates the foetus in her own body, the female who makes the milk to suckle it when it is born, the female who bears the brunt of the load of bringing it up and protecting it. The female sex is exploited, and the fundamental evolutionary basis for the exploitation is the fact that eggs are larger than sperms.

Of course in many species the father does work hard and faithfully at looking after the young. But even so, we must expect that there will normally be some evolutionary pressure on males to invest a little bit less in each child, and to try to have more children by different wives. By this I simply mean that there will be a tendency for genes that say 'Body, if you are male leave your mate a little bit earlier than my rival allele would have you do, and look for another female', to be successful in the gene pool. The extent to which this evolutionary pressure actually prevails in practice varies greatly from species to species. In many, for example in the birds of paradise, the female receives no help at all from any male, and she rears her children on her own. Other species such as kittiwakes form monogamous pairbonds of exemplary fidelity, and both partners cooperate in the work of bringing up children. Here we must suppose that some evolutionary counter-pressure has been at work: there must be a penalty attached to the selfish mate-exploitation strategy as well as a benefit, and in kittiwakes the penalty outweighs the benefit. It will in any case only pay a father to desert his wife and child if the wife has a reasonable chance of rearing the child on her own.

Trivers has considered the possible courses of action open to a mother who has been deserted by her mate. Best of all for her would be to try to deceive another male into adopting her child, 'thinking' it is his own. This might not be too difficult if it is still a foetus, not yet born. Of course, while the child bears half her genes, it bears no genes at all from the gullible step-father. Natural selection would severely penalize such gullibility in males and indeed would favour males who took active steps to kill any potential step-children as soon as they mated with a new wife. This is very probably the explanation of the so-called Bruce effect: male mice secrete a chemical which when smelt by a pregnant female can cause her to abort. She only aborts if the smell is different from that of her former mate. In this way a male mouse destroys his potential step-children, and renders his new wife receptive to his own sexual advances. Ardrey, incidentally, sees the Bruce effect as a population control mechanism! A similar example is that of male lions, who, when newly arrived in a pride, sometimes murder existing cubs, presumably because these are not their own children.

A male can achieve the same result without necessarily killing step-children. He can enforce a period of prolonged courtship before he copulates with a female, driving away all other males who approach her, and preventing her from escaping. In this way he can wait and see whether she is harbouring any little step-children in her womb, and desert her if so. We shall see below a reason why a female might want a long 'engagement' period before copulation. Here we have a reason why a male might want one too. Provided he can isolate her from all contact with other males, it helps to avoid being the unwitting benefactor of another male's children.

Assuming then that a deserted female cannot fool a new male into adopting her child, what else can she do? Much may depend on how old the child is. If it is only just conceived, it is true that she has invested the whole of one egg in it and perhaps more, but it may still pay her to abort it and find a new mate as quickly as possible. In these circumstances it would be to the mutual advantage both of her and of the potential new husband that she should abort—since we are assuming she has no hope of fooling him into adopting the child. This could explain why the Bruce effect works from the female's point of view.

Another option open to a deserted female is to stick it out, and try and rear the child on her own. This will especially pay her if the child is already quite old. The older he is the more has already been invested in him, and the less it will take out of her to finish the job of rearing him. Even if he is still quite young, it might yet pay her to try to salvage something from her initial investment, even if she has to work twice as hard to feed the child, now that the male has gone. It is no comfort to her that the child contains half the male's genes too, and that she could spite him by abandoning it. There is no point in spite for its own sake. The child carries half her genes, and the dilemma is now hers alone.

Paradoxically, a reasonable policy for a female who is in danger of being deserted might be to walk out on the male *before* he walks out on her. This could pay her, even if she has already invested more in the child than the male has. The unpleasant truth is that in some circumstances an advantage accrues to the partner who deserts *first*, whether it is the father or the mother. As Trivers puts it, the partner who is left behind is placed in a cruel bind. It is a rather horrible but very subtle argument. A parent may be expected to desert the moment it is possible for him or her to say the following: 'This child is now far enough developed that either of us *could* finish off rearing it on our own. Therefore it would pay me to desert now, provided I could be sure my partner would not desert as well. If I did desert now, my partner would do whatever is best for her/his genes. He/she would be forced into making a more drastic decision than I am making now, because I would have already left. My partner would "know" that if he/she left as well, the child would surely die. Therefore, assuming that my partner will take the decision that is best for his/her own selfish genes, I conclude that my own best course of action is to desert first. This is especially so, since my partner may be "thinking" along exactly the same lines, and may seize the initiative at any minute by deserting me!' As always, the subjective soliloquy is intended for illustration only. The point is that genes for deserting *first* could be favourably selected simply because genes for deserting *second* would not be.

We have looked at some of the things that a female might do if she has been deserted by her mate. But these all have the air of making the best of a bad job. Is there anything a female can do to reduce the extent to which her mate exploits her in the first place? She has a strong card in her hand. She can refuse to copulate. She is in demand, in a seller's market. This is because she brings the dowry of a large, nutritious egg. A male who successfully copulates gains a valuable food reserve for his offspring. The female is potentially in a position to drive a hard bargain before she copulates. Once she has copulated she has played her ace—her egg has been committed to the male. It is all very well to talk about driving hard bargains, but we know very well it is not really like that. Is there any realistic way in which something equivalent to driving a hard bargain could evolve by natural selection? I shall consider two main possibilities, called the domestic-bliss strategy and the he-man strategy.

The simplest version of the domestic-bliss strategy is this. The female looks the males over, and tries to spot signs of fidelity and domesticity in advance. There is bound to be variation in the population of males in their predisposition to be faithful husbands. If females could recognize such qualities in advance, they could benefit themselves by choosing males possessing them. One way for a female to do this is to play hard to get for a long time, to be coy. Any male who is not patient enough to wait until the female eventually consents to copulate is not likely to be a good bet as a faithful husband. By insisting on a long engagement period, a female weeds out casual suitors, and only finally copulates with a male who has proved his qualities of fidelity and perseverance in advance. Feminine coyness is in fact very common among animals, and so are prolonged courtship or engagement periods. As we have already seen, a long engagement can also benefit a male where there is a danger of his being duped into caring for another male's child.

Courtship rituals often include considerable pre-copulation investment by the male. The female may refuse to copulate until the male has built her a nest. Or the male may have to feed her quite substantial amounts of food. This, of course, is very good from the female's point of view, but it also suggests another possible version of the domestic-bliss strategy. Could females force males to invest so heavily in their offspring *before* they allow copulation that it would no longer pay the males to desert *after* copulation? The idea is appealing. A male who waits for a coy female eventually to copulate with him is paying a cost: he is forgoing the chance to copulate with other females, and he is spending a lot of time and energy in courting her. By the time he is finally allowed to copulate with a particular female, he will inevitably be heavily 'committed' to her. There will be little temptation for him to desert her, if he knows that any future female he approaches will also procrastinate in the same manner before she will get down to business.

As I showed in a paper, there is a mistake in Trivers's reasoning here. He thought that prior investment in itself committed an individual to future investment. This is fallacious economics. A business man should never say 'I have already invested so much in the Concorde airliner (for instance) that I cannot afford to scrap it now.' He should always ask instead whether it would pay him *in the future*, to cut his losses, and abandon the project now, even though he has already invested heavily in it. Similarly, it is no use a female forcing a male to invest heavily in her in the hope that this, on its own, will deter the male from subsequently deserting. This version of the domestic-bliss strategy depends upon one further crucial assumption. This is that a majority of the females can be relied upon to play the same game. If there are loose females in the population, prepared to welcome males who have deserted their wives, then it could pay a male to desert his wife, no matter how much he has already invested in her children.

Much therefore depends on how the majority of females behave. If we were allowed to think in terms of a conspiracy of females there would be no problem. But a conspiracy of females can no more evolve than the conspiracy of doves which we considered in Chapter 5. Instead, we must look for evolutionarily stable strategies. Let us take Maynard Smith's method of analysing aggressive contests, and apply it to sex.* It will be a little bit more complicated than the case of the hawks and doves, because we shall have two female strategies and two male strategies.

As in Maynard Smith's studies, the word 'strategy' refers to a blind unconscious behaviour program. Our two female strategies will be called *coy* and *fast*, and the two male strategies will be called *faithful* and *philanderer*. The behavioural rules of the four types are as follows. Coy females will not copulate with a male until he has gone through a long and expensive courtship period lasting several weeks. Fast females will copulate immediately with anybody. Faithful males are prepared to go on courting for a long time, and after copulation they stay with the female and help her to rear the young. Philanderer males lose patience quickly if a female will not copulate with them straight away: they go off and look for another female; after copulation too they do not stay and act as good fathers, but go off in search of fresh females. As in the case of the hawks and doves, these are not the only possible strategies, but it is illuminating to study their fates nevertheless.

Like Maynard Smith, we shall use some arbitrary hypothetical values for the various costs and benefits. To be more general it can be done with algebraic symbols, but numbers are easier to understand. Suppose that the genetic pay-off gained by each parent when a child is reared successfully is +15 units. The cost of rearing one child, the cost of all its food, all the time spent looking after it, and all the risks taken on its behalf, is −20 units. The cost is expressed as negative, because it is 'paid out' by the parents. Also negative is the cost of wasting time in prolonged courtship. Let this cost be −3 units.

Imagine we have a population in which all the females are coy, and all the males are faithful. It is an ideal monogamous society. In each couple, the male and the female both get the same average pay-off. They get +15 for each child reared; they share the cost of rearing it (−20) equally between the two of them, an average of −10 each. They both pay the −3 point penalty for wasting time in prolonged courtship. The average pay-off for each is therefore $+15 − 10 − 3 = +2$.

Now suppose a single fast female enters the population. She does very well. She does not pay the cost of delay, because she does not indulge in prolonged courtship. Since all the males in the population are faithful, she can reckon on finding a good father for her children whomever she mates with. Her average pay-off per child is $+15 − 10 = +5$. She is 3 units better off than her coy rivals. Therefore fast genes will start to spread.

If the success of fast females is so great that they come to predominate in the population, things will start to change in the male camp too. So far, faithful males have had a monopoly. But now if a philanderer male arises in the population, he starts to do better than his faithful rivals. In a population where all the females are fast, the pickings for a philanderer male are rich indeed. He gets the +15 points if a child is successfully reared, and he pays neither of the two costs. What this lack of cost mainly means to him is that he is free to go off and mate with new females. Each of his unfortunate wives struggles on alone with the child, paying the entire −20 point cost, although she does not pay anything for wasting time in courting. The net pay-off for a fast female when she encounters a philanderer male is $+15 - 20 = -5$; the pay-off to the philanderer himself is $+15$. In a population in which all the females are fast, philanderer genes will spread like wildfire.

If the philanderers increase so successfully that they come to dominate the male part of the population, the fast females will be in dire straits. Any coy female would have a strong advantage. If a coy female encounters a philanderer male, no business results. She insists on prolonged courtship; he refuses and goes off in search of another female. Neither partner pays the cost of wasting time. Neither gains anything either, since no child is produced. This gives a net pay-off of zero for a coy female in a population where all the males are philanderers. Zero may not seem much, but it is better than the -5 which is the average score for a fast female. Even if a fast female decided to leave her young after being deserted by a philanderer, she would still have paid the considerable cost of an egg. So, coy genes start to spread through the population again.

To complete the hypothetical cycle, when coy females increase in numbers so much that they predominate, the philanderer males, who had such an easy time with the fast females, start to feel the pinch. Female after female insists on a long and arduous courtship. The philanderers flit from female to female, and always the story is the same. The net pay-off for a philanderer male when all the females are coy is zero. Now if a single faithful male should turn up, he is the only one with whom the coy females will mate. His net pay-off is $+2$, better than that of the philanderers. So, faithful genes start to increase, and we come full circle.

As in the case of the aggression analysis, I have told the story as though it was an endless oscillation. But, as in that case, it can be shown that really there would be no oscillation. The system would converge to a stable state.* If you do the sums, it turns out that a population in which $\frac{5}{6}$ of the females are coy, and $\frac{5}{8}$ of the males are faithful, is evolutionarily stable. This is, of course, just for the particular arbitrary numbers that we started out with, but it is easy to work out what the stable ratios would be for any other arbitrary assumptions.

As in Maynard Smith's analyses, we do not have to think of there being two different sorts of male and two different sorts of female. The ESS could equally well be achieved if each male spends $\frac{5}{8}$ of his time being faithful and the rest of his time philandering; and each female spends $\frac{5}{6}$ of her time being coy and $\frac{1}{6}$ of her time being fast. Whichever way we think of the ESS, what it means is this. Any tendency for members of either sex to deviate from their appropriate stable ratio will be penalized by a consequent change in the ratio of strategies of the other sex, which is, in turn, to the disadvantage of the original deviant. Therefore the ESS will be preserved.

We can conclude that it is certainly possible for a population consisting largely of coy females and faithful males to evolve. In these circumstances the domestic-bliss strategy for females really does seem to work. We do not have to think in terms of a conspiracy of coy females. Coyness can actually pay a female's selfish genes.

There are various ways in which females can put this type of strategy into practice. I have already suggested that a female might refuse to copulate with a male who has not already built her a nest, or at least helped her to build a nest. It is indeed the case that in many monogamous birds copulation does not take place until after the nest is built. The effect of this is that at the moment of conception the male has invested a good deal more in the child than just his cheap sperms.

Demanding that a prospective mate should build a nest is one effective way for a female to trap him. It might be thought that almost anything that costs the male a great deal would do in theory, even if that cost is not directly paid in the form of benefit to the unborn children. If all females of a population forced males to do some difficult and costly deed, like slaying a dragon or climbing a mountain, before they would consent to copulate with them, they could in theory be reducing the temptation for the males to desert after copulation. Any male tempted to desert his mate and try to spread more of his genes by another female would be put off by the thought that he would have to kill another dragon. In practice, however, it is unlikely that females would impose such arbitrary tasks as dragon-killing, or Holy-Grail-seeking on their suitors. The reason is that a rival female who imposed a task no less arduous, but more useful to her and her children, would have an advantage over more romantically minded females who demanded a pointless labour of love. Building a nest may be less romantic than slaying a dragon or swimming the Hellespont, but it is much more useful.

Also useful to the female is the practice I have already mentioned of courtship feeding by the male. In birds this has usually been regarded as a kind of regression to juvenile behaviour on the part of the female. She begs from the male, using the same gestures as a young bird would use. It has been supposed that this is automatically attractive to the male, in the same way as a man finds a lisp or pouting lips attractive in an adult woman. The female bird at this time needs all the extra food she can get, for she is building up her reserves for the effort of manufacturing her enormous eggs. Courtship feeding by the male probably represents direct investment by him in the eggs themselves. It therefore has the effect of reducing the disparity between the two parents in their initial investment in the young.

Several insects and spiders also demonstrate the phenomenon of courtship feeding. Here an alternative interpretation has sometimes been only too obvious. Since, as in the case of the praying mantis, the male may be in danger of being eaten by the larger female, anything that he can do to reduce her appetite may be to his advantage. There is a macabre sense in which the unfortunate male mantis can be said to invest in his children. He is used as food to help make the eggs which will then be fertilized, posthumously, by his own stored sperms.

A female, playing the domestic-bliss strategy, who simply looks the males over and tries to *recognize* qualities of fidelity in advance, lays herself open to deception. Any male who can pass himself off as a good loyal domestic type, but who in reality is concealing a strong tendency towards desertion and unfaithfulness, could have a great advantage. As long as his deserted former wives have any chance of bringing up some of the children, the philanderer stands to pass on more genes than a rival male who is an honest husband and father. Genes for effective deception by males will tend to be favoured in the gene pool.

Conversely, natural selection will tend to favour females who become good at seeing through such deception. One way they can do this is to play especially hard to get when they are courted by a new male, but in successive breeding seasons to be increasingly ready to accept quickly the advances of last year's mate. This will automatically penalize young males embarking on their first breeding season, whether they are deceivers or not. The brood of naïve first year females would tend to contain a relatively high proportion of genes from unfaithful fathers, but faithful fathers have the advantage in the second and subsequent years of a mother's life, for they do not have to go through the same prolonged energy-wasting and time-consuming courtship rituals. If the majority of individuals in a population are the children of experienced rather than naïve mothers—a reasonable assumption in any long-lived species—genes for honest, good fatherhood will come to prevail in the gene pool.

For simplicity, I have talked as though a male were either purely honest or thoroughly deceitful. In reality it is more probable that all males, indeed all individuals, are a little bit deceitful, in that they are programmed to take advantage of opportunities to exploit their mates. Natural selection, by sharpening up the ability of each partner to detect dishonesty in the other, has kept large-scale deceit down to a fairly low level. Males have more to gain from dishonesty than females, and we must expect that, even in those species where males show considerable parental altruism, they will usually tend to do a bit less work than the females, and to be a bit more ready to abscond. In birds and mammals this is certainly normally the case.

There are species, however, in which the male actually does more work in caring for the children than the female does. Among birds and mammals these cases of paternal devotion are exceptionally rare, but they are common among fish. Why?* This is a challenge for the selfish gene theory which has puzzled me for a long time. An ingenious solution was recently suggested to me in a tutorial by Miss T. R. Carlisle. She makes use of Trivers's 'cruel bind' idea, referred to on page 193, as follows.

Many fish do not copulate, but instead simply spew out their sex cells into the water. Fertilization takes place in the open water, not inside the body of one of the partners. This is probably how sexual reproduction first began. Land animals like birds, mammals and reptiles, on the other hand, cannot afford this kind of external fertilization, because their sex cells are too vulnerable to drying-up. The gametes of one sex—the male, since sperms are mobile—are introduced into the wet interior of a member of the other sex—the female. So much is just fact. Now comes the idea. After copulation, the land-dwelling female is left in physical possession of the embryo. It is inside her body. Even if she lays the fertilized egg almost immediately, the male still has time to vanish, thereby forcing the female into Trivers's 'cruel bind'. The male is inevitably provided with an opportunity to take the prior decision to desert, closing the female's options, and forcing her to decide whether to leave the young to certain death, or whether to stay with it and rear it. Therefore, maternal care is more common among land animals than paternal care.

But for fish and other water-dwelling animals things are very different. If the male does not physically introduce his sperms into the female's body there is no necessary sense in which the female is left 'holding the baby'. Either partner might make a quick getaway and leave the other one in possession of the newly fertilized eggs. But there is even a possible reason why it might often be the male who is most vulnerable to being deserted. It seems probable that an evolutionary battle will develop over who sheds their sex cells first. The partner who does so has the advantage that he or she can then leave the other one in possession of the new embryos. On the other hand, the partner who spawns first runs the risk that his prospective partner may subsequently fail to follow suit. Now the male is more vulnerable here, if only because sperms are lighter and more likely to diffuse than eggs. If a female spawns too early, i.e. before the male is ready, it will not greatly matter because the eggs, being relatively large and heavy, are likely to stay together as a coherent clutch for some time. Therefore a female fish can afford to take the 'risk' of spawning early. The male dare not take this risk, since if he spawns too early his sperms will have diffused away before the female is ready, and she will then not spawn herself, because it will not be worth her while to do so. Because of the diffusion problem, the male must wait until the female spawns, and then he must shed his sperms over the eggs. But she has had a precious few seconds in which to disappear, leaving the male in possession, and forcing him on to the horns of Trivers's dilemma. So this theory neatly explains why paternal care is common in water but rare on dry land.

Leaving fish, I now turn to the other main female strategy, the he-man strategy. In species where this policy is adopted the females, in effect, resign themselves to getting no help from the father of their children, and go all-out for good genes instead. Once again they use their weapon of withholding copulation. They refuse to mate with just any male, but exercise the utmost care and discrimination before they will allow a male to copulate with them. Some males undoubtedly do contain a larger number of good genes than other males, genes that would benefit the survival prospects of both sons and daughters. If a female can somehow detect good genes in males, using externally visible clues, she can benefit her own genes by allying them with good paternal genes. To use our analogy of the rowing crews, a female can minimize the chance that her genes will be dragged down through getting into bad company. She can try to hand-pick good crew-mates for her own genes.

The chances are that most of the females will agree with each other on which are the best males, since they all have the same information to go on. Therefore these few lucky males will do most of the copulating. This they are quite capable of doing, since all they must give to each female is some cheap sperms. This is presumably what has happened in elephant seals and in birds of paradise. The females are allowing just a few males to get away with the ideal selfish-exploitation strategy which all males aspire to, but they are making sure that only the best males are allowed this luxury.

From the point of view of a female trying to pick good genes with which to ally her own, what is she looking for? One thing she wants is evidence of ability to survive. Obviously any potential mate who is courting her has proved his ability to survive at least into adulthood, but he has not necessarily proved that he can survive much longer. Quite a good policy for a female might be to go for old men. Whatever their shortcomings, they have at least proved they can survive, and she is likely to be allying her genes with genes for longevity. However, there is no point in ensuring that her children live long lives if they do not also give her lots of grandchildren. Longevity is not prima facie evidence of virility. Indeed a long-lived male may have survived precisely *because* he does not take risks in order to reproduce. A female who selects an old male is not necessarily going to have more descendants than a rival female who chooses a young one who shows some other evidence of good genes.

What other evidence? There are many possibilities. Perhaps strong muscles as evidence of ability to catch food, perhaps long legs as evidence of ability to run away from predators. A female might benefit her genes by allying them with such traits, since they might be useful qualities in both her sons and her daughters. To begin with, then, we have to imagine females choosing males on the basis of perfectly genuine labels or indicators which tend to be evidence of good underlying genes. But now here is a very interesting point realized by Darwin, and clearly enunciated by Fisher. In a society where males compete with each other to be chosen as he-men by females, one of the best things a mother can do for her genes is to make a son who will turn out in his turn to be an attractive he-man. If she can ensure that her son is one of the fortunate few males who wins most of the copulations in the society when he grows up, she will have an enormous number of grandchildren. The result of this is that one of the most desirable qualities a male can have in the eyes of a female is, quite simply, sexual attractiveness itself. A female who mates with a super-attractive he-man is more likely to have sons who are attractive to females of the next generation, and who will make lots of grandchildren for her. Originally, then, females may be thought of as selecting males on the basis of obviously useful qualities like big muscles, but once such qualities became widely accepted as attractive among the females of the species, natural selection would continue to favour them simply because they were attractive.

Extravagances such as the tails of male birds of paradise may therefore have evolved by a kind of unstable, runaway process.* In the early days, a slightly longer tail than usual may have been selected by females as a desirable quality in males, perhaps because it betokened a fit and healthy constitution. A short tail on a male might have been an indicator of some vitamin deficiency—evidence of poor food-getting ability. Or perhaps short-tailed males were not very good at running away from predators, and so had had their tails bitten off. Notice that we don't have to assume that the short tail was in itself genetically inherited, only that it served as an indicator of some genetic inferiority. Anyway, for whatever reason, let us suppose that females in the ancestral bird of paradise species preferentially went for males with longer than average tails. Provided there was *some* genetic contribution to the natural variation in male tail-length, this would in time cause the average tail-length of males in the population to increase. Females followed a simple rule: look all the males over, and go for the one with the longest tail. Any female who departed from this rule was penalized, *even if* tails had already become so long that they actually encumbered males possessing them. This was because any female who did not produce long-tailed sons had little chance of one of her sons being regarded as attractive. Like a fashion in women's clothes, or in American car design, the trend toward longer tails took off and gathered its own momentum. It was stopped only when tails became so grotesquely long that their manifest disadvantages started to outweigh the advantage of sexual attractiveness.

This is a hard idea to swallow, and it has attracted its sceptics ever since Darwin first proposed it, under the name of sexual selection. One person who does not believe it is A. Zahavi, whose 'Fox, fox' theory we have already met. He puts forward his own maddeningly contrary 'handicap principle' as a rival explanation.* He points out that the very fact that females are trying to select for good genes among males opens the door to deception by the males. Strong muscles may be a genuinely good quality for a female to select, but then what is to stop males from growing dummy muscles with no more real substance than human padded shoulders? If it costs a male less to grow false muscles than real ones, sexual selection should favour genes for producing false muscles. It will not be long, however, before counter-selection leads to the evolution of females capable of seeing through the deception. Zahavi's basic premise is that false sexual advertisement will eventually be seen through by females. He therefore concludes that really successful males will be those who do not advertise falsely, those who palpably demonstrate that they are not deceiving. If it is strong muscles we are talking about, then males who merely assume the visual *appearance* of strong muscles will soon be detected by the females. But a male who demonstrates, by the equivalent of lifting weights or ostentatiously doing press-ups, that he really has strong muscles, will succeed in convincing the females. In other words Zahavi believes that a he-man must not only *seem* to be a good quality male: he must really *be* a good quality male; otherwise, he will not be accepted as such by sceptical females. Displays will therefore evolve that only a genuine he-man is capable of doing.

So far so good. Now comes the part of Zahavi's theory that really sticks in the throat. He suggests that the tails of birds of paradise and peacocks, the huge antlers of deer, and the other sexually-selected features which have always seemed paradoxical because they appear to be handicaps to their possessors evolve precisely *because* they are handicaps. A male bird with a long and cumbersome tail is showing off to females that he is such a strong he-man that he can survive *in spite of* his tail. Think of a woman watching two men run a race. If both arrive at the finishing post at the same time, but one has deliberately encumbered himself with a sack of coal on his back, the women will naturally draw the conclusion that the man with the burden is really the faster runner.

I do not believe this theory, although I am not quite so confident in my scepticism as I was when I first heard it. I pointed out then that the logical conclusion to it should be the evolution of males with only one leg and only one eye. Zahavi, who comes from Israel, instantly retorted: 'Some of our best generals have only one eye!' Nevertheless, the problem remains that the handicap theory seems to contain a basic contradiction. If the handicap is a genuine one—and it is of the essence of the theory that it has to be a genuine one—then the handicap itself will penalize the offspring just as surely as it may attract females. It is, in any case, important that the handicap must not be passed on to daughters.

If we rephrase the handicap theory in terms of genes, we have something like this. A gene that makes males develop a handicap, such as a long tail, becomes more numerous in the gene pool because females choose males who have handicaps. Females choose males who have handicaps, because genes that make females so choose also become frequent in the gene pool. This is because females with a taste for handicapped males will automatically tend to be selecting males with good genes in other respects, since those males have survived to adulthood in spite of the handicap. These good 'other' genes will benefit the bodies of the children, which therefore survive to propagate the genes

for the handicap itself, and also the genes for choosing handicapped males. Provided the genes for the handicap itself exert
their effect only in sons, just as the genes for a sexual preference for the handicap affect only daughters, the theory just might
be made to work. So long as it is formulated only in words,
we cannot be sure whether it will work or not. We get a better
idea of how feasible such a theory is when it is rephrased in
terms of a mathematical model. So far mathematical geneticists who have tried to make the handicap principle into a workable model have failed. This may be because it is not a workable
principle, or it may be because they are not clever enough. One
of them is Maynard Smith, and my hunch favours the former
possibility.

If a male can demonstrate his superiority over other males in
a way that does not involve deliberately handicapping himself,
nobody would doubt that he could increase his genetic success
in that way. Thus elephant seals win and hold on to their harems,
not by being aesthetically attractive to females, but by the simple
expedient of beating up any male who tries to move in on the
harem. Harem-holders tend to win these fights against would-be
usurpers, if only for the obvious reason that that is why they are
harem-holders. Usurpers do not often win fights, because if they
were capable of winning they would have done so before! Any
female who mates only with a harem-holder is therefore allying
her genes with a male who is strong enough to beat off successive challenges from the large surplus of desperate bachelor
males. With luck her sons will inherit their father's ability to hold
a harem. In practice a female elephant seal does not have much
option, because the harem-owner beats *her* up if she tries to
stray. The principle remains, however, that females who choose
to mate with males who win fights may benefit their genes by so
doing. As we have seen, there are examples of females preferring

to mate with males who hold territories and with males who have high status in the dominance hierarchy.

To sum up this chapter so far, the various different kinds of breeding system that we find among animals—monogamy, promiscuity, harems, and so on—can be understood in terms of conflicting interests between males and females. Individuals of either sex 'want' to maximize their total reproductive output during their lives. Because of a fundamental difference between the size and numbers of sperms and eggs, males are in general likely to be biased towards promiscuity and lack of paternal care. Females have two main available counter-ploys, which I have called the he-man and the domestic-bliss strategies. The ecological circumstances of a species will determine whether the females are biased towards one or the other of these counter-ploys, and will also determine how the males respond. In practice all intermediates between he-man and domestic-bliss are found and, as we have seen, there are cases in which the father does even more child-care than the mother. This book is not concerned with the details of particular animal species, so I will not discuss what might predispose a species towards one form of breeding system rather than another. Instead I will consider the differences that are commonly observed between males and females in general, and show how these may be interpreted. I shall therefore not be emphasizing those species in which the differences between the sexes are slight, these being in general the ones whose females have favoured the domestic-bliss strategy.

Firstly, it tends to be the males who go in for sexually attractive, gaudy colours, and the females who tend to be more drab. Individuals of both sexes want to avoid being eaten by predators, and there will be some evolutionary pressure on both sexes to be drably coloured. Bright colours attract predators no less than they attract sexual partners. In gene terms, this means that genes for bright colours are more likely to meet their end in the stomachs of predators than are genes for drab colours. On the other hand, genes for drab colours may be less likely than genes for bright colours to find themselves in the next generation, because drab individuals have difficulty in attracting a mate. There are therefore two conflicting selection pressures: predators tending to remove bright-colour genes from the gene pool, and sexual partners tending to remove genes for drabness. As in so many other cases, efficient survival machines can be regarded as a compromise between conflicting selection pressures. What interests us at the moment is that the optimal compromise for a male seems to be different from the optimal compromise for a female. This is of course fully compatible with our view of males as high-risk, high-reward gamblers. Because a male produces many millions of sperms to every egg produced by a female, sperms heavily outnumber eggs in the population. Any given egg is therefore much more likely to enter into sexual fusion than any given sperm is. Eggs are a relatively valuable resource, and therefore a female does not need to be so sexually attractive as a male does in order to ensure that her eggs are fertilized. A male is perfectly capable of siring all the children born to a large population of females. Even if a male has a short life because his gaudy tail attracts predators, or gets tangled in the bushes, he may have fathered a very large number of children before he dies. An unattractive or drab male may live even as long as a female, but he has few children, and his genes are not passed on. What shall it profit a male if he shall gain the whole world, and lose his immortal genes?

Another common sexual difference is that females are more fussy than males about whom they mate with. One of the reasons for fussiness by an individual of either sex is the need to avoid mating with a member of another species. Such hybridizations are a bad thing for a variety of reasons. Sometimes, as in the case of a man copulating with a sheep, the copulation does not lead to an embryo being formed, so not much is lost. When more closely related species like horses and donkeys cross-breed, however, the cost, at least to the female partner, can be considerable. An embryo mule is likely to be formed and it then clutters up her womb for eleven months. It takes a large quantity of her total parental investment, not only in the form of food absorbed through the placenta, and then later in the form of milk, but above all in time which could have been spent in rearing other children. Then when the mule reaches adulthood it turns out to be sterile. This is presumably because, although horse chromosomes and donkey chromosomes are sufficiently similar to cooperate in the building of a good strong mule body, they are not similar enough to work together properly in meiosis. Whatever the exact reason, the very considerable investment by the mother in the rearing of a mule is totally wasted from the point of view of her genes. Female horses should be very, very careful that the individual they copulate with is another horse, and not a donkey. In gene terms, any horse gene that says, 'Body, if you are female, copulate with any old male, whether he is a donkey or a horse', is a gene which may next find itself in the dead-end body of a mule, and the mother's parental investment in that baby mule detracts heavily from her capacity to rear fertile horses. A male, on the other hand, has less to lose if he mates with a member of the wrong species, and, although he may have nothing to gain either, we should expect males to be less fussy in their choice of sexual partners. Where this has been looked at, it has been found to be true.

Even within a species, there may be reasons for fussiness. Incestuous mating, like hybridization, is likely to have damaging genetic consequences, in this case because lethal and semi-lethal recessive genes are brought out into the open. Once again, females have more to lose than males, since their investment in any particular child tends to be greater. Where incest taboos exist, we should expect females to be more rigid in their adherence to the taboos than males. If we assume that the older partner in an incestuous relationship is relatively likely to be the active initiator, we should expect that incestuous unions in which the male is older than the female should be more common than unions in which the female is older. For instance father/daughter incest should be commoner than mother/son. Brother/sister incest should be intermediate in commonness.

In general, males should tend to be more promiscuous than females. Since a female produces a limited number of eggs at a relatively slow rate, she has little to gain from having a large number of copulations with different males. A male on the other hand, who can produce millions of sperms every day, has everything to gain from as many promiscuous matings as he can snatch. Excess copulations may not actually cost a female much, other than a little lost time and energy, but they do not do her positive good. A male on the other hand can never get enough copulations with as many different females as possible: the word excess has no meaning for a male.

I have not explicitly talked about man but inevitably, when we think about evolutionary arguments such as those in this chapter, we cannot help reflecting about our own species and our own experience. Notions of females withholding copulation until a male shows some evidence of long-term fidelity may strike a familiar chord. This might suggest that human females play the domestic-bliss rather than the he-man strategy. Many human societies are indeed monogamous. In our own society, parental investment by both parents is large and not obviously unbalanced. Mothers certainly do more direct work for children than fathers do, but fathers often work hard in a more indirect sense to provide the material resources that are poured into the children. On the other hand, some human societies are promiscuous, and many are harem-based. What this astonishing variety suggests is that man's way of life is largely determined by culture rather than by genes. However, it is still possible that human males in general have a tendency towards promiscuity, and females a tendency towards monogamy, as we would predict on evolutionary grounds. Which of these two tendencies wins in particular societies depends on details of cultural circumstance, just as in different animal species it depends on ecological details.

One feature of our own society that seems decidedly anomalous is the matter of sexual advertisement. As we have seen, it is strongly to be expected on evolutionary grounds that, where the sexes differ, it should be the males that advertise and the females that are drab. Modern western man is undoubtedly exceptional in this respect. It is of course true that some men dress flamboyantly and some women dress drably but, on average, there can be no doubt that in our society the equivalent of the peacock's tail is exhibited by the female, not by the male. Women paint their faces and glue on false eyelashes. Apart from special cases, like actors, men do not. Women seem to be interested in their own personal appearance and they are encouraged in this by their magazines and journals. Men's magazines are less preoccupied with male sexual attractiveness, and a man who is unusually interested in his own dress and appearance is apt to arouse suspicion, both among men and among women. When a woman is described in conversation, it is quite likely that her sexual attractiveness, or lack of it, will be prominently mentioned. This is true, whether the speaker is a man or a woman. When a man is described, the adjectives used are much more likely to have nothing to do with sex.

Faced with these facts, a biologist would be forced to suspect that he was looking at a society in which females compete for males, rather than vice versa. In the case of birds of paradise, we decided that females are drab because they do not need to compete for males. Males are bright and ostentatious because females are in demand and can afford to be choosy. The reason female birds of paradise are in demand is that eggs are a more scarce resource than sperms. What has happened in modern western man? Has the male really become the sought-after sex, the one that is in demand, the sex that can afford to be choosy? If so, why?

YOU SCRATCH MY BACK, I'LL RIDE ON YOURS

We have considered parental, sexual, and aggressive interactions between survival machines belonging to the same species. There are striking aspects of animal interactions which do not seem to be obviously covered by any of these headings. One of these is the propensity that so many animals have for living in groups. Birds flock, insects swarm, fish and whales school, plains-dwelling mammals herd together or hunt in packs. These aggregations usually consist of members of a single species only, but there are exceptions. Zebras often herd together with gnus, and mixed-species flocks of birds are sometimes seen.

The suggested benefits that a selfish individual can wrest from living in a group constitute rather a miscellaneous list. I am not going to trot out the catalogue, but will mention just a few suggestions. In the course of this I shall return to the remaining examples of apparently altruistic behaviour that I gave in Chapter 1, and which I promised to explain. This will lead into a consideration of the social insects, without which no account of animal altruism would be complete. Finally in this rather miscellaneous chapter, I shall mention the important idea of reciprocal altruism, the principle of 'You scratch my back, I'll scratch yours'.

If animals live together in groups their genes must get more benefit out of the association than they put in. A pack of hyenas can catch prey so much larger than a lone hyena can bring down that it pays each selfish individual to hunt in a pack, even though this involves sharing food. It is probably for similar reasons that some spiders cooperate in building a huge communal web. Emperor penguins conserve heat by huddling together. Each one gains by presenting a smaller surface area to the elements than he would on his own. A fish who swims obliquely behind another fish may gain a hydrodynamic advantage from the turbulence produced by the fish in front. This could be partly why fish school. A related trick concerned with air turbulence is known to racing cyclists, and it may account for the V-formation of flying birds. There is probably competition to avoid the disadvantageous position at the head of the flock. Possibly the birds take turns as unwilling leader—a form of the delayed reciprocal altruism to be discussed at the end of the chapter.

Many of the suggested benefits of group living have been concerned with avoiding being eaten by predators. An elegant formulation of such a theory was given by W. D. Hamilton, in a paper called 'Geometry for the Selfish Herd'. Lest this lead to misunderstanding, I must stress that by 'selfish herd' he meant 'herd of selfish individuals'.

Once again we start with a simple 'model' which, though abstract, helps us to understand the real world. Suppose a species of animal is hunted by a predator that always tends to attack the nearest prey individual. From the predator's point of view this is a reasonable strategy, since it tends to cut down energy expenditure. From the prey's point of view it has an interesting consequence. It means that each prey individual will constantly try to avoid being the nearest to a predator. If the prey can detect the predator at a distance, it will simply run away. But if the predator is apt to turn up suddenly without warning, say it lurks concealed in long grass, then each prey individual can still take steps to minimize its chance of being the nearest to a predator. We can picture each prey individual as being surrounded by a 'domain of danger'. This is defined as that area of ground in which any point is nearer to that individual than it is to any other individual. For instance, if the prey individuals march spaced out in a regular geometric formation, the domain of danger round each one (unless he is on the edge) might be roughly hexagonal in shape. If a predator happens to be lurking in the hexagonal domain of danger surrounding individual A, then individual A is likely to be eaten. Individuals on the edge of the herd are especially vulnerable, since their domain of danger is not a relatively small hexagon, but includes a wide area on the open side.

Now clearly a sensible individual will try to keep his domain of danger as small as possible. In particular, he will try to avoid being on the edge of the herd. If he finds himself on the edge he will take immediate steps to move towards the centre. Unfortunately somebody has to be on the edge, but as far as each individual is concerned it is not going to be him! There will be a ceaseless migration in from the edges of an aggregation towards the centre. If the herd was previously loose and straggling, it will soon become tightly bunched as a result of the inward migration. Even if we start our model with no tendency towards aggregation at all, and the prey animals start by being randomly dispersed, the selfish urge of each individual will be to reduce his domain of danger by trying to position himself in a gap between other individuals. This will quickly lead to the formation of aggregations which will become ever more densely bunched.

Obviously, in real life the bunching tendency will be limited by opposing pressures: otherwise all individuals would collapse in a writhing heap! But still, the model is interesting as it shows us that even very simple assumptions can predict aggregation. Other, more elaborate models have been proposed. The fact that they are more realistic does not detract from the value of the simpler Hamilton model in helping us to think about the problem of animal aggregation.

The selfish-herd model in itself has no place for cooperative interactions. There is no altruism here, only selfish exploitation by each individual of every other individual. But in real life there are cases where individuals seem to take active steps to preserve fellow members of the group from predators. Bird alarm calls spring to mind. These certainly function as alarm signals in that they cause individuals who hear them to take immediate evasive action. There is no suggestion that the caller is 'trying to draw the predator's fire' away from his colleagues. He is simply informing them of the predator's existence—warning them. Nevertheless the act of calling seems, at least at first sight, to be altruistic, because it has the *effect* of calling the predator's attention to the caller. We can infer this indirectly from a fact which was noticed by P. R. Marler. The physical characteristics of the calls seem to be ideally shaped to be difficult to locate. If an acoustic engineer were asked to design a sound that a predator would find it hard to approach, he would produce something very like the real alarm calls of many small song-birds. Now in nature this shaping of the calls must have been produced by natural selection, and we know what that means. It means that large numbers of individuals have died because their alarm calls were not quite perfect. Therefore there seems to be danger attached to giving alarm calls. The selfish gene theory has to come up with a convincing advantage of giving alarm calls which is big enough to counteract this danger.

In fact this is not very difficult. Bird alarm calls have been held up so many times as 'awkward' for the Darwinian theory that it has become a kind of sport to dream up explanations for them.

As a result, we now have so many good explanations that it is hard to remember what all the fuss was about. Obviously, if there is a chance that the flock contains some close relatives, a gene for giving an alarm call can prosper in the gene pool because it has a good chance of being in the bodies of some of the individuals saved. This is true, even if the caller pays dearly for his altruism by attracting the predator's attention to himself.

If you are not satisfied with this kin selection idea, there are plenty of other theories to choose from. There are many ways in which the caller could gain selfish benefit from warning his fellows. Trivers reels off five good ideas, but I find the following two of my own rather more convincing.

The first I call the *cave* theory, from the Latin for 'beware', still used (pronounced 'kay-vee') by schoolboys to warn of approaching authority. This theory is suitable for camouflaged birds that crouch frozen in the undergrowth when danger threatens. Suppose a flock of such birds is feeding in a field. A hawk flies past in the distance. He has not yet seen the flock and he is not flying directly towards them, but there is a danger that his keen eyes will spot them at any moment and he will race into the attack. Suppose one member of the flock sees the hawk, but the rest have not yet done so. This one sharp-eyed individual could immediately freeze and crouch in the grass. But this would do him little good, because his companions are still walking around conspicuously and noisily. Any one of them could attract the hawk's attention and then the whole flock is in peril. From a purely selfish point of view the best policy for the individual who spots the hawk first is to hiss a quick warning to his companions, and so shut them up and reduce the chance that they will inadvertently summon the hawk into his own vicinity.

The other theory I want to mention may be called the 'never break ranks' theory. This one is suitable for species of birds that fly off when a predator approaches, perhaps up into a tree. Once again, imagine that one individual in a flock of feeding birds has spotted a predator. What is he to do? He could simply fly off himself, without warning his colleagues. But now he would be a bird on his own, no longer part of a relatively anonymous flock, but an odd man out. Hawks are actually known to go for odd pigeons out, but even if this were not so there are plenty of theoretical reasons for thinking that breaking ranks might be a suicidal policy. Even if his companions eventually follow him, the individual who first flies up off the ground temporarily increases his domain of danger. Whether Hamilton's particular theory is right or wrong, there must be some important advantage in living in flocks; otherwise, the birds would not do it. Whatever that advantage may be, the individual who leaves the flock ahead of the others will, at least in part, forfeit that advantage. If he must not break ranks, then, what is the observant bird to do? Perhaps he should just carry on as if nothing had happened and rely on the protection afforded by his membership of the flock. But this too carries grave risks. He is still out in the open, highly vulnerable. He would be much safer up in a tree. The best policy is indeed to fly up into a tree, *but to make sure everybody else does too*. That way, he will not become an odd man out and he will not forfeit the advantages of being part of a crowd, but he will gain the advantage of flying off into cover. Once again, uttering a warning call is seen to have a purely selfish advantage. E. L. Charnov and J. R. Krebs have proposed a similar theory in which they go so far as to use the word 'manipulation' to describe what the calling bird does to the rest of his flock. We have come a long way from pure, disinterested altruism!

Superficially, these theories may seem incompatible with the statement that the individual who gives the alarm call endangers himself. Really there is no incompatibility. He would endanger himself even more by not calling. Some individuals have died because they gave alarm calls, especially the ones whose calls were easy to locate. Other individuals have died because they did not give alarm calls. The *cave* theory and the 'never break ranks' theory are just two out of many ways of explaining why.

What of the stotting Thomson's gazelle, which I mentioned in Chapter 1, and whose apparently suicidal altruism moved Ardrey to state categorically that it could be explained only by group selection? Here the selfish gene theory has a more exacting challenge. Alarm calls in birds do work, but they are clearly designed to be as inconspicuous and discreet as possible. Not so the stotting high jumps. They are ostentatious to the point of downright provocation. The gazelles look as if they are deliberately inviting the predator's attention, almost as if they are teasing the predator. This observation has led to a delightfully daring theory. The theory was originally foreshadowed by N. Smythe but, pushed to its logical conclusion, it bears the unmistakeable signature of A. Zahavi.

Zahavi's theory can be put like this. The crucial bit of lateral thinking is the idea that stotting, far from being a signal to the other gazelles, is really aimed at the predators. It is noticed by the other gazelles and it affects their behaviour, but this is incidental, for it is primarily selected as a signal to the predator. Translated roughly into English it means: 'Look how high I can jump, I am obviously such a fit and healthy gazelle, you can't catch me, you would be much wiser to try and catch my neighbour who is not jumping so high!' In less anthropomorphic terms, genes for jumping high and ostentatiously are unlikely to be eaten by predators because predators tend to choose prey who look easy to catch. In particular, many mammal predators are known to go for the old and the unhealthy. An individual who jumps high is advertising, in an exaggerated way, the fact that he is neither old nor unhealthy. According to this theory, the display is far from altruistic. If anything it is selfish, since its object is to persuade the predator to chase somebody else. In a way there is a competition to see who can jump the highest, the loser being the one chosen by the predator.

The other example that I said I would return to is the case of the kamikáze bees, who sting honey-raiders but commit almost certain suicide in the process. The honey bee is just one example of a highly *social* insect. Others are wasps, ants, and termites or 'white ants'. I want to discuss social insects generally, not just suicidal bees. The exploits of the social insects are legendary, in particular their astonishing feats of cooperation and apparent altruism. Suicidal stinging missions typify their prodigies of self-abnegation. In the 'honey-pot' ants there is a caste of workers with grotesquely swollen, food-packed abdomens, whose sole function in life is to hang motionless from the ceiling like bloated light-bulbs, being used as food stores by the other workers. In the human sense they do not live as individuals at all; their individuality is subjugated, apparently to the welfare of the community. A society of ants, bees, or termites achieves a kind of individuality at a higher level. Food is shared to such an extent that one may speak of a communal stomach. Information is shared so efficiently by chemical signals and by the famous 'dance' of the bees that the community behaves almost as if it were a unit with a nervous system and sense organs of its own. Foreign intruders are recognized and repelled with something of the selectivity of a body's immune reaction system. The rather high temperature inside a beehive is regulated nearly as precisely as that of the human body, even though an individual bee is not a 'warm blooded' animal. Finally and most importantly, the analogy extends to reproduction. The majority of individuals in a social insect colony are sterile workers. The 'germ line'—the line of immortal gene continuity—flows through the bodies of a minority of individuals, the reproductives. These are the analogues of our own reproductive cells in our testes and ovaries. The sterile workers are the analogy of our liver, muscle, and nerve cells.

Kamikaze behaviour and other forms of altruism and cooperation by workers are not astonishing once we accept the fact that they are sterile. The body of a normal animal is manipulated to ensure the survival of its genes both through bearing offspring and through caring for other individuals containing the same genes. Suicide in the interests of caring for other individuals is incompatible with future bearing of one's own offspring. Suicidal self-sacrifice therefore seldom evolves. But a worker bee never bears offspring of its own. All its efforts are directed to preserving its genes by caring for relatives other than its own offspring. The death of a single sterile worker bee is no more serious to its genes than is the shedding of a leaf in autumn to the genes of a tree.

There is a temptation to wax mystical about the social insects, but there is really no need for this. It is worth looking in some detail at how the selfish gene theory deals with them, and in particular at how it explains the evolutionary origin of that extraordinary phenomenon of worker sterility from which so much seems to follow.

A social insect colony is a huge family, usually all descended from the same mother. The workers, who seldom or never reproduce themselves, are often divided into a number of distinct castes, including small workers, large workers, soldiers, and highly specialized castes like the honey-pots. Reproductive females are called queens. Reproductive males are sometimes called drones or kings. In the more advanced societies, the reproductives never work at anything except procreation, but at this one task they are extremely good. They rely on the workers for their food and protection, and the workers are also responsible for looking after the brood. In some ant and termite species the queen has swollen into a gigantic egg factory, scarcely recognizable as an insect at all, hundreds of times the size of a worker and quite incapable of moving. She is constantly tended by workers who groom her, feed her, and transport her ceaseless flow of eggs to the communal nurseries. If such a monstrous queen ever has to move from the royal cell she rides in state on the backs of squadrons of toiling workers.

In Chapter 7 I introduced the distinction between bearing and caring. I said that mixed strategies, combining bearing and caring, would normally evolve. In Chapter 5 we saw that mixed evolutionarily stable strategies could be of two general types. Either each individual in the population could behave in a mixed way: thus individuals usually achieve a judicious mixture of bearing and caring; *or* the population may be divided into two different types of individual: this was how we first pictured the balance between hawks and doves. Now it is theoretically possible for an evolutionarily stable balance between bearing and caring to be achieved in the latter kind of way: the population could be divided into bearers and carers. But this can only be evolutionarily stable if the carers are close kin to the individuals for whom they care, at least as close as they would be to their own offspring if they had any. Although it is theoretically possible for evolution to proceed in this direction, it seems to be only in the social insects that it has actually happened.*

Social insect individuals are divided into two main classes, bearers and carers. The bearers are the reproductive males and females. The carers are the workers—infertile males and females in the termites, infertile females in all other social insects. Both types do their job more efficiently because they do not have to cope with the other. But from whose point of view is it efficient?

The question which will be hurled at the Darwinian theory is the familiar cry: 'What's in it for the workers?'

Some people have answered, 'Nothing.' They feel that the queen is having it all her own way, manipulating the workers by chemical means to her own selfish ends, making them care for her own teeming brood. This is a version of Alexander's 'parental manipulation' theory which we met in Chapter 8. The opposite idea is that the workers 'farm' the reproductives, manipulating them to increase their productivity in propagating replicas of the workers' genes. To be sure, the survival machines that the queen makes are not offspring to the workers, but they are close relatives nevertheless. It was Hamilton who brilliantly realized that, at least in the ants, bees, and wasps, the workers may actually be more closely related to the brood than the queen herself is! This led him, and later Trivers and Hare, on to one of the most spectacular triumphs of the selfish gene theory. The reasoning goes like this.

Insects of the group known as the Hymenoptera, including ants, bees, and wasps, have a very odd system of sex determination. Termites do not belong to this group and they do not share the same peculiarity. A hymenopteran nest typically has only one mature queen. She made one mating flight when young and stored up the sperms for the rest of her long life—ten years or even longer. She rations the sperms out to her eggs over the years, allowing the eggs to be fertilized as they pass out through her tubes. But not all the eggs are fertilized. The unfertilized ones develop into males. A male therefore has no father, and all the cells of his body contain just a single set of chromosomes (all obtained from his mother) instead of a double set (one from the father and one from the mother) as in ourselves. In terms of the analogy of Chapter 3, a male hymenopteran has only one copy of each 'volume' in each of his cells, instead of the usual two.

A female hymenopteran, on the other hand, is normal in that she does have a father, and she has the usual double set of chromosomes in each of her body cells. Whether a female develops into a worker or a queen depends not on her genes but on how she is brought up. That is to say, each female has a complete set of queen-making genes, and a complete set of worker-making genes (or, rather, sets of genes for making each specialized caste of worker, soldier, etc.). Which set of genes is 'turned on' depends on how the female is reared, in particular on the food she receives.

Although there are many complications, this is essentially how things are. We do not know why this extraordinary system of sexual reproduction evolved. No doubt there were good reasons, but for the moment we must just treat it as a curious fact about the Hymenoptera. Whatever the original reason for the oddity, it plays havoc with Chapter 6's neat rules for calculating relatedness. It means that the sperms of a single male, instead of all being different as they are in ourselves, are all exactly the same. A male has only a single set of genes in each of his body cells, not a double set. Every sperm must therefore receive the full set of genes rather than a 50 per cent sample, and all sperms from a given male are therefore identical. Let us now try to calculate the relatedness between a mother and son. If a male is known to possess a gene A, what are the chances that his mother shares it? The answer must be 100 per cent, since the male had no father and obtained all his genes from his mother. But now suppose a queen is known to have the gene B. The chance that her son shares the gene is only 50 per cent, since he contains only half her genes. This sounds like a contradiction, but it is not. A male gets *all* his genes from his mother, but a mother only gives *half* her genes to her son. The solution to the apparent paradox lies in the fact that a male has only half the usual number of genes. There is no point in puzzling over whether the 'true' index of relatedness is $\frac{1}{2}$ or 1. The index is only a man-made measure, and if it leads to difficulties in particular cases, we may have to abandon it and go back to first principles. From the point of view of a gene A in the body of a queen, the chance that the gene is shared by a son is $\frac{1}{2}$, just as it is for a daughter. From a queen's point of view therefore, her offspring, of either sex, are as closely related to her as human children are to their mother.

Things start to get intriguing when we come to sisters. Full sisters not only share the same father: the two sperms that conceived them were identical in every gene. The sisters are therefore equivalent to identical twins as far as their paternal genes are concerned. If one female has a gene A, she must have got it from either her father or her mother. If she got it from her mother then there is a 50 per cent chance that her sister shares it. But if she got it from her father, the chances are 100 per cent that her sister shares it. Therefore the relatedness between hymenopteran full sisters is not $\frac{1}{2}$ as it would be for normal sexual animals, but $\frac{3}{4}$.

It follows that a hymenopteran female is more closely related to her full sisters than she is to her offspring of either sex.* As Hamilton realized (though he did not put it in quite the same way) this might well predispose a female to farm her own mother as an efficient sister-making machine. A gene for vicariously making sisters replicates itself more rapidly than a gene for making offspring directly. Hence worker sterility evolved. It is presumably no accident that true sociality, with worker sterility, seems to have evolved no fewer than eleven times *independently* in the Hymenoptera and only once in the whole of the rest of the animal kingdom, namely in the termites.

However, there is a catch. If the workers are successfully to farm their mother as a sister-producing machine, they must somehow curb her natural tendency to give them an equal number of little brothers as well. From the point of view of a worker, the chance of any one brother containing a particular one of her genes is only $\frac{1}{4}$. Therefore, if the queen were allowed to produce male and female reproductive offspring in equal proportions, the farm would not show a profit as far as the workers are concerned. They would not be maximizing the propagation of their precious genes.

Trivers and Hare realized that the workers must try to bias the sex ratio in favour of females. They took the Fisher calculations on optimal sex ratios (which we looked at in the previous chapter) and re-worked them for the special case of the Hymenoptera. It turned out that the stable ratio of investment for a mother is, as usual, 1:1. But the stable ratio for a sister is 3:1 in favour of sisters rather than brothers. If you are a hymenopteran female, the most efficient way for you to propagate your genes is to refrain from breeding yourself, and to make your mother provide you with reproductive sisters and brothers in the ratio 3:1. But if you *must* have offspring of your own, you can benefit your genes best by having reproductive sons and daughters in equal proportions.

As we have seen, the difference between queens and workers is not a genetic one. As far as her genes are concerned, an embryo female might be destined to become either a worker, who 'wants' a 3:1 sex ratio, or a queen, who 'wants' a 1:1 ratio. So what does this 'wanting' mean? It means that a gene that finds itself in a queen's body can propagate itself best if that body invests equally in reproductive sons and daughters. But the same gene finding itself in a worker's body can propagate itself best by making the mother of that body have more daughters than sons. There is no real paradox here. A gene must take best advantage of the levers of power that happen to be at its disposal. If it finds itself in a position to influence the development of a body that is destined to turn into a queen, its optimal strategy to exploit that control is one thing. If it finds itself in a position to influence the way a worker's body develops, its optimal strategy to exploit that power is different.

This means there is a conflict of interests down on the farm. The queen is 'trying' to invest equally in males and females. The workers are trying to shift the ratio of reproductives in the direction of three females to every one male. If we are right to picture the workers as the farmers and the queen as their brood mare, presumably the workers will be successful in achieving their 3:1 ratio. If not, if the queen really lives up to her name and the workers are her slaves and the obedient tenders of the royal nurseries, then we should expect the 1:1 ratio which the queen 'prefers' to prevail. Who wins in this special case of a battle of the generations? This is a matter that can be put to the test and that is what Trivers and Hare did, using a large number of species of ants.

The sex ratio that is of interest is the ratio of male to female reproductives. These are the large winged forms which emerge from the ants' nest in periodic bursts for mating flights, after which the young queens may try to found new colonies. It is these winged forms that have to be counted to obtain an estimate of the sex ratio. Now the male and female reproductives are, in many species, very unequal in size. This complicates things since, as we saw in the previous chapter, the Fisher calculations about optimal sex ratio strictly apply, not to *numbers* of males and females, but to *quantity of investment* in males and females. Trivers and Hare made allowance for this by weighing them. They took 20 species of ant and estimated the sex ratio in terms of investment in reproductives. They found a rather convincingly close fit to the 3:1 female to male ratio predicted by the theory that the workers are running the show for their own benefit.*

It seems then that in the ants studied, the conflict of interests is 'won' by the workers. This is not too surprising since worker bodies, being the guardians of the nurseries, have more power in practical terms than queen bodies. Genes trying to manipulate the world through queen bodies are outmanœuvred by genes manipulating the world through worker bodies. It is interesting to look around for some special circumstances in which we might expect queens to have more practical power than workers. Trivers and Hare realized that there was just such a circumstance which could be used as a critical test of the theory.

This arises from the fact that there are some species of ant that take slaves. The workers of a slave-making species either do no ordinary work at all or are rather bad at it. What they are good at is going on slaving raids. True warfare in which large rival armies fight to the death is known only in man and in social insects. In many species of ants the specialized caste of workers known as soldiers have formidable fighting jaws, and devote their time to fighting for the colony against other ant armies. Slaving raids are just a particular kind of war effort. The slavers mount an attack on a nest of ants belonging to a different species, attempt to kill the defending workers or soldiers, and carry off the unhatched young. These young ones hatch out in the nest of their captors. They do not 'realize' that they are slaves and they set to work following their built-in nervous programs, doing all the duties that they would normally perform in their own nest. The slave-making workers or soldiers go on further slaving expeditions while the slaves stay at home and get on with the everyday business of running an ants' nest, cleaning, foraging, and caring for the brood.

The slaves are, of course, blissfully ignorant of the fact that they are unrelated to the queen and to the brood that they are tending. Unwittingly they are rearing new platoons of slave-makers. No doubt natural selection, acting on the genes of the slave species, tends to favour anti-slavery adaptations. However, these are evidently not fully effective because slavery is a wide spread phenomenon.

The consequence of slavery that is interesting from our present point of view is this. The queen of the slave-making species is now in a position to bend the sex ratio in the direction she 'prefers'. This is because her own true-born children, the slavers, no longer hold the practical power in the nurseries. This power is now held by the slaves. The slaves 'think' they are looking after their own siblings and they are presumably doing whatever *would be appropriate in their own nests* to achieve their desired 3:1 bias in favour of sisters. But the queen of the slave-making species is able to get away with counter-measures and there is no selection operating on the slaves to neutralize these counter-measures, since the slaves are totally unrelated to the brood.

For example, suppose that in any ant species, queens 'attempt' to disguise male eggs by making them smell like female ones. Natural selection will normally favour any tendency by workers to 'see through' the disguise. We may picture an evolutionary battle in which queens continually 'change the code', and workers 'break the code'. The war will be won by whoever manages to get more of her genes into the next generation, via the bodies of the reproductives. This will normally be the workers, as we have seen. But when the queen of a *slave-making* species changes the code, the slave workers cannot evolve any ability to break the code. This is because any gene in a slave worker 'for breaking the code' is not represented in the body of any reproductive individual, and so is not passed on. The reproductives all belong to the slave-making species, and are kin to the queen but not to the slaves. If the genes of the slaves find their way into any reproductives at all, it will be into the reproductives that emerge from the original nest from which they were kidnapped. The slave workers will, if anything, be busy breaking the wrong code! Therefore, queens of a slave-making species can get away with changing their code freely, without there being any danger that genes for breaking the code will be propagated into the next generation.

The upshot of this involved argument is that we should expect in slave-making species that the ratio of investment in reproductives of the two sexes should approach 1:1 rather than 3:1. For once, the queen will have it all her own way. This is just what Trivers and Hare found, although they only looked at two slave-making species.

I must stress that I have told the story in an idealized way. Real life is not so neat and tidy. For instance, the most familiar social insect species of all, the honey bee, seems to do entirely the 'wrong' thing. There is a large surplus of investment in males over queens—something that does not appear to make sense from either the workers' or the mother queen's point of view. Hamilton has offered a possible solution to this puzzle. He points out that when a queen bee leaves the hive she goes with a large swarm of attendant workers, who help her to start a new colony. These workers are lost to the parent hive, and the cost of making them must be reckoned as part of the cost of reproduction: for every queen who leaves, many *extra* workers have to be made. Investment in these extra workers should be counted as part of the investment in reproductive females. The extra workers should be weighed in the balance against the males when the sex ratio is computed. So this was not a serious difficulty for the theory after all.

A more awkward spanner in the elegant works of the theory is the fact that, in some species, the young queen on her mating flight mates with several males instead of one. This means that the average relatedness among her daughters is less than $\frac{3}{4}$, and may even approach $\frac{1}{4}$ in extreme cases. It is tempting, though probably not very logical, to regard this as a cunning blow struck by queens against workers! Incidentally, this might seem to suggest that workers should chaperone a queen on her mating flight, to prevent her from mating more than once. But this would in no way help the workers' own genes—only the genes of the coming generation of workers. There is no trade-union spirit among the workers as a class. All that each one of them 'cares' about is her own genes. A worker might have 'liked' to have chaperoned her own mother, but she lacked the opportunity, not having been conceived in those days. A young queen on her mating flight is the sister of the present generation of workers, not the mother. Therefore they are on *her* side rather than on the side of the next generation of workers, who are merely their nieces. My head is now spinning, and it is high time to bring this topic to a close.

I have used the analogy of farming for what hymenopteran workers do to their mothers. The farm is a gene farm. The workers use their mother as a more efficient manufacturer of copies of their own genes than they would be themselves. The genes come off the production line in packages called reproductive individuals. This farming analogy should not be confused with a quite different sense in which the social insects may be said to farm. Social insects discovered, as man did long after, that settled cultivation of food can be more efficient than hunting and gathering.

For example, several species of ants in the New World, and, quite independently, termites in Africa, cultivate 'fungus gardens'. The best known are the so-called parasol ants of South America. These are immensely successful. Single colonies with more than two million individuals have been found. Their nests consist of huge spreading underground complexes of passages and galleries going down to a depth of ten feet or more, made by the excavation of as much as 40 tons of soil. The underground chambers contain the fungus gardens. The ants deliberately sow fungus of a particular species in special compost beds which they prepare by chewing leaves into fragments. Instead of foraging directly for their own food, the workers forage for leaves to make compost. The 'appetite' of a colony of parasol ants for leaves is gargantuan. This makes them a major economic pest, but the leaves are not food for themselves but food for their fungi. The ants eventually harvest and eat the fungi and feed them to their brood. The fungi are more efficient at breaking down leaf material than the ants' own stomachs would be, which is how the ants benefit by the arrangement. It is possible that the fungi benefit too, even though they are cropped: the ants propagate them more efficiently than their own spore dispersal mechanism might achieve. Furthermore, the ants 'weed' the fungus gardens, keeping them clear of alien species of fungi. By removing competition, this may benefit the ants' own domestic fungi. A kind of relationship of mutual altruism could be said to exist between ants and fungi. It is remarkable that a very similar system of fungus-farming has evolved independently, among the quite unrelated termites.

Ants have their own domestic animals as well as their crop plants. Aphids—greenfly and similar bugs—are highly specialized for sucking the juice out of plants. They pump the sap up out of the plants' veins more efficiently than they subsequently digest it. The result is that they excrete a liquid that has had only some of its nutritious value extracted. Droplets of sugar-rich 'honeydew' pass out of the back end at a great rate, in some cases more than the insect's own body-weight every hour. The honeydew normally rains down on to the ground—it may well have been the providential food known as 'manna' in the Old Testament. But ants of several species intercept it as soon as it leaves the bug. The ants 'milk' the aphids by stroking their hind-quarters with their feelers and legs. Aphids respond to this, in some cases apparently holding back their droplets until an ant strokes them, and even withdrawing a droplet if an ant is not ready to accept it.

It has been suggested that some aphids have evolved a backside that looks and feels like an ant's face, the better to attract ants. What the aphids have to gain from the relationship is apparently protection from their natural enemies. Like our own dairy cattle they lead a sheltered life, and aphid species that are much cultivated by ants have lost their normal defensive mechanisms. In some cases ants care for the aphid eggs inside their own underground nests, feed the young aphids, and finally, when they are grown, gently carry them up to the protected grazing grounds.

A relationship of mutual benefit between members of different species is called mutualism or symbiosis. Members of different species often have much to offer each other because they can bring different 'skills' to the partnership. This kind of fundamental asymmetry can lead to evolutionarily stable strategies of mutual cooperation. Aphids have the right sort of mouthparts for pumping up plant sap, but such sucking mouthparts are no good for self-defence. Ants are no good at sucking sap from plants, but they are good at fighting. Ant genes for cultivating and protecting aphids have been favoured in ant gene pools. Aphid genes for cooperating with the ants have been favoured in aphid gene pools.

Symbiotic relationships of mutual benefit are common among animals and plants. A lichen appears superficially to be an individual plant like any other. But it is really an intimate symbiotic union between a fungus and a green alga. Neither partner could live without the other. If their union had become just a bit more intimate we would no longer have been able to tell that a lichen was a double organism at all. Perhaps then there are other double or multiple organisms which we have not recognized as such. Perhaps even we ourselves?

Within each one of our cells there are numerous tiny bodies called mitochondria. The mitochondria are chemical factories, responsible for providing most of the energy we need. If we lost our mitochondria we would be dead within seconds. Recently it has been plausibly argued that mitochondria are, in origin, symbiotic bacteria who joined forces with our type of cell very early in evolution. Similar suggestions have been made for other small bodies within our cells. This is one of those revolutionary ideas which it takes time to get used to, but it is an idea whose time has come. I speculate that we shall come to accept the more radical idea that each one of our genes is a symbiotic unit. We are gigantic colonies of symbiotic genes. One cannot really speak of 'evidence' for this idea, but, as I tried to suggest in earlier chapters, it is really inherent in the very way we think about how genes work in sexual species. The other side of this coin is that viruses may be genes who have broken loose from 'colonies' such as ourselves. Viruses consist of pure DNA (or a related self-replicating molecule) surrounded by a protein jacket. They are all parasitic. The suggestion is that they have evolved from 'rebel' genes who escaped, and now travel from body to body directly through the air, rather than via the more conventional vehicles—sperms and eggs. If this is true, we might just as well regard ourselves as colonies of viruses! Some of them cooperate symbiotically, and travel from body to body in sperms and eggs. These are the conventional 'genes'. Others live parasitically, and travel by whatever means they can. If the parasitic DNA travels in sperms and eggs, it perhaps forms the 'paradoxical' surplus of DNA which I mentioned in Chapter 3. If it travels through the air, or by other direct means, it is called 'virus' in the usual sense.

But these are speculations for the future. At present we are concerned with symbiosis at the higher level of relationships between many-celled organisms, rather than within them. The word symbiosis is conventionally used for associations between members of different species. But, now that we have eschewed the 'good of the species' view of evolution, there seems no logical reason to distinguish associations between members of different species as things apart from associations between members of the same species. In general, associations of mutual benefit will evolve if each partner can get more out than he puts in. This is true whether we are speaking of members of the same hyena pack, or of widely distinct creatures such as ants and aphids, or bees and flowers. In practice it may be difficult to distinguish cases of genuine two-way mutual benefit from cases of one-sided exploitation.

The evolution of associations of mutual benefit is theoretically easy to imagine if the favours are given and received simultaneously, as in the case of the partners who make up a lichen. But problems arise if there is a delay between the giving of a favour and its repayment. This is because the first recipient of a favour may be tempted to cheat and refuse to pay it back when his turn comes. The resolution of this problem is interesting and is worth discussing in detail. I can do this best in terms of a hypothetical example.

Suppose a species of bird is parasitized by a particularly nasty kind of tick which carries a dangerous disease. It is very important that these ticks should be removed as soon as possible. Normally an individual bird can pull off its owns ticks when preening itself. There is one place, however—the top of the head—which it cannot reach with its own bill. The solution to the problem quickly occurs to any human. An individual may not be able to reach his own head, but nothing is easier than for a friend to do it for him. Later, when the friend is parasitized himself, the good deed can be paid back. Mutual grooming is in fact very common in both birds and mammals.

This makes immediate intuitive sense. Anybody with conscious foresight can see that it is sensible to enter into mutual back-scratching arrangements. But we have learnt to beware of what seems intuitively sensible. The gene has no foresight. Can the theory of selfish genes account for mutual back-scratching, or 'reciprocal altruism', where there is a delay between good deed and repayment? Williams briefly discussed the problem in his 1966 book, to which I have already referred. He concluded, as had Darwin, that delayed reciprocal altruism can evolve in species that are capable of recognizing and remembering each other as individuals. Trivers, in 1971, took the matter further. When he wrote, he did not have available to him Maynard Smith's concept of the evolutionarily stable strategy. If he had, my guess is that he would have made use of it, for it provides a natural way to express his ideas. His reference to the 'Prisoner's Dilemma'—a favourite puzzle in game theory—shows that he was already thinking along the same lines.

Suppose B has a parasite on the top of his head. A pulls it off him. Later, the time comes when A has a parasite on his head. He naturally seeks out B in order that B may pay back his good deed. B simply turns up his nose and walks off. B is a cheat, an individual who accepts the benefit of other individuals' altruism, but who does not pay it back, or who pays it back insufficiently. Cheats do better than indiscriminate altruists because they gain the benefits without paying the costs. To be sure, the cost of grooming another individual's head seems small compared with the benefit of having a dangerous parasite removed, but it is not negligible. Some valuable energy and time has to be spent.

Let the population consist of individuals who adopt one of two strategies. As in Maynard Smith's analyses, we are not talking about conscious strategies, but about unconscious behaviour programs laid down by genes. Call the two strategies Sucker and Cheat. Suckers groom anybody who needs it, indiscriminately. Cheats accept altruism from suckers, but they never groom anybody else, not even somebody who has previously groomed them. As in the case of the hawks and doves, we arbitrarily assign pay-off points. It does not matter what the exact values are, so long as the benefit of being groomed exceeds the cost of grooming. If the incidence of parasites is high, any individual sucker in a population of suckers can reckon on being groomed about as often as he grooms. The average pay-off for a sucker among suckers is therefore positive. They all do quite nicely in fact, and the word sucker seems inappropriate. But now suppose a cheat arises in the population. Being the only cheat, he can count on being groomed by everybody else, but he pays nothing in return. His average pay-off is better than the average for a sucker. Cheat genes will therefore start to spread through the population. Sucker genes will soon be driven to extinction. This is because, no matter what the ratio in the population, cheats will always do better than suckers. For instance, consider the case when the population consists of 50 per cent suckers and 50 per cent cheats. The average pay-off for both suckers and cheats will be less than that for any individual in a population of 100 per cent suckers. But still, cheats will be doing better than suckers because they are getting all the benefits—such as they are—and paying nothing back. When the proportion of cheats reaches 90 per cent, the average pay-off for all individuals will be very low: many of both types may by now be dying of the infection carried by the ticks. But still the cheats will be doing better than the suckers. Even if the whole population declines toward extinction, there will never be any time when suckers do better than cheats. Therefore, as long as we consider only these two strategies, nothing can stop the extinction of the suckers and, very probably, the extinction of the whole population too.

But now, suppose there is a third strategy called Grudger. Grudgers groom strangers and individuals who have previously groomed them. However, if any individual cheats them, they remember the incident and bear a grudge: they refuse to groom that individual in the future. In a population of grudgers and suckers it is impossible to tell which is which. Both types behave altruistically towards everybody else, and both earn an equal and high average pay-off. In a population consisting largely of cheats, a single grudger would not be very successful. He would expend a great deal of energy grooming most of the individuals he met— for it would take time for him to build up grudges against all of them. On the other hand, nobody would groom him in return. If grudgers are rare in comparison with cheats, the grudger gene will go extinct. Once the grudgers manage to build up in numbers so that they reach a critical proportion, however, their chance of meeting each other becomes sufficiently great to off-set their wasted effort in grooming cheats. When this critical proportion is reached they will start to average a higher pay-off than cheats, and the cheats will be driven at an accelerating rate towards extinction. When the cheats are nearly extinct their rate of decline will become slower, and they may survive as a minor-ity for quite a long time. This is because for any one rare cheat there is only a small chance of his encountering the same grudger twice: therefore the proportion of individuals in the population who bear a grudge against any given cheat will be small.

I have told the story of these strategies as though it were intui-tively obvious what would happen. In fact it is not all that obvi-ous, and I did take the precaution of simulating it on a computer to check that intuition was right. Grudger does indeed turn out to be an evolutionarily stable strategy against sucker and cheat, in the sense that, in a population consisting largely of grudgers, neither cheat nor sucker will invade. Cheat is also an ESS, how-ever, because a population consisting largely of cheats will not be invaded by either grudger or sucker. A population could sit at either of these two ESSs. In the long term it might flip from one to the other. Depending on the exact values of the pay-offs—the assumptions in the simulation were of course completely arbitrary—one or other of the two stable states will have a larger 'zone of attraction' and will be more likely to be attained. Note incidentally that, although a population of cheats may be more likely to go extinct than a population of grudgers, this in no way affects its status as an ESS. If a population arrives at an ESS that drives it extinct, then it goes extinct, and that is just too bad.*

It is quite entertaining to watch a computer simulation that starts with a strong majority of suckers, a minority of grudgers that is just above the critical frequency, and about the same-sized minority of cheats. The first thing that happens is a dramatic crash in the population of suckers as the cheats ruthlessly exploit them. The cheats enjoy a soaring population explosion, reaching their peak just as the last sucker perishes. But the cheats still have the grudgers to reckon with. During the precipitous decline of the suckers, the grudgers have been slowly decreasing in numbers, taking a battering from the prospering cheats, but just managing to hold their own. After the last sucker has gone and the cheats can no longer get away with selfish exploitation so easily, the grudgers slowly begin to increase at the cheats' expense. Steadily their population rise gathers momentum. It accelerates steeply, the cheat population crashes to near extinction, then levels out as they enjoy the privileges of rarity and the comparative freedom from grudges which this brings. However, slowly and inexorably the cheats are driven out of existence, and the grudgers are left in sole possession. Paradoxically, the presence of the suckers actually endangered the grudgers early on in the story because they were responsible for the temporary prosperity of the cheats.

By the way, my hypothetical example about the dangers of not being groomed is quite plausible. Mice kept in isolation tend to develop unpleasant sores on those parts of their heads that they cannot reach. In one study, mice kept in groups did not suffer in this way, because they licked each others' heads. It would be interesting to test the theory of reciprocal altruism experimentally and it seems that mice might be suitable subjects for the work.

Trivers discusses the remarkable symbiosis of the cleaner-fish. Some fifty species, including small fish and shrimps, are known to make their living by picking parasites off the surface of larger fish of other species. The large fish obviously benefit from being cleaned, and the cleaners get a good supply of food. The relationship is symbiotic. In many cases the large fish open their mouths and allow cleaners right inside to pick their teeth, and then to swim out through the gills which they also clean. One might expect that a large fish would craftily wait until he had been thoroughly cleaned, and then gobble up the cleaner. Yet instead he usually lets the cleaner swim off unmolested. This is a considerable feat of apparent altruism because in many cases the cleaner is of the same size as the large fish's normal prey.

Cleaner-fish have special stripy patterns and special dancing displays which label them as cleaners. Large fish tend to refrain from eating small fish who have the right kind of stripes, and who approach them with the right kind of dance. Instead they go into a trance-like state and allow the cleaner free access to their exterior and interior. Selfish genes being what they are, it is not surprising that ruthless, exploiting cheats have cashed in. There are species of small fish that look just like cleaners and dance in the same kind of way in order to secure safe conduct into the vicinity of large fish. When the large fish has gone into its expectant trance the cheat, instead of pulling off a parasite, bites a chunk out of the large fish's fin and beats a hasty retreat. But in spite of the cheats, the relationship between fish cleaners and their clients is mainly amicable and stable. The profession of cleaner plays an important part in the daily life of the coral reef community. Each cleaner has his own territory, and large fish have been seen queuing up for attention like customers at a barber's shop. It is probably this site-tenacity that makes possible the evolution of delayed reciprocal altruism in this case. The benefit to a large fish of being able to return repeatedly to the same 'barber's shop', rather than continually searching for a new one, must outweigh the cost of refraining from eating the cleaner. Since cleaners are small, this is not hard to believe. The presence of cheating cleaner-mimics probably indirectly endangers the bonafide cleaners by setting up a minor pressure on large fish to eat stripy dancers. Site-tenacity on the part of genuine cleaners enables customers to find them and to avoid cheats.

A long memory and a capacity for individual recognition are well developed in man. We might therefore expect reciprocal altruism to have played an important part in human evolution. Trivers goes so far as to suggest that many of our psychological characteristics—envy, guilt, gratitude, sympathy, etc.—have been shaped by natural selection for improved ability to cheat, to detect cheats, and to avoid being thought to be a cheat. Of particular interest are 'subtle cheats' who appear to be reciprocating, but who consistently pay back slightly less than they receive. It is even possible that man's swollen brain, and his predisposition to reason mathematically, evolved as a mechanism of ever more devious cheating, and ever more penetrating detection of cheating in others. Money is a formal token of delayed reciprocal altruism.

There is no end to the fascinating speculation that the idea of reciprocal altruism engenders when we apply it to our own species. Tempting as it is, I am no better at such speculation than the next man, and I leave the reader to entertain himself.

MEMES
The new replicators

So far, I have not talked much about man in particular, though
I have not deliberately excluded him either. Part of the reason
I have used the term 'survival machine' is that 'animal' would
have left out plants and, in some people's minds, humans. The
arguments I have put forward should, prima facie, apply to any
evolved being. If a species is to be excepted, it must be for good
particular reasons. Are there any good reasons for supposing
our own species to be unique? I believe the answer is yes.

Most of what is unusual about man can be summed up in one
word: 'culture'. I use the word not in its snobbish sense, but as a
scientist uses it. Cultural transmission is analogous to genetic
transmission in that, although basically conservative, it can give
rise to a form of evolution. Geoffrey Chaucer could not hold
a conversation with a modern Englishman, even though they
are linked to each other by an unbroken chain of some twenty
generations of Englishmen, each of whom could speak to his
immediate neighbours in the chain as a son speaks to his father.
Language seems to 'evolve' by non-genetic means, and at a rate
which is orders of magnitude faster than genetic evolution.

Cultural transmission is not unique to man. The best non-
human example that I know has recently been described by P. F.

Jenkins in the song of a bird called the saddleback which lives on islands off New Zealand. On the island where he worked there was a total repertoire of about nine distinct songs. Any given male sang only one or a few of these songs. The males could be classified into dialect groups. For example, one group of eight males with neighbouring territories sang a particular song called the CC song. Other dialect groups sang different songs. Sometimes the members of a dialect group shared more than one distinct song. By comparing the songs of fathers and sons, Jenkins showed that song patterns were not inherited genetically. Each young male was likely to adopt songs from his territorial neighbours by imitation, in an analogous way to human language. During most of the time Jenkins was there, there was a fixed number of songs on the island, a kind of 'song pool' from which each young male drew his own small repertoire. But occasionally Jenkins was privileged to witness the 'invention' of a new song, which occurred by a mistake in the imitation of an old one. He writes: 'New song forms have been shown to arise variously by change of pitch of a note, repetition of a note, the elision of notes and the combination of parts of other existing songs... The appearance of the new form was an abrupt event and the product was quite stable over a period of years. Further, in a number of cases the variant was transmitted accurately in its new form to younger recruits so that a recognizably coherent group of like singers developed.' Jenkins refers to the origins of new songs as 'cultural mutations'.

Song in the saddleback truly evolves by non-genetic means. There are other examples of cultural evolution in birds and monkeys, but these are just interesting oddities. It is our own species that really shows what cultural evolution can do. Language is only one example out of many. Fashions in dress and diet, ceremonies and customs, art and architecture, engineering and technology, all evolve in historical time in a way that looks like highly speeded up genetic evolution, but has really nothing to do with genetic evolution. As in genetic evolution though, the change may be progressive. There is a sense in which modern science is actually better than ancient science. Not only does our under-standing of the universe change as the centuries go by: it improves. Admittedly the current burst of improvement dates back only to the Renaissance, which was preceded by a dismal period of stag-nation, in which European scientific culture was frozen at the level achieved by the Greeks. But, as we saw in Chapter 5, genetic evolution too may proceed as a series of brief spurts between sta-ble plateaux.

The analogy between cultural and genetic evolution has fre-quently been pointed out, sometimes in the context of quite unnecessary mystical overtones. The analogy between scientific progress and genetic evolution by natural selection has been illu-minated especially by Sir Karl Popper. I want to go even further into directions which are also being explored by, for example, the geneticist L. L. Cavalli-Sforza, the anthropologist F. T. Cloak, and the ethologist J. M. Cullen.

As an enthusiastic Darwinian, I have been dissatisfied with explanations that my fellow enthusiasts have offered for human behaviour. They have tried to look for 'biological advantages' in various attributes of human civilization. For instance, tribal reli-gion has been seen as a mechanism for solidifying group iden-tity, valuable for a pack-hunting species whose individuals rely on cooperation to catch large and fast prey. Frequently the evolu-tionary preconception in terms of which such theories are framed is implicitly group selectionist, but it is possible to rephrase the theories in terms of orthodox gene selection. Man may well have spent large portions of the last several million years living in small kin groups. Kin selection and selection in favour of recipro-cal altruism may have acted on human genes to produce many of our basic psychological attributes and tendencies. These ideas are plausible as far as they go, but I find that they do not begin to square up to the formidable challenge of explaining culture, cultural evolution, and the immense differences between human cultures around the world, from the utter selfishness of the Ik of Uganda, as described by Colin Turnbull, to the gentle altruism of Margaret Mead's Arapesh. I think we have got to start again and go right back to first principles. The argument I shall advance, surprising as it may seem coming from the author of the earlier chapters, is that, for an understanding of the evolution of modern man, we must begin by throwing out the gene as the sole basis of our ideas on evolution. I am an enthusiastic Darwinian, but I think Darwinism is too big a theory to be confined to the narrow context of the gene. The gene will enter my thesis as an analogy, nothing more.

What, after all, is so special about genes? The answer is that they are replicators. The laws of physics are supposed to be true all over the accessible universe. Are there any principles of biology that are likely to have similar universal validity? When astronauts voyage to distant planets and look for life, they can expect to find creatures too strange and unearthly for us to imagine. But is there anything that must be true of all life, wherever it is found, and whatever the basis of its chemistry? If forms of life exist whose chemistry is based on silicon rather than carbon, or ammonia rather than water, if creatures are discovered that boil to death at −100 degrees centigrade, if a form of life is found that is not based on chemistry at all but on electronic reverberating circuits, will there still be any general principle that is true of all life? Obviously I do not know but, if I had to bet, I would put my money on one fundamental principle. This is the law that all life evolves by the differential survival of replicating entities.* The gene, the DNA molecule, happens to be the replicating entity that prevails on our own planet. There may be others. If there are, provided certain other conditions are met, they will almost inevitably tend to become the basis for an evolutionary process.

But do we have to go to distant worlds to find other kinds of replicator and other, consequent, kinds of evolution? I think that a new kind of replicator has recently emerged on this very planet. It is staring us in the face. It is still in its infancy, still drifting clumsily about in its primeval soup, but already it is achieving evolutionary change at a rate that leaves the old gene panting far behind.

The new soup is the soup of human culture. We need a name for the new replicator, a noun that conveys the idea of a unit of cultural transmission, or a unit of *imitation*. 'Mimeme' comes from a suitable Greek root, but I want a monosyllable that sounds a bit like 'gene'. I hope my classicist friends will forgive me if I abbreviate mimeme to *meme*.* If it is any consolation, it could alternatively be thought of as being related to 'memory', or to the French word *même*. It should be pronounced to rhyme with 'cream'.

Examples of memes are tunes, ideas, catch-phrases, clothes fashions, ways of making pots or of building arches. Just as genes propagate themselves in the gene pool by leaping from body to body via sperms or eggs, so memes propagate themselves in the meme pool by leaping from brain to brain via a process which, in the broad sense, can be called imitation. If a scientist hears, or reads about, a good idea, he passes it on to his colleagues and students. He mentions it in his articles and his lectures. If the idea catchs on, it can be said to propagate itself, spreading from brain to brain. As my colleague N. K. Humphrey neatly summed up an earlier draft of this chapter: '... memes should be regarded as living structures, not just metaphorically but technically.* When you plant a fertile meme in my mind you literally parasitize my brain, turning it into a vehicle for the meme's propagation in just the way that a virus may parasitize the genetic mechanism of a host cell. And this isn't just a way of talking—the meme for, say, "belief in life after death" is actually realized physically, millions of times over, as a structure in the nervous systems of individual men the world over.'

Consider the idea of God. We do not know how it arose in the meme pool. Probably it originated many times by independent 'mutation'. In any case, it is very old indeed. How does it replicate itself? By the spoken and written word, aided by great music and great art. Why does it have such high survival value? Remember that 'survival value' here does not mean value for a gene in a gene pool, but value for a meme in a meme pool. The question really means: What is it about the idea of a god that gives it its stability and penetrance in the cultural environment? The survival value of the god meme in the meme pool results from its great psychological appeal. It provides a superficially plausible answer to deep and troubling questions about existence. It suggests that injustices in this world may be rectified in the next. The 'everlasting arms' hold out a cushion against our own inadequacies which, like a doctor's placebo, is none the less effective for being imaginary. These are some of the reasons why the idea of God is copied so readily by successive generations of individual brains. God exists, if only in the form of a meme with high survival value, or infective power, in the environment provided by human culture.

Some of my colleagues have suggested to me that this account of the survival value of the god meme begs the question. In the last analysis they wish always to go back to 'biological advantage'. To them it is not good enough to say that the idea of a god has 'great psychological appeal'. They want to know *why* it has great psychological appeal. Psychological appeal means appeal to brains, and brains are shaped by natural selection of genes in gene pools. They want to find some way in which having a brain like that improves gene survival.

I have a lot of sympathy with this attitude, and I do not doubt that there are genetic advantages in our having brains of the kind that we have. But nevertheless I think that these colleagues, if they look carefully at the fundamentals of their own assumptions, will find that they are begging just as many questions as I am. Fundamentally, the reason why it is good policy for us to try to explain biological phenomena in terms of gene advantage is that genes are replicators. As soon as the primeval soup provided conditions in which molecules could make copies of themselves, the replicators themselves took over. For more than three thousand million years, DNA has been the only replicator worth talking about in the world. But it does not necessarily hold these monopoly rights for all time. Whenever conditions arise in which a new kind of replicator *can* make copies of itself, the new replicators *will* tend to take over, and start a new kind of evolution of their own. Once this new evolution begins, it will in no necessary sense be subservient to the old. The old gene-selected evolution, by making brains, provided the soup in which the first memes arose. Once self-copying memes had arisen, their own, much faster, kind of evolution took off. We biologists have assimilated the idea of genetic evolution so deeply that we tend to forget that it is only one of many possible kinds of evolution.

Imitation, in the broad sense, is how memes *can* replicate. But just as not all genes that can replicate do so successfully, so some memes are more successful in the meme pool than others. This is the analogue of natural selection. I have mentioned particular examples of qualities that make for high survival value among memes. But in general they must be the same as those discussed for the replicators of Chapter 2: longevity, fecundity, and copying-fidelity. The longevity of any one copy of a meme is probably relatively unimportant, as it is for any one copy of a gene. The copy of the tune 'Auld Lang Syne' that exists in my brain will last only for the rest of my life.* The copy of the same tune that is printed in my volume of *The Scottish Student's Song Book* is unlikely to last much longer. But I expect there will be copies of the same tune on paper and in peoples' brains for centuries to come. As in the case of genes, fecundity is much more important than longevity of particular copies. If the meme is a scientific idea, its spread will depend on how acceptable it is to the population of individual scientists; a rough measure of its survival value could be obtained by counting the number of times it is referred to in successive years in scientific journals.* If it is a popular tune, its spread through the meme pool may be gauged by the number of people heard whistling it in the streets. If it is a style of women's shoe, the population memeticist may use sales statistics from shoe shops. Some memes, like some genes, achieve brilliant short-term success in spreading rapidly, but do not last long in the meme pool. Popular songs and stiletto heels are examples. Others, such as the Jewish religious laws, may continue to propagate themselves for thousands of years, usually because of the great potential permanence of written records.

This brings me to the third general quality of successful replicators: copying-fidelity. Here I must admit that I am on shaky ground. At first sight it looks as if memes are not high-fidelity replicators at all. Every time a scientist hears an idea and passes it on to somebody else, he is likely to change it somewhat. I have made no secret of my debt in this book to the ideas of R. L. Trivers. Yet I have not repeated them in his own words. I have twisted them round for my own purposes, changing the emphasis, blending them with ideas of my own and of other people. The memes are being passed on to you in altered form. This looks quite unlike the particulate, all-or-none quality of gene transmission.

It looks as though meme transmission is subject to continuous mutation, and also to blending.

It is possible that this appearance of non-particulateness is illusory, and that the analogy with genes does not break down. After all, if we look at the inheritance of many genetic characters such as human height or skin-colouring, it does not look like the work of indivisible and unblendable genes. If a black and a white person mate, their children do not come out either black or white: they are intermediate. This does not mean the genes concerned are not particulate. It is just that there are so many of them concerned with skin colour, each one having such a small effect, that they *seem* to blend. So far I have talked of memes as though it was obvious what a single unit-meme consisted of. But of course it is far from obvious. I have said a tune is one meme, but what about a symphony: how many memes is that? Is each movement one meme, each recognizable phrase of melody, each bar, each chord, or what?

I appeal to the same verbal trick as I used in Chapter 3. There I divided the 'gene complex' into large and small genetic units, and units within units. The 'gene' was defined, not in a rigid all-or-none way, but as a unit of convenience, a length of chromosome with just sufficient copying-fidelity to serve as a viable unit of natural selection. If a single phrase of Beethoven's ninth symphony is sufficiently distinctive and memorable to be abstracted from the context of the whole symphony, and used as the call-sign of a maddeningly intrusive European broadcasting station, then to that extent it deserves to be called one meme. It has, incidentally, materially diminished my capacity to enjoy the original symphony.

Similarly, when we say that all biologists nowadays believe in Darwin's theory, we do not mean that every biologist has, graven in his brain, an identical copy of the exact words of Charles

Darwin himself. Each individual has his own way of interpreting Darwin's ideas. He probably learned them not from Darwin's own writings, but from more recent authors. Much of what Darwin said is, in detail, wrong. Darwin if he read this book would scarcely recognize his own original theory in it, though I hope he would like the way I put it. Yet, in spite of all this, there is something, some essence of Darwinism, which is present in the head of every individual who understands the theory. If this were not so, then almost any statement about two people agreeing with each other would be meaningless. An 'idea-meme' might be defined as an entity that is capable of being transmitted from one brain to another. The meme of Darwin's theory is therefore that essential basis of the idea which is held in common by all brains that understand the theory. The *differences* in the ways that people represent the theory are then, by definition, not part of the meme. If Darwin's theory can be subdivided into components, such that some people believe component A but not component B, while others believe B but not A, then A and B should be regarded as separate memes. If almost everybody who believes in A also believes in B—if the memes are closely 'linked' to use the genetic term—then it is convenient to lump them together as one meme.

Let us pursue the analogy between memes and genes further. Throughout this book, I have emphasized that we must not think of genes as conscious, purposeful agents. Blind natural selection, however, makes them behave rather as if they were purposeful, and it has been convenient, as a shorthand, to refer to genes in the language of purpose. For example, when we say 'genes are trying to increase their numbers in future gene pools', what we really mean is 'those genes that behave in such a way as to increase their numbers in future gene pools tend to be the genes whose effects we see in the world'. Just as we have found it convenient to think of genes as active agents, working purpose-fully for their own survival, perhaps it might be convenient to think of memes in the same way. In neither case must we get mystical about it. In both cases the idea of purpose is only a metaphor, but we have already seen what a fruitful metaphor it is in the case of genes. We have even used words like 'selfish' and 'ruthless' of genes, knowing full well it is only a figure of speech. Can we, in exactly the same spirit, look for selfish or ruthless memes?

There is a problem here concerning the nature of competition. Where there is sexual reproduction, each gene is competing par-ticularly with its own alleles—rivals for the same chromosomal slot. Memes seem to have nothing equivalent to chromosomes, and nothing equivalent to alleles. I suppose there is a trivial sense in which many ideas can be said to have 'opposites'. But in gen-eral memes resemble the early replicating molecules, floating chaotically free in the primeval soup, rather than modern genes in their neatly paired, chromosomal regiments. In what sense then are memes competing with each other? Should we expect them to be 'selfish' or 'ruthless', if they have no alleles? The answer is that we might, because there is a sense in which they must indulge in a kind of competition with each other.

Any user of a digital computer knows how precious computer time and memory storage space are. At many large computer centres they are literally costed in money; or each user may be allotted a ration of time, measured in seconds, and a ration of space, measured in 'words'. The computers in which memes live are human brains.* Time is possibly a more important limiting factor than storage space, and it is the subject of heavy competi-tion. The human brain, and the body that it controls, cannot do more than one or a few things at once. If a meme is to dominate the attention of a human brain, it must do so at the expense of 'rival' memes. Other commodities for which memes compete are radio and television time, billboard space, newspaper column-inches, and library shelf-space.

In the case of genes, we saw in Chapter 3 that co-adapted gene complexes may arise in the gene pool. A large set of genes concerned with mimicry in butterflies became tightly linked together on the same chromosome, so tightly that they can be treated as one gene. In Chapter 5 we met the more sophisticated idea of the evolutionarily stable set of genes. Mutually suitable teeth, claws, guts, and sense organs evolved in carnivore gene pools, while a different stable set of characteristics emerged from herbivore gene pools. Does anything analogous occur in meme pools? Has the god meme, say, become associated with any other particular memes, and does this association assist the survival of each of the participating memes? Perhaps we could regard an organized church, with its architecture, rituals, laws, music, art, and written tradition, as a co-adapted stable set of mutually-assisting memes.

To take a particular example, an aspect of doctrine that has been very effective in enforcing religious observance is the threat of hell fire. Many children and even some adults believe that they will suffer ghastly torments after death if they do not obey the priestly rules. This is a peculiarly nasty technique of persuasion, causing great psychological anguish throughout the Middle Ages and even today. But it is highly effective. It might almost have been planned deliberately by a Machiavellian priesthood trained in deep psychological indoctrination techniques. However, I doubt if the priests were that clever. Much more probably, unconscious memes have ensured their own survival by virtue of those same qualities of pseudo-ruthlessness that successful genes display. The idea of hell fire is, quite simply, *self perpetuating*, because of its own deep psychological impact. It has become linked with the god meme because the two reinforce each other, and assist each other's survival in the meme pool.

Another member of the religious meme complex is called faith. It means blind trust, in the absence of evidence, even in the teeth of evidence. The story of Doubting Thomas is told, not so that we shall admire Thomas, but so that we can admire the other apostles in comparison. Thomas demanded evidence. Nothing is more lethal for certain kinds of meme than a tendency to look for evidence. The other apostles, whose faith was so strong that they did not need evidence, are held up to us as worthy of imitation. The meme for blind faith secures its own perpetuation by the simple unconscious expedient of discouraging rational inquiry.

Blind faith can justify anything.* If a man believes in a different god, or even if he uses a different ritual for worshipping the same god, blind faith can decree that he should die—on the cross, at the stake, skewered on a Crusader's sword, shot in a Beirut street, or blown up in a bar in Belfast. Memes for blind faith have their own ruthless ways of propagating themselves. This is true of patriotic and political as well as religious blind faith.

Memes and genes may often reinforce each other, but they sometimes come into opposition. For example, the habit of celibacy is presumably not inherited genetically. A gene for celibacy is doomed to failure in the gene pool, except under very special circumstances such as we find in the social insects. But still, a *meme* for celibacy can be successful in the meme pool. For example, suppose the success of a meme depends critically on how much time people spend in actively transmitting it to other people. Any time spent in doing other things than attempting to transmit the meme may be regarded as time wasted from the meme's point of view. The meme for celibacy is transmitted by priests to young boys who have not yet decided what they want to do with their lives. The medium of transmission is human influence of various kinds, the spoken and written word, personal example, and so on. Suppose, for the sake of argument, it happened to be the case that marriage weakened the power of a priest to influence his flock, say because it occupied a large proportion of his time and attention. This has, indeed, been advanced as an official reason for the enforcement of celibacy among priests. If this were the case, it would follow that the meme for celibacy could have greater survival value than the meme for marriage. Of course, exactly the opposite would be true for a *gene* for celibacy. If a priest is a survival machine for memes, celibacy is a useful attribute to build into him. Celibacy is just a minor partner in a large complex of mutually-assisting religious memes.

I conjecture that co-adapted meme-complexes evolve in the same kind of way as co-adapted gene-complexes. Selection favours memes that exploit their cultural environment to their own advantage. This cultural environment consists of other memes which are also being selected. The meme pool therefore comes to have the attributes of an evolutionarily stable set, which new memes find it hard to invade.

I have been a bit negative about memes, but they have their cheerful side as well. When we die there are two things we can leave behind us: genes and memes. We were built as gene machines, created to pass on our genes. But that aspect of us will be forgotten in three generations. Your child, even your grandchild, may bear a resemblance to you, perhaps in facial features, in a talent for music, in the colour of her hair. But as each generation passes, the contribution of your genes is halved. It does not take long to reach negligible proportions. Our genes may be immortal but the *collection* of genes that is any one of us is bound to crumble away. Elizabeth II is a direct descendant of William the Conqueror. Yet it is quite probable that she bears not a single one of the old king's genes. We should not seek immortality in reproduction.

But if you contribute to the world's culture, if you have a good idea, compose a tune, invent a sparking plug, write a poem, it may live on, intact, long after your genes have dissolved in the common pool. Socrates may or may not have a gene or two alive in the world today, as G. C. Williams has remarked, but who cares? The meme-complexes of Socrates, Leonardo, Copernicus, and Marconi are still going strong.

However speculative my development of the theory of memes may be, there is one serious point which I would like to emphasize once again. This is that when we look at the evolution of cultural traits and at their survival value, we must be clear *whose* survival we are talking about. Biologists, as we have seen, are accustomed to looking for advantages at the gene level (or the individual, the group, or the species level according to taste). What we have not previously considered is that a cultural trait may have evolved in the way that it has, simply because it is *advantageous to itself*.

We do not have to look for conventional biological survival values of traits like religion, music, and ritual dancing, though these may also be present. Once the genes have provided their survival machines with brains that are capable of rapid imitation, the memes will automatically take over. We do not even have to posit a genetic advantage in imitation, though that would certainly help. All that is necessary is that the brain should be *capable* of imitation: memes will then evolve that exploit the capability to the full.

I now close the topic of the new replicators, and end the chapter on a note of qualified hope. One unique feature of man, which may or may not have evolved memically, is his capacity for conscious foresight. Selfish genes (and, if you allow the speculation of this chapter, memes too) have no foresight. They are unconscious, blind, replicators. The fact that they replicate, together with certain further conditions, means, willy-nilly, that they will tend towards the evolution of qualities which, in the special sense of this book, can be called selfish. A simple replicator, whether gene or meme, cannot be expected to forgo short-term selfish advantage even if it would really pay it, in the long term, to do so. We saw this in the chapter on aggression. Even though a 'conspiracy of doves' would be better for *every single individual* than the evolutionarily stable strategy, natural selection is bound to favour the ESS.

It is possible that yet another unique quality of man is a capacity for genuine, disinterested, true altruism. I hope so, but I am not going to argue the case one way or the other, nor to speculate over its possible memic evolution. The point I am making now is that, even if we look on the dark side and assume that individual man is fundamentally selfish, our conscious foresight—our capacity to simulate the future in imagination—could save us from the worst selfish excesses of the blind replicators. We have at least the mental equipment to foster our long-term selfish interests rather than merely our short-term selfish interests. We can see the long-term benefits of participating in a 'conspiracy of doves', and we can sit down together to discuss ways of making the conspiracy work. We have the power to defy the selfish genes of our birth and, if necessary, the selfish memes of our indoctrination. We can even discuss ways of deliberately cultivating and nurturing pure, disinterested altruism—something that has no place in nature, something that has never existed before in the whole history of the world. We are built as gene machines and cultured as meme machines, but we have the power to turn against our creators. We, alone on earth, can rebel against the tyranny of the selfish replicators.*

NICE GUYS FINISH FIRST

Nice guys finish last. The phrase seems to have originated in the world of baseball, although some authorities claim priority for an alternative connotation. The American biologist Garrett Hardin used it to summarize the message of what may be called 'sociobiology' or 'selfish genery'. It is easy to see its aptness. If we translate the colloquial meaning of 'nice guy' into its Darwinian equivalent, a nice guy is an individual that assists other members of its species, at its own expense, to pass their genes on to the next generation. Nice guys, then, seem bound to decrease in numbers: niceness dies a Darwinian death. But there is another, technical, interpretation of the colloquial word 'nice'. If we adopt this definition, which is not too far from the colloquial meaning, nice guys can finish *first*. This more optimistic conclusion is what this chapter is about.

Remember the grudgers of Chapter 10. These were birds that helped each other in an apparently altruistic way, but refused to help—bore a grudge against—individuals that had previously refused to help them. Grudgers came to dominate the population because they passed on more genes to future generations than either suckers (who helped others indiscriminately and were exploited) or cheats (who tried ruthlessly to exploit everybody and ended up doing each other down). The story of the grudgers illustrated an important general principle, which Robert Trivers called 'reciprocal altruism'. As we saw in the example of the cleaner fish (pages 243–4), reciprocal altruism is not confined to members of a single species. It is at work in all relationships that are called symbiotic—for instance the ants milking their aphid 'cattle' (pages 235–6). Since Chapter 10 was written, the American political scientist Robert Axelrod (working partly in collaboration with W. D. Hamilton, whose name has cropped up on so many pages of this book) has taken the idea of reciprocal altruism on in exciting new directions. It was Axelrod who coined the technical meaning of the word 'nice' to which I alluded in my opening paragraph.

Axelrod, like many political scientists, economists, mathematicians, and psychologists, was fascinated by a simple gambling game called Prisoner's Dilemma. It is so simple that I have known clever men misunderstand it completely, thinking that there must be more to it! But its simplicity is deceptive. Whole shelves in libraries are devoted to the ramifications of this beguiling game. Many influential people think it holds the key to strategic defence planning, and that we should study it to prevent a third world war. As a biologist, I agree with Axelrod and Hamilton that many wild animals and plants are engaged in ceaseless games of Prisoner's Dilemma, played out in evolutionary time.

In its original, human, version, here is how the game is played. There is a 'banker', who adjudicates and pays out winnings to the two players. Suppose that I am playing against you (though, as we shall see, 'against' is precisely what we don't have to be). There are only two cards in each of our hands, labelled COOPERATE and DEFECT. To play, we each choose one of our cards and lay it face down on the table. Face down so that neither of us can be influenced by the other's move: in effect, we move simultaneously. We now wait in suspense for the banker to turn the cards over. The suspense is because our winnings depend not just on which card we have played (which we each know), but on the other player's card too (which we don't know until the banker reveals it).

Since there are 2 × 2 cards, there are four possible outcomes. For each outcome, our winnings are as follows (quoted in dollars in deference to the North American origins of the game):

Outcome I: We have both played COOPERATE. The banker pays each of us $300. This respectable sum is called the reward for mutual cooperation.

Outcome II: We have both played DEFECT. The banker fines each of us $10. This is called the punishment for mutual defection.

Outcome III: You have played COOPERATE; I have played DEFECT. The banker pays me $500 (the temptation to defect) and fines you (the sucker) $100.

Outcome IV: You have played DEFECT; I have played COOPER-ATE. The banker pays you the temptation pay-off of $500 and fines me, the sucker, $100.

Outcomes III and IV are obviously mirror images: one player does very well and the other does very badly. In outcomes I and II we do as well as one another, but I is better for *both* of us than II. The exact quantities of money don't matter. It doesn't even matter how many of them are positive (payments) and how many of them, if any, are negative (fines). What matters, for the game to qualify as a true Prisoner's Dilemma, is their rank order. The temptation to defect must be better than the reward for mutual cooperation, which must be better than the punishment for mutual defection, which must be better than the sucker's pay-off. (Strictly

What you do

	Cooperate	Defect
Cooperate	Fairly good **REWARD** (for mutual cooperation) e.g. $300	Very bad **SUCKER'S PAY-OFF** e.g. $100 fine
Defect	Very good **TEMPTATION** (to defect) e.g. $500	Fairly bad **PUNISHMENT** (for mutual defection) e.g. $10 fine

What I do (row label, positioned at left between Cooperate and Defect rows)

Figure A. Pay-offs to me from various outcomes of the Prisoner's Dilemma game

speaking, there is one further condition for the game to qualify as a true Prisoner's Dilemma: the average of the temptation and the sucker pay-offs must not exceed the reward. The reason for this additional condition will emerge later.) The four outcomes are summarized in the pay-off matrix in Figure A.

Now, why the 'dilemma'? To see this, look at the pay-off matrix and imagine the thoughts that might go through my head as I play against you. I know that there are only two cards you can play, COOPERATE and DEFECT. Let's consider them in order. If you have played DEFECT (this means we have to look at the right-hand column), the best card I could have played would have been DEFECT too. Admittedly I'd have suffered the penalty for mutual defection, but if I'd cooperated I'd have got the sucker's pay-off which is even worse. Now let's turn to the other thing you could have done (look at the left-hand column), play the COOP-ERATE card. Once again DEFECT is the best thing I could have done. If I had cooperated we'd both have got the rather high score of $300. But if I'd defected I'd have got even more—$500. The conclusion is that, regardless of which card you play, my best move is *Always Defect*.

So I have worked out by impeccable logic that, regardless of what you do, I must defect. And you, with no less impeccable logic, will work out just the same thing. So when two rational players meet, they will both defect, and both will end up with a fine or a low pay-off. Yet each knows perfectly well that, if only they had *both* played COOPERATE, both would have obtained the relatively high reward for mutual cooperation ($300 in our example). That is why the game is called a dilemma, why it seems so maddeningly paradoxical, and why it has even been proposed that there ought to be a law against it.

'Prisoner' comes from one particular imaginary example. The currency in this case is not money but prison sentences. Two men—call them Peterson and Moriarty—are in jail, suspected of collaborating in a crime. Each prisoner, in his separate cell, is invited to betray his colleague (DEFECT) by turning King's evidence against him. What happens depends upon what both prisoners do, and neither knows what the other has done. If Peterson throws the blame entirely on Moriarty, and Moriarty renders the story plausible by remaining silent (cooperating with his erstwhile and, as it turns out, treacherous friend), Moriarty gets a heavy jail sentence while Peterson gets off scot-free, having yielded to the temptation to defect. If each betrays the other, both are convicted of the crime, but receive some credit for giving evidence and get a somewhat reduced, though still stiff, sentence, the punishment for mutual defection. If both cooperate (with each other, not with the authorities) by refusing to speak, there is not enough evidence to convict either of them of the main crime, and they receive a small sentence for a lesser offence, the reward for mutual cooperation. Although it may seem odd to call a jail sentence a 'reward', that is how the men would see it if the alternative was a longer spell behind bars. You will notice that, although the 'pay-offs' are not in dollars but in jail sentences, the essential features of the game are preserved (look at the rank order of desirability of the four outcomes). If you put yourself in each prisoner's place, assuming both to be motivated by rational self-interest and remembering that they cannot talk to one another to make a pact, you will see that neither has any choice but to betray the other, thereby condemning both to heavy sentences.

Is there any way out of the dilemma? Both players know that, whatever their opponent does, they themselves cannot do better than DEFECT; yet both also know that, if only *both* had cooperated, *each* one would have done better. If only...if only...if only there could be some way of reaching agreement, some way of reassuring each player that the other can be trusted not to go for the selfish jackpot, some way of policing the agreement.

In the simple game of Prisoner's Dilemma, there is no way of ensuring trust. Unless at least one of the players is a really saintly sucker, too good for this world, the game is doomed to end in mutual defection with its paradoxically poor result for both players. But there is another version of the game. It is called the 'Iterated' or 'Repeated' Prisoner's Dilemma. The iterated game is more complicated, and in its complication lies hope.

The iterated game is simply the ordinary game repeated an indefinite number of times with the same players. Once again you and I face each other, with a banker sitting between. Once again we each have a hand of just two cards, labelled COOPER-ATE and DEFECT. Once again we move by each playing one or other of these cards and the banker shells out, or levies fines, according to the rules given on page 263. But now, instead of that being the end of the game, we pick up our cards and prepare for another round. The successive rounds of the game give us the opportunity to build up trust or mistrust, to reciprocate or placate, forgive or avenge. In an indefinitely long game, the important point is that we can both win at the expense of the banker, rather than at the expense of one another.

After ten rounds of the game, I could theoretically have won as much as $5,000, but only if you have been extraordinarily silly (or saintly) and played COOPERATE every time, in spite of the fact that I was consistently defecting. More realistically, it is easy for each of us to pick up $3,000 of the banker's money by both playing COOPERATE on all ten rounds of the game. For this we don't have to be particularly saintly, because we can both see, from the other's past moves, that the other is to be trusted. We can, in effect, police each other's behaviour. Another thing that is quite likely to happen is that neither of us trusts the other: we both play DEFECT for all ten rounds of the game, and the banker gains $100 in fines from each of us. Most likely of all is that we partially trust one another, and each play some mixed sequence of COOPERATE and DEFECT, ending up with some intermediate sum of money.

The birds in Chapter 10 who removed ticks from each other's feathers were playing an Iterated Prisoner's Dilemma game. How is this so? It is important, you remember, for a bird to pull off his own ticks, but he cannot reach the top of his own head and needs a companion to do that for him. It would seem only fair that he should return the favour later. But this service costs a bird time and energy, albeit not much. If a bird can get away with cheating—with having his own ticks removed but then refusing to reciprocate—he gains all the benefits without paying the costs. Rank the outcomes, and you'll find that indeed we have a true game of Prisoner's Dilemma. Both cooperating (pulling each other's ticks off) is pretty good, but there is still a temptation to do even better by refusing to pay the costs of reciprocating. Both defecting

What you do

	Cooperate	Defect
	Fairly good	Very bad
Cooperate	**REWARD**	**SUCKER'S PAY-OFF**
	I get my ticks removed, although I also pay the costs of removing yours.	I keep my ticks, while also paying the costs of removing yours.
	Very good	Fairly bad
Defect	**TEMPTATION**	**PUNISHMENT**
	I get my ticks removed, and I don't pay the costs of removing yours.	I keep my ticks with the small consolation of not removing yours.

(leftmost label: **What I do**)

Figure B. The bird tick-removing game: pay-offs to me from various outcomes

(refusing to pull ticks off) is pretty bad, but not so bad as putting effort into pulling another's ticks off and still ending up infested with ticks oneself. The pay-off matrix is Figure B.

But this is only one example. The more you think about it, the more you realize that life is riddled with Iterated Prisoner's Dilemma games, not just human life but animal and plant life too. Plant life? Yes, why not? Remember that we are not talking about conscious strategies (though at times we might be), but about strategies in the 'Maynard Smithian' sense, strategies of the kind that genes might preprogram. Later we shall meet plants, various animals, and even bacteria, all playing the game of Iterated Prisoner's Dilemma. Meanwhile, let's explore more fully what is so important about iteration.

Unlike the simple game, which is rather predictable in that DEFECT is the only rational strategy, the iterated version offers plenty of strategic scope. In the simple game there are only two possible strategies, COOPERATE and DEFECT. Iteration, however, allows lots of conceivable strategies, and it is by no means obvious which one is best. The following, for instance, is just one among thousands: 'cooperate most of the time, but on a random 10 per cent of rounds throw in a defect'. Or strategies might be conditional upon the past history of the game. My 'grudger' is an example of this; it has a good memory for faces, and although fundamentally cooperative it defects if the other player has ever defected before. Other strategies might be more forgiving and have shorter memories.

Clearly the strategies available in the iterated game are limited only by our ingenuity. Can we work out which is best? This was the task that Axelrod set himself. He had the entertaining idea of running a competition, and he advertised for experts in games theory to submit strategies. Strategies, in this sense, are preprogrammed rules for action, so it was appropriate for contestants to send in their entries in computer language. Fourteen strategies were submitted. For good measure Axelrod added a fifteenth, called Random, which simply played COOPERATE and DEFECT randomly, and served as a kind of baseline 'non-strategy': if a strategy can't do better than Random, it must be pretty bad.

Axelrod translated all 15 strategies into one common programming language, and set them against one another in one big computer. Each strategy was paired off in turn with every other one (including a copy of itself) to play Iterated Prisoner's Dilemma. Since there were 15 strategies, there were 15 × 15, or 225 separate games going on in the computer. When each pairing had gone through 200 moves of the game, the winnings were totalled up and the winner declared.

We are not concerned with which strategy won against any particular opponent. What matters is which strategy accumulated the most 'money', summed over all its 15 pairings. 'Money'

		What you do	
		Cooperate	**Defect**
What I do	**Cooperate**	Fairly good **REWARD** for mutual cooperation **3 points**	Very bad **SUCKER'S PAY-OFF** **0 points**
	Defect	Very good **TEMPTATION** to defect **5 points**	Fairly bad **PUNISHMENT** for mutual defection **1 points**

Figure C. Axelrod's computer tournament: pay-offs to me from various outcomes

means simply 'points', awarded according to the following scheme: mutual cooperation, 3 points; temptation to defect, 5 points; punishment for mutual defection, 1 point (equivalent to a light fine in our earlier game); sucker's pay-off, 0 points (equivalent to a heavy fine in our earlier game).

The maximum possible score that any strategy could achieve was 15,000 (200 rounds at 5 points per round, for each of 15 opponents). The minimum possible score was 0. Needless to say, neither of these two extremes was realized. The most that a strategy can realistically hope to win in an average one of its 15 pairings cannot be much more than 600 points. This is what two players would each receive if they both consistently cooperated, scoring 3 points for each of the 200 rounds of the game. If one of them succumbed to the temptation to defect, it would very probably end up with fewer points than 600 because of retaliation by the other player (most of the submitted strategies had some kind of retaliatory behaviour built into them). We can use 600 as a kind of benchmark for a game, and express all scores as a percentage of this benchmark. On this scale it is theoretically possible to score up to 166 per cent (1,000 points), but in practice no strategy's average score exceeded 600.

Remember that the 'players' in the tournament were not humans but computer programs, preprogrammed strategies. Their human authors played the same role as genes programming bodies (think of Chapter 4's computer chess and the Andromeda computer). You can think of the strategies as miniature 'proxies' for their authors. Indeed, one author could have submitted more than one strategy (although it would have been cheating—and Axelrod would presumably not have allowed it—for an author to 'pack' the competition with strategies, one of which received the benefits of sacrificial cooperation from the others).

Some ingenious strategies were submitted, though they were, of course, far less ingenious than their authors. The winning strategy, remarkably, was the simplest and superficially least ingenious of all. It was called Tit for Tat, and was submitted by Professor Anatol Rapoport, a well-known psychologist and games theorist from Toronto. Tit for Tat begins by cooperating on the first move and thereafter simply copies the previous move of the other player.

How might a game involving Tit for Tat proceed? As ever, what happens depends upon the other player. Suppose, first, that the other player is also Tit for Tat (remember that each strategy played against copies of itself as well as against the other 14). Both Tit for Tats begin by cooperating. In the next move, each player copies the other's previous move, which was COOPERATE. Both continue to COOPERATE until the end of the game, and both end up with the full 100 per cent 'benchmark' score of 600 points.

Now suppose Tit for Tat plays against a strategy called Naive Prober. Naive Prober wasn't actually entered in Axelrod's competition, but it is instructive nevertheless. It is basically identical to

Tit for Tat except that, once in a while, say on a random one in ten moves, it throws in a gratuitous defection and claims the high temptation score. Until Naive Prober tries one of its probing defections the players might as well be two Tit for Tats. A long and mutually profitable sequence of cooperation seems set to run its course, with a comfortable 100 per cent benchmark score for both players. But suddenly, without warning, say on the eighth move, Naive Prober defects. Tit for Tat, of course, has played COOPERATE on this move, and so is landed with the sucker's pay-off of 0 points. Naive Prober appears to have done well, since it has obtained 5 points from that move. But in the next move Tit for Tat 'retaliates'. It plays DEFECT, simply following its rule of imitating the opponent's previous move. Naive Prober meanwhile, blindly following its own built-in copying rule, has copied its opponent's COOPERATE move. So it now collects the sucker's pay-off of 0 points, while Tit for Tat gets the high score of 5. In the next move, Naive Prober—rather unjustly one might think— 'retaliates' against Tit for Tat's defection. And so the alternation continues. During these alternating runs both players receive on average 2.5 points per move (the average of 5 and 0). This is lower than the steady 3 points per move that both players can amass in a run of mutual cooperation (and, by the way, this is the reason for the 'additional condition' left unexplained on page 264). So, when Naive Prober plays against Tit for Tat, both do worse than when Tit for Tat plays against another Tit for Tat. And when Naive Prober plays against another Naive Prober, both tend to do, if anything, even worse still, since runs of reverberating defection tend to get started earlier.

Now consider another strategy, called Remorseful Prober. Remorseful Prober is like Naive Prober, except that it takes active steps to break out of runs of alternating recrimination. To do this it needs a slightly longer 'memory' than either Tit for Tat or Naive

Prober. Remorseful Prober remembers whether it has just spontaneously defected, and whether the result was prompt retaliation. If so, it 'remorsefully' allows its opponent 'one free hit' without retaliating. This means that runs of mutual recrimination are nipped in the bud. If you now work through an imaginary game between Remorseful Prober and Tit for Tat, you'll find that runs of would-be mutual retaliation are promptly scotched. Most of the game is spent in mutual cooperation, with both players enjoying the consequent generous score. Remorseful Prober does better against Tit for Tat than Naive Prober does, though not as well as Tit for Tat does against itself.

Some of the strategies entered in Axelrod's tournament were much more sophisticated than either Remorseful Prober or Naive Prober, but they too ended up with fewer points, on average, than simple Tit for Tat. Indeed the least successful of all the strategies (except Random) was the most elaborate. It was submitted by 'Name withheld'—a spur to pleasing speculation: Some *eminence grise* in the Pentagon? The head of the CIA? Henry Kissinger? Axelrod himself? I suppose we shall never know.

It isn't all that interesting to examine the details of the particular strategies that were submitted. This isn't a book about the ingenuity of computer programmers. It is more interesting to classify strategies according to certain categories, and examine the success of these broader divisions. The most important category that Axelrod recognizes is 'nice'. A nice strategy is defined as one that is never the first to defect. Tit for Tat is an example. It is capable of defecting, but it does so only in retaliation. Both Naive Prober and Remorseful Prober are nasty strategies because they sometimes defect, however rarely, when not provoked. Of the 15 strategies entered in the tournament, 8 were nice. Significantly, the 8 top-scoring strategies were the very same 8 nice strategies, the 7 nasties trailing well behind. Tit for Tat obtained an average of 504.5 points: 84 per cent of our benchmark of 600, and a good score. The other nice strategies scored only slightly less, with scores ranging from 83.4 per cent down to 78.6 per cent. There is a big gap between this score and the 66.8 per cent obtained by Graaskamp, the most successful of all the nasty strategies. It seems pretty convincing that nice guys do well in this game.

Another of Axelrod's technical terms is 'forgiving'. A forgiving strategy is one that, although it may retaliate, has a short memory. It is swift to overlook old misdeeds. Tit for Tat is a forgiving strategy. It raps a defector over the knuckles instantly but, after that, lets bygones be bygones. Chapter 10's grudger is totally unforgiving. Its memory lasts the entire game. It never forgets a grudge against a player who has ever defected against it, even once. A strategy formally identical to Grudger was entered in Axelrod's tournament under the name of Friedman, and it didn't do particularly well. Of all the nice strategies (note that it is technically nice, although it is totally unforgiving), grudger/Friedman did next to worst. The reason unforgiving strategies don't do very well is that they can't break out of runs of mutual recrimination, even when their opponent is 'remorseful'.

It is possible to be even more forgiving than Tit for Tat. Tit for Two Tats allows its opponents two defections in a row before it eventually retaliates. This might seem excessively saintly and magnanimous. Nevertheless Axelrod worked out that, if only somebody had submitted Tit for Two Tats, it would have won the tournament. This is because it is so good at avoiding runs of mutual recrimination.

So, we have identified two characteristics of winning strategies: niceness and forgivingness. This almost utopian-sounding conclusion—that niceness and forgivingness pay—came as a surprise to many of the experts, who had tried to be too cunning by submitting subtly nasty strategies; while even those who had submitted nice strategies had not dared anything so forgiving as Tit for Two Tats.

Axelrod announced a second tournament. He received 62 entries and again added Random, making 63 in all. This time, the exact number of moves per game was not fixed at 200 but was left open, for a good reason that I shall come to later. We can still express scores as a percentage of the 'benchmark', or 'always cooperate' score, even though that benchmark needs more complicated calculation and is no longer a fixed 600 points.

Programmers in the second tournament had all been provided with the results of the first, including Axelrod's analysis of why Tit for Tat and other nice and forgiving strategies had done so well. It was only to be expected that the contestants would take note of this background information, in one way or another. In fact, they split into two schools of thought. Some reasoned that niceness and forgivingness were evidently winning qualities, and they accordingly submitted nice, forgiving strategies. John Maynard Smith went so far as to submit the super-forgiving Tit for Two Tats. The other school of thought reasoned that lots of their colleagues, having read Axelrod's analysis, would now submit nice, forgiving strategies. They therefore submitted nasty strategies, trying to exploit these anticipated softies!

But once again nastiness didn't pay. Once again, Tit for Tat, submitted by Anatol Rapoport, was the winner, and it scored a massive 96 per cent of the benchmark score. And again nice strategies, in general, did better than nasty ones. All but one of the top 15 strategies were nice, and all but one of the bottom 15 were nasty. But although the saintly Tit for Two Tats would have won the first tournament if it had been submitted, it did not win the second. This was because the field now included more subtle nasty strategies capable of preying ruthlessly upon such an out-and-out softy.

This underlines an important point about these tournaments. Success for a strategy depends upon which other strategies happen to be submitted. This is the only way to account for the difference between the second tournament, in which Tit for Two Tats was ranked well down the list, and the first tournament, which Tit for Two Tats would have won. But, as I said before, this is not a book about the ingenuity of computer programmers. Is there an objective way in which we can judge which is the truly best strategy, in a more general and less arbitrary sense? Readers of earlier chapters will already be prepared to find the answer in the theory of evolutionarily stable strategies.

I was one of those to whom Axelrod circulated his early results, with an invitation to submit a strategy for the second tournament. I didn't do so, but I did make another suggestion. Axelrod had already begun to think in ESS terms, but I felt that this tendency was so important that I wrote to him suggesting that he should get in touch with W. D. Hamilton, who was then, though Axelrod didn't know it, in a different department of the same university, the University of Michigan. He did indeed immediately contact Hamilton, and the result of their subsequent collaboration was a brilliant joint paper published in the journal *Science* in 1981, a paper that won the Newcomb Cleveland Prize of the American Association for the Advancement of Science. In addition to discussing some delightfully way-out biological examples of Iterated Prisoner's Dilemmas, Axelrod and Hamilton gave what I regard as due recognition to the ESS approach.

Contrast the ESS approach with the 'round-robin' system that Axelrod's two tournaments followed. A round-robin is like a football league. Each strategy was matched against each other strategy an equal number of times. The final score of a strategy was the sum of the points it gained against all the other strategies. To be successful in a round-robin tournament, therefore, a strategy has to be a good competitor against all the other strategies that people *happen* to have submitted. Axelrod's name for a strategy that is good against a wide variety of other strategies is 'robust'. Tit for Tat turned out to be a robust strategy. But the set of strategies that people happen to have submitted is an arbitrary set. This was the point that worried us above. It just so happened that in Axelrod's original tournament about half the entries were nice. Tit for Tat won in this climate, and Tit for Two Tats would have won in this climate if it had been submitted. But suppose that nearly all the entries had just happened to be nasty. This could very easily have occurred. After all, 6 out of the 14 strategies submitted were nasty. If 13 of them had been nasty, Tit for Tat wouldn't have won. The 'climate' would have been wrong for it. Not only the money won, but the rank order of success among strategies depends upon which strategies happen to have been submitted; depends, in other words, upon something as arbitrary as human whim. How can we reduce this arbitrariness? By 'thinking ESS'.

The important characteristic of an evolutionarily stable strategy, you will remember from earlier chapters, is that it carries on doing well when it is already numerous in the population of strategies. To say that Tit for Tat, say, is an ESS would be to say that Tit for Tat does well in a climate dominated by Tit for Tat. This could be seen as a special kind of 'robustness'. As evolutionists we are tempted to see it as the only kind of robustness that matters. Why does it matter so much? Because, in the world of Darwinism, winnings are not paid out as money; they are paid out as offspring. To a Darwinian, a successful strategy is one that has become numerous in the population of strategies. For a strategy to remain successful, it must do well specifically when it is numerous, that is in a climate dominated by copies of itself.

Axelrod did, as a matter of fact, run a third round of his tournament as natural selection might have run it, looking for an ESS. Actually he didn't call it a third round, since he didn't solicit new entries but used the same 63 as for Round 2. I find it convenient to treat it as Round 3, because I think it differs from the two 'round-robin' tournaments more fundamentally than the two round-robin tournaments differ from each other.

Axelrod took the 63 strategies and threw them again into the computer to make 'generation 1' of an evolutionary succession. In 'generation 1', therefore, the 'climate' consisted of an equal representation of all 63 strategies. At the end of generation 1, winnings to each strategy were paid out, not as 'money' or 'points', but as *offspring*, identical to their (asexual) parents. As generations went by, some strategies became scarcer and eventually went extinct. Other strategies became more numerous. As the proportions changed, so, consequently, did the 'climate' in which future moves of the game took place.

Eventually, after about 1,000 generations, there were no further changes in proportions, no further changes in climate. Stability was reached. Before this, the fortunes of the various strategies rose and fell, just as in my computer simulation of the cheats, suckers, and grudgers. Some of the strategies started going extinct from the start, and most were extinct by generation 200. Of the nasty strategies, one or two of them began by increasing in frequency, but their prosperity, like that of cheat in my simulation, was short-lived. The only nasty strategy to survive beyond generation 200 was one called Harrington. Harrington's fortunes rose steeply for about the first 150 generations. Thereafter it declined rather gradually, approaching extinction around generation 1,000. Harrington did well temporarily for the same reason as my original cheat did. It exploited softies like Tit for Two Tats (too forgiving) while these were still around. Then, as the softies were driven extinct, Harrington followed them, having no easy prey left. The field was free for 'nice' but 'provocable' strategies like Tit for Tat.

Tit for Tat itself, indeed, came out top in five out of six runs of Round 3, just as it had in Rounds 1 and 2. Five other nice but provocable strategies ended up nearly as successful (frequent in the population) as Tit for Tat; indeed, one of them won the sixth run. When all the nasties had been driven extinct, there was no way in which any of the nice strategies could be distinguished from Tit for Tat or from each other, because they all, being nice, simply played COOPERATE against each other.

A consequence of this indistinguishability is that, although Tit for Tat seems like an ESS, it is strictly not a true ESS. To be an ESS, remember, a strategy must not be invadable, when it is common, by a rare, mutant strategy. Now it is true that Tit for Tat cannot be invaded by any nasty strategy, but another nice strategy is a different matter. As we have just seen, in a population of nice strategies they will all look and behave exactly like one another: they will all COOPERATE all the time. So any other nice strategy, like the totally saintly Always Cooperate, although admittedly it will not enjoy a positive selective advantage over Tit for Tat, can nevertheless drift into the population without being noticed. So technically Tit for Tat is not an ESS.

You might think that since the world stays just as nice, we could as well regard Tit for Tat as an ESS. But alas, look what happens next. Unlike Tit for Tat, Always Cooperate is not stable against invasion by nasty strategies such as Always Defect. Always Defect does well against Always Cooperate, since it gets the high temptation score every time. Nasty strategies like

Always Defect will come in to keep down the numbers of too nice strategies like Always Cooperate.

But although Tit for Tat is strictly speaking not a true ESS, it is probably fair to treat some sort of mixture of basically nice but retaliatory 'Tit for Tat-like' strategies as roughly equivalent to an ESS in practice. Such a mixture might include a small admixture of nastiness. Robert Boyd and Jeffrey Lorberbaum, in one of the more interesting follow-ups to Axelrod's work, looked at a mixture of Tit for Two Tats and a strategy called Suspicious Tit for Tat. Suspicious Tit for Tat is technically nasty, but it is not *very* nasty. It behaves just like Tit for Tat itself after the first move, but—this is what makes it technically nasty—it does defect on the very first move of the game. In a climate entirely dominated by Tit for Tat, Suspicious Tit for Tat does not prosper, because its initial defection triggers an unbroken run of mutual recrimination. When it meets a Tit for Two Tats player, on the other hand, Tit for Two Tats's greater forgivingness nips this recrimination in the bud. Both players end the game with at least the 'benchmark', all C, score and with Suspicious Tit for Tat scoring a bonus for its initial defection. Boyd and Lorberbaum showed that a population of Tit for Tat could be invaded, evolutionarily speaking, by a *mixture* of Tit for Two Tats and Suspicious Tit for Tat, the two prospering in each other's company. This combination is almost certainly not the only combination that could invade in this kind of way. There are probably lots of mixtures of slightly nasty strategies with nice and very forgiving strategies that are together capable of invading. Some might see this as a mirror for familiar aspects of human life.

Axelrod recognized that Tit for Tat is not strictly an ESS, and he therefore coined the phrase 'collectively stable strategy' to describe it. As in the case of true ESSs, it is possible for more than one strategy to be collectively stable at the same time. And again, it is a matter of luck which one comes to dominate a population. Always Defect is also stable, as well as Tit for Tat. In a population that has already come to be dominated by Always Defect, no other strategy does better. We can treat the system as bistable, with Always Defect being one of the stable points, Tit for Tat (or some mixture of mostly nice, retaliatory strategies) the other stable point. Whichever stable point comes to dominate the population first will tend to stay dominant.

But what does 'dominate' mean, in quantitative terms? How many Tit for Tats must there be in order for Tit for Tat to do better than Always Defect? That depends upon the detailed pay-offs that the banker has agreed to shell out in this particular game. All we can say in general is that there is a critical frequency, a knife-edge. On one side of the knife-edge the critical frequency of Tit for Tat is exceeded, and selection will favour more and more Tit for Tats. On the other side of the knife-edge the critical frequency of Always Defect is exceeded, and selection will favour more and more Always Defects. We met the equivalent of this knife-edge, you will remember, in the story of the grudgers and cheats in Chapter 10.

It obviously matters, therefore, on which side of the knife-edge a population happens to *start*. And we need to know how it might happen that a population could occasionally cross from one side of the knife-edge to the other. Suppose we start with a population already sitting on the Always Defect side. The few Tit for Tat individuals don't meet each other often enough to be of mutual benefit. So natural selection pushes the population even further towards the Always Defect extreme. If only the population could just manage, by random drift, to get itself over the knife-edge, it could coast down the slope to the Tit for Tat side, and everyone would do much better at the banker's (or 'nature's') expense. But of course populations have no group will, no group intention or purpose. They cannot strive to leap the knife-edge. They will cross it only if the undirected forces of nature happen to lead them across.

How could this happen? One way to express the answer is that it might happen by 'chance'. But 'chance' is just a word expressing ignorance. It means 'determined by some as yet unknown, or unspecified, means'. We can do a little better than 'chance'. We can try to think of practical ways in which a minority of Tit for Tat individuals might happen to increase to the critical mass. This amounts to a quest for possible ways in which Tit for Tat individuals might happen to cluster together in sufficient numbers that they can all benefit at the banker's expense.

This line of thought seems to be promising, but it is rather vague. How exactly might mutually resembling individuals find themselves clustered together, in local aggregations? In nature, the obvious way is through genetic relatedness—kinship. Animals of most species are likely to find themselves living close to their sisters, brothers, and cousins, rather than to random members of the population. This is not necessarily through choice. It follows automatically from 'viscosity' in the population. Viscosity means any tendency for individuals to continue living close to the place where they were born. For instance, through most of history, and in most parts of the world (though not, as it happens, in our modern world), individual humans have seldom strayed more than a few miles from their birthplace. As a result, local clusters of genetic relatives tend to build up. I remember visiting a remote island off the west coast of Ireland, and being struck by the fact that almost everyone on the island had the most enormous jug-handle ears. This could hardly have been because large ears suited the climate (there are strong offshore winds). It was because most of the inhabitants of the island were close kin of one another.

Genetic relatives will tend to be alike not just in facial features but in all sorts of other respects as well. For instance, they will tend to resemble each other with respect to genetic tendencies to play—or not to play—Tit for Tat. So even if Tit for Tat is rare in the population as a whole, it may still be locally common. In a local area, Tit for Tat individuals may meet each other often enough to prosper from mutual cooperation, even though calculations that take into account only the global frequency in the total population might suggest that they are below the 'knife-edge' critical frequency.

If this happens, Tit for Tat individuals, cooperating with one another in cosy little local enclaves, may prosper so well that they grow from small local clusters into larger local clusters. These local clusters may grow so large that they spread out into other areas, areas that had hitherto been dominated, numerically, by individuals playing Always Defect. In thinking of these local enclaves, my Irish island is a misleading parallel because it is physically cut off. Think, instead, of a large population in which there is not much movement, so that individuals tend to resemble their immediate neighbours more than their more distant neighbours, even though there is continuous interbreeding all over the whole area.

Coming back to our knife-edge, then, Tit for Tat could surmount it. All that is required is a little local clustering, of a sort that will naturally tend to arise in natural populations. Tit for Tat has a built-in gift, even when rare, for crossing the knife-edge over to its own side. It is as though there were a secret passage underneath the knife-edge. But that secret passage contains a one-way valve: there is an asymmetry. Unlike Tit for Tat, Always Defect, though a true ESS, cannot use local clustering to cross the knife-edge. On the contrary. Local clusters of Always Defect individuals, far from prospering by each other's presence, do especially *badly* in each other's presence. Far from quietly helping one another at the expense of the banker, they do one another down. Always Defect, then, unlike Tit for Tat, gets no help from kinship or viscosity in the population.

So, although Tit for Tat may be only dubiously an ESS, it has a sort of higher-order stability. What can this mean? Surely, stable is stable. Well, here we are taking a longer view. Always Defect resists invasion for a long time. But if we wait long enough, perhaps thousands of years, Tit for Tat will eventually muster the numbers required to tip it over the knife-edge, and the population will flip. But the reverse will not happen. Always Defect, as we have seen, cannot benefit from clustering, and so does not enjoy this higher-order stability.

Tit for Tat, as we have seen, is 'nice', meaning never the first to defect, and 'forgiving', meaning that it has a short memory for past misdeeds. I now introduce another of Axelrod's evocative technical terms. Tit for Tat is also 'not envious'. To be *envious*, in Axelrod's terminology, means to strive for more money than the other player, rather than for an absolutely large quantity of the banker's money. To be non-envious means to be quite happy if the other player wins just as much money as you do, so long as you both thereby win more from the banker. Tit for Tat never actually 'wins' a game. Think about it and you'll see that it *cannot* score more than its 'opponent' in any particular game because it never defects except in retaliation. The most it can do is draw with its opponent. But it tends to achieve each draw with a high, shared score. Where Tit for Tat and other nice strategies are concerned, the very word 'opponent' is inappropriate. Sadly, however, when psychologists set up games of Iterated Prisoner's Dilemma between real humans, nearly all players succumb to envy and therefore do relatively poorly in terms of money. It seems that many people, perhaps without even thinking about it, would rather do down the other player than cooperate with the other player to do down the banker. Axelrod's work has shown what a mistake this is.

It is only a mistake in certain kinds of game. Games theorists divide games into 'zero sum' and 'nonzero sum'. A zero sum game is one in which a win for one player is a loss for the other. Chess is zero sum, because the aim of each player is to win, and this means to make the other player lose. Prisoner's Dilemma, however, is a nonzero sum game. There is a banker paying out money, and it is possible for the two players to link arms and laugh all the way to the bank.

This talk of laughing all the way to the bank reminds me of a delightful line from Shakespeare:

> The first thing we do, let's kill all the lawyers.
>
> 2 *Henry VI*

In what are called civil 'disputes' there is often in fact great scope for cooperation. What looks like a zero sum confrontation can, with a little goodwill, be transformed into a mutually beneficial nonzero sum game. Consider divorce. A good marriage is obviously a nonzero sum game, brimming with mutual cooperation. But even when it breaks down there are all sorts of reasons why a couple could benefit by continuing to cooperate, and treating their divorce, too, as nonzero sum. As if child welfare were not a sufficient reason, the fees of two lawyers will make a nasty dent in the family finances. So obviously a sensible and civilized couple begin by going *together* to see one lawyer, don't they?

Well, actually no. At least in England and, until recently, in all fifty states of the USA, the law, or more strictly—and significantly—the lawyers' own professional code, doesn't allow them to. Lawyers must accept only one member of a couple as a client. The other person is turned from the door, and either has no legal advice at all or is forced to go to another lawyer. And that is when the fun begins. In separate chambers but with one voice, the two lawyers immediately start referring to 'us' and 'them'. 'Us', you understand, doesn't mean me and my wife; it means me and my lawyer against her and her lawyer. When the case comes to court, it is actually listed as 'Smith *versus* Smith'! It is *assumed* to be adversarial, whether the couple feel adversarial or not, whether or not they have specifically agreed that they want to be sensibly amicable. And who benefits from treating it as an 'I win, you lose' tussle? The chances are, only the lawyers.

The hapless couple have been dragged into a zero sum game. For the lawyers, however, the case of *Smith v. Smith* is a nice fat *non*zero sum game, with the Smiths providing the pay-offs and the two professionals milking their clients' joint account in elaborately coded cooperation. One way in which they cooperate is to make proposals that they both know the other side will not accept. This prompts a counter proposal that, again, both know is unacceptable. And so it goes on. Every letter, every telephone call exchanged between the cooperating 'adversaries' adds another wad to the bill. With luck, this procedure can be dragged out for months or even years, with costs mounting in parallel. The lawyers don't get together to work all this out. On the contrary, it is ironically their scrupulous separateness that is the chief instrument of their cooperation at the expense of the clients. The lawyers may not even be aware of what they are doing. Like the vampire bats that we shall meet in a moment, they are playing to well-ritualized rules. The system works without any conscious overseeing or organizing. It is all geared to forcing us into zero sum games. Zero sum for the clients, but very much *non*zero sum for the lawyers.

What is to be done? The Shakespeare option is messy. It would be cleaner to get the law changed. But most parliamentarians are drawn from the legal profession, and have a zero sum mentality. It is hard to imagine a more adversarial atmosphere than the British House of Commons. (The law courts at least preserve the decencies of debate. As well they might, since 'my learned friend and I' are cooperating very nicely all the way to the bank.) Perhaps well-meaning legislators and, indeed, contrite lawyers should be taught a little game theory. It is only fair to add that some lawyers play exactly the opposite role, persuading clients who are itching for a zero sum fight that they would do better to reach a nonzero sum settlement out of court.

What about other games in human life? Which are zero sum and which nonzero sum? And—because this is not the same thing—which aspects of life do we *perceive* as zero or nonzero sum? Which aspects of human life foster 'envy', and which foster cooperation against a 'banker'? Think, for instance, about wage-bargaining and 'differentials'. When we negotiate our pay-rises, are we motivated by 'envy', or do we cooperate to maximize our real income? Do we assume, in real life as well as in psychological experiments, that we are playing a zero sum game when we are not? I simply pose these difficult questions. To answer them would go beyond the scope of this book.

Football is a zero sum game. At least, it usually is. Occasionally it can become a nonzero sum game. This happened in 1977 in the English Football League (Association Football or 'Soccer'; the other games called football—Rugby Football, Australian Football, American Football, Irish Football, etc., are also normally zero sum games). Teams in the Football League are split into four divisions. Clubs play against other clubs within their own division, accumulating points for each win or draw throughout the season. To be in the First Division is prestigious, and also lucrative for a club since it ensures large crowds. At the end of each season, the bottom three clubs in the First Division are relegated to the Second Division for the next season. Relegation seems to be regarded as a terrible fate, worth going to great efforts to avoid.

May 18, 1977 was the last day of that year's football season. Two of the three relegations from the First Division had already been determined, but the third relegation was still in contention. It would definitely be one of three teams, Sunderland, Bristol, or Coventry. These three teams, then, had everything to play for on that Saturday. Sunderland were playing against a fourth team (whose tenure in the First Division was not in doubt). Bristol and Coventry happened to be playing against each other. It was known that, if Sunderland lost their game, then Bristol and Coventry needed only to draw against each other in order to stay in the First Division. But if Sunderland won, then the team relegated would be either Bristol or Coventry, depending on the outcome of their game against each other. The two crucial games were theoretically simultaneous. As a matter of fact, however, the Bristol–Coventry game happened to be running five minutes late. Because of this, the result of the Sunderland game became known before the end of the Bristol–Coventry game. Thereby hangs this whole complicated tale.

For most of the game between Bristol and Coventry the play was, to quote one contemporary news report, 'fast and often furious', an exciting (if you like that sort of thing) ding-dong battle. Some brilliant goals from both sides had seen to it that the score was 2-all by the eightieth minute of the match. Then, two minutes before the end of the game, the news came through from the other ground that Sunderland had lost. Immediately, the Coventry team manager had the news flashed up on the giant electronic message board at the end of the ground. Apparently all 22 players could read, and they all realized that they needn't bother to play hard any more. A draw was all that either team needed in order to avoid relegation. Indeed, to put effort into scoring goals was now positively bad policy since, by taking players away from defence, it carried the risk of actually losing—and being relegated after all. Both sides became intent on securing a draw. To quote the same news report: 'Supporters who had been fierce rivals seconds before when Don Gillies fired in an 80th minute equaliser for Bristol, suddenly joined in a combined celebration. Referee Ron Challis watched helpless as the players pushed the ball around with little or no challenge to the man in possession.' What had previously been a zero sum game had suddenly, because of a piece of news from the outside world, become a nonzero sum game. In the terms of our earlier discussion, it is as if an external 'banker' had magically appeared, making it possible for both Bristol and Coventry to benefit from the same outcome, a draw.

Spectator sports like football are normally zero sum games for a good reason. It is more exciting for crowds to watch players striving mightily against one another than to watch them conniving amicably. But real life, both human life and plant and animal life, is not set up for the benefit of spectators. Many situations in real life are, as a matter of fact, equivalent to nonzero sum games. Nature often plays the role of 'banker', and individuals can therefore benefit from one another's success. They do not have to do down rivals in order to benefit themselves. Without departing from the fundamental laws of the selfish gene, we can see how cooperation and mutual assistance can flourish even in a basically selfish world. We can see how, in Axelrod's meaning of the term, nice guys may finish first.

But none of this works unless the game is *iterated*. The players must know (or 'know') that the present game is not the last one between them. In Axelrod's haunting phrase, the 'shadow of the future' must be long. But how long must it be? It can't be infinitely long. From a theoretical point of view it doesn't matter how long the game is; the important thing is that neither player should *know* when the game is going to end. Suppose you and I were playing against each other, and suppose we both knew that the number of rounds in the game was to be exactly 100. Now we both understand that the 100th round, being the last, will be equivalent to a simple one-off game of Prisoner's Dilemma. Therefore the only rational strategy for either of us to play on the 100th round will be DEFECT, and we can each assume that the other player will work that out and be fully resolved to defect on the last round. The last round can therefore be written off as predictable. But now the 99th round will be the equivalent of a one-off game, and the only rational choice for each player on this last but one game is also DEFECT. The 98th round succumbs to the same reasoning, and so on back. Two strictly rational players, each of whom assumes that the other is strictly rational, can do nothing but defect if they both know how many rounds the game is destined to run. For this reason, when games theorists talk about the Iterated or Repeated Prisoner's Dilemma game, they always assume that the end of the game is unpredictable, or known only to the banker.

Even if the exact number of rounds in the game is not known for certain, in real life it is often possible to make a statistical guess as to how much longer the game is *likely* to last. This assessment may become an important part of strategy. If I notice the banker fidget and look at his watch, I may well conjecture that the game is about to be brought to an end, and I may therefore feel tempted to defect. If I suspect that you too have noticed the banker fidgeting, I may fear that you too may be contemplating defection. I will probably be anxious to get my defection in first. Especially since I may fear that you are fearing that I...

The mathematician's simple distinction between the one-off Prisoner's Dilemma game and the Iterated Prisoner's Dilemma game is too simple. Each player can be expected to behave as if he possessed a continuously updated estimate of how long the game is likely to go on. The longer his estimate, the more he will play according to the mathematician's expectations for the true iterated game: in other words, the nicer, more forgiving, less envious he will be. The shorter his estimate of the future of the game, the more he will be inclined to play according to the mathematician's expectations for the one-off game: the nastier, and less forgiving will he be.

Axelrod draws a moving illustration of the importance of the shadow of the future from a remarkable phenomenon that grew up during the First World War, the so-called live-and-let-live system. His source is the research of the historian and sociologist Tony Ashworth. It is quite well known that at Christmas in 1914 British and German troops briefly fraternized and drank together in no man's land. Less well known, but in my opinion more interesting, is the fact that unofficial and unspoken nonaggression pacts, a 'live-and-let-live' system, flourished all up and down the front lines for at least two years starting in 1914. A senior British officer, on a visit to the trenches, is quoted as being astonished to observe German soldiers walking about within rifle range behind their own line. 'Our men appeared to take no notice. I privately made up my mind to do away with that sort of thing when we took over; such things should not be allowed. These people evidently did not know there was a war on. Both sides apparently believed in the policy of "live-and-let-live".'

The theory of games and the Prisoner's Dilemma had not been invented in those days but, with hindsight, we can see pretty clearly what was going on, and Axelrod provides a fascinating analysis. In the entrenched warfare of those times, the shadow of the future for each platoon was long. That is to say, each dug-in group of British soldiers could expect to be facing the same dug-in group of Germans for many months. Moreover, the ordinary soldiers never knew when, if ever, they were going to be moved; army orders are notoriously arbitrary, capricious, and incomprehensible to those receiving them. The shadow of the future was quite long enough, and indeterminate enough, to foster the development of a Tit for Tat type of cooperation. Provided, that is, that the situation was equivalent to a game of Prisoner's Dilemma.

To qualify as a true Prisoner's Dilemma, remember, the payoffs have to follow a particular rank order. Both sides must see mutual cooperation (CC) as preferable to mutual defection. Defection while the other side cooperates (DC) is even better if you can get away with it. Cooperation while the other side defects (CD) is worst of all. Mutual defection (DD) is what the general staff would like to see. They want to see their own chaps, keen as mustard, potting Jerries (or Tommies) whenever the opportunity arises.

Mutual cooperation was undesirable from the generals' point of view, because it wasn't helping them to win the war. But it was highly desirable from the point of view of the individual soldiers on both sides. They didn't want to be shot. Admittedly—and this takes care of the other pay-off conditions needed to make the situation a true Prisoner's Dilemma—they probably agreed with the generals in preferring to win the war rather than lose it. But that is not the choice that faces an individual soldier. The outcome of the entire war is unlikely to be materially affected by what he, as an individual, does. Mutual cooperation with the particular enemy soldiers facing you across no man's land most definitely does affect your own fate, and is greatly preferable to mutual defection, even though you might, for patriotic or disciplinary reasons, marginally prefer to defect (DC) if you could get away with it. It seems that the situation was a true Prisoner's Dilemma. Something like Tit for Tat could be expected to grow up, and it did.

The locally stable strategy in any particular part of the trench lines was not necessarily Tit for Tat itself. Tit for Tat is one of a family of nice, retaliatory but forgiving strategies, all of which are, if not technically stable, at least difficult to invade once they arise. Three Tits for a Tat, for instance, grew up in one local area according to a contemporary account.

> We go out at night in front of the trenches...The German working parties are also out, so it is not considered etiquette to fire. The really nasty things are rifle grenades...They can kill as many as eight or nine men if they do fall into a trench...But we never use ours unless the Germans get particularly noisy, as on their system of retaliation three for every one of ours come back.

It is important, for any member of the Tit for Tat family of strategies, that the players are punished for defection. The threat of retaliation must always be there. Displays of retaliatory capability were a notable feature of the live-and-let-live system. Crack shots on both sides would display their deadly virtuosity by firing, not at enemy soldiers, but at inanimate targets close to the enemy soldiers, a technique also used in Western films (like shooting out candle flames). It does not seem ever to have been satisfactorily answered why the two first operational atomic bombs were used—against the strongly voiced wishes of the leading physicists responsible for developing them—to destroy two cities instead of being deployed in the equivalent of spectacularly shooting out candles.

An important feature of Tit for Tat-like strategies is that they are forgiving. This, as we have seen, helps to damp down what might otherwise become long and damaging runs of mutual recrimination. The importance of damping down retaliation is dramatized by the following memoir by a British (as if the first sentence left us in any doubt) officer:

I was having tea with A company when we heard a lot of shouting and went to investigate. We found our men and the Germans standing on their respective parapets. Suddenly a salvo arrived but did no damage. Naturally both sides got down and our men started swearing at the Germans, when all at once a brave German got on to his parapet and shouted out 'We are very sorry about that; we hope no one was hurt. It is not our fault, it is that damned Prussian artillery.'

Axelrod comments that this apology 'goes well beyond a merely instrumental effort to prevent retaliation. It reflects moral regret for having violated a situation of trust, and it shows concern that someone might have been hurt.' Certainly an admirable and very brave German.

Axelrod also emphasizes the importance of predictability and ritual in maintaining a stable pattern of mutual trust. A pleasing example of this was the 'evening gun' fired by British artillery with clockwork regularity at a certain part of the line. In the words of a German soldier:

At seven it came—so regularly that you could set your watch by it...It always had the same objective, its range was accurate, it never varied laterally or went beyond or fell short of the mark...There were even some inquisitive fellows who crawled out...a little before seven, in order to see it burst.

The German artillery did just the same thing, as the following account from the British side shows:

So regular were they [the Germans] in their choice of targets, times of shooting, and number of rounds fired, that...Colonel Jones...knew to a minute where the next shell would fall. His calculations were very accurate, and he was able to take what seemed to uninitiated Staff Officers big risks, knowing that the shelling would stop before he reached the place being shelled.

Axelrod remarks that such 'rituals of perfunctory and routine firing sent a double message. To the high command they conveyed aggression, but to the enemy they conveyed peace.'

The live-and-let-live system could have been worked out by verbal negotiation, by conscious strategists bargaining round a table. In fact it was not. It grew up as a series of local conventions, through people responding to one another's *behaviour*; the individual soldiers were probably hardly aware that the growing up was going on. This need not surprise us. The strategies in Axelrod's computer were definitely unconscious. It was their behaviour that defined them as nice or nasty, as forgiving or unforgiving, envious or the reverse. The programmers who designed them may have been any of these things, but that is irrelevant. A nice, forgiving, non-envious strategy could easily be programmed into a computer by a very nasty man. And vice versa. A strategy's niceness is recognized by its behaviour, not by its motives (for it has none) nor by the personality of its author (who has faded into the background by the time the program is running in the computer). A computer program can behave in a strategic manner, without being aware of its strategy or, indeed, of anything at all.

We are, of course, entirely familiar with the idea of unconscious strategists, or at least of strategists whose consciousness, if any, is irrelevant. Unconscious strategists abound in the pages of this book. Axelrod's programs are an exellent model for the way we, throughout the book, have been thinking of animals and plants, and indeed of genes. So it is natural to ask whether his optimistic conclusions—about the success of non-envious, forgiving niceness—also apply in the world of nature. The answer is yes, of course they do. The only conditions are that nature should sometimes set up games of Prisoner's Dilemma, that the shadow of the future should be long, and that the games should be nonzero sum games. These conditions are certainly met, all round the living kingdoms.

Nobody would ever claim that a bacterium was a conscious strategist, yet bacterial parasites are probably engaged in ceaseless games of Prisoner's Dilemma with their hosts and there is no reason why we should not attribute Axelrodian adjectives—forgiving, non-envious, and so on—to their strategies. Axelrod and Hamilton point out that normally harmless or beneficial bacteria can turn nasty, even causing lethal sepsis, in a person who is injured. A doctor might say that the person's 'natural resistance' is lowered by the injury. But perhaps the real reason is to do with games of Prisoner's Dilemma. Do the bacteria, perhaps, have something to gain, but usually keep themselves in check? In the game between human and bacteria, the 'shadow of the future' is normally long since a typical human can be expected to live for years from any given starting-point. A seriously wounded human, on the other hand, may present a potentially much shorter shadow of the future to his bacterial guests. The 'temptation to defect' correspondingly starts to look like a more attractive option than the 'reward for mutual cooperation'. Needless to say, there is no suggestion that the bacteria work all this out in their nasty little heads! Selection on generations of bacteria has presumably built into them an unconscious rule of thumb which works by purely biochemical means.

Plants, according to Axelrod and Hamilton, may even take revenge, again obviously unconsciously. Fig trees and fig wasps share an intimate cooperative relationship. The fig that you eat is not really a fruit. There is a tiny hole at the end, and if you go into this hole (you'd have to be as small as a fig wasp to do so, and they are minute: thankfully too small to notice when you eat a fig), you find hundreds of tiny flowers lining the walls. The fig is a dark indoor hothouse for flowers, an indoor pollination chamber.

And the only agents that can do the pollinating are fig wasps. The tree, then, benefits from harbouring the wasps. But what is in it for the wasps? They lay their eggs in some of the tiny flowers, which the larvae then eat. They pollinate other flowers within the same fig. 'Defecting', for a wasp, would mean laying eggs in too many of the flowers in a fig and pollinating too few of them. But how could a fig tree 'retaliate'? According to Axelrod and Hamilton, 'It turns out in many cases that if a fig wasp entering a young fig does not pollinate enough flowers for seeds and instead lays eggs in almost all, the tree cuts off the developing fig at an early stage. All progeny of the wasp then perish.'

A bizarre example of what appears to be a Tit for Tat arrangement in nature was discovered by Eric Fischer in a hermaphrodite fish, the sea bass. Unlike us, these fish don't have their sex determined at conception by their chromosomes. Instead, every individual is capable of performing both female and male functions. In any one spawning episode they shed either eggs or sperm. They form monogamous pairs and, within the pair, take turns to play the male and female roles. Now, we may surmise that any individual fish, if it could get away with it, would 'prefer' to play the male role all the time, because the male role is cheaper. Putting it another way, an individual that succeeded in persuading its partner to play the female most of the time would gain all the benefits of 'her' economic investment in eggs, while 'he' has resources left over to spend on other things, for instance on mating with other fish.

In fact, what Fischer observed was that the fishes operate a system of pretty strict alternation. This is just what we should expect if they are playing Tit for Tat. And it is plausible that they should, because it does appear that the game is a true Prisoner's Dilemma, albeit a somewhat complicated one. To play the COOPERATE card means to play the female role when it is your turn to do so.

Attempting to play the male role when it is your turn to play the female is equivalent to playing the DEFECT card. Defection is vulnerable to retaliation: the partner can refuse to play the female role next time it is 'her' (his?) turn to do so, or 'she' can simply terminate the whole relationship. Fischer did indeed observe that pairs with an uneven sharing of sex roles tended to break up.

A question that sociologists and psychologists sometimes ask is why blood donors (in countries, such as Britain, where they are not paid) give blood. I find it hard to believe that the answer lies in reciprocity or disguised selfishness in any simple sense. It is not as though regular blood donors receive preferential treatment when they come to need a transfusion. They are not even issued with little gold stars to wear. Maybe I am naïve, but I find myself tempted to see it as a genuine case of pure, disinterested altruism. Be that as it may, blood-sharing in vampire bats seems to fit the Axelrod model well. We learn this from the work of G. S. Wilkinson.

Vampires, as is well known, feed on blood at night. It is not easy for them to get a meal, but if they do it is likely to be a big one. When dawn comes, some individuals will have been unlucky and return completely empty, while those individuals that have managed to find a victim are likely to have sucked a surplus of blood. On a subsequent night the luck may run the other way. So, it looks like a promising case for a bit of reciprocal altruism. Wilkinson found that those individuals who struck lucky on any one night did indeed sometimes donate blood, by regurgitation, to their less fortunate comrades. Out of 110 regurgitations that Wilkinson witnessed, 77 could easily be understood as cases of mothers feeding their children, and many other instances of blood-sharing involved other kinds of genetic relatives. There still remained, however, some examples of blood-sharing among unrelated bats, cases where the 'blood is thicker than water' explanation would not fit the facts. Significantly the individuals involved here tended to be frequent roostmates—they had every opportunity to interact with one another repeatedly, as is required for an Iterated Prisoner's Dilemma. But were the other requirements for a Prisoner's Dilemma met? The pay-off matrix in Figure D is what we should expect if they were.

Do vampire economics really conform to this table? Wilkinson looked at the rate at which starved vampires lose weight. From this he calculated the time it would take a sated bat to starve to death, the time it would take an empty bat to starve to death, and all intermediates. This enabled him to cash out blood in the currency of hours of prolonged life. He found, not really surprisingly, that the exchange rate is different, depending upon how starved a bat is. A given amount of blood adds more hours to the life of a highly starved bat than to a less starved one. In other

	What you do	
	Cooperate	Defect
Cooperate (What I do)	Fairly good **REWARD** I get blood on my unlucky nights, which saves me from starving. I have to give blood on my lucky nights, which doesn't cost me too much.	Very bad **SUCKER'S PAY-OFF** I pay the cost of saving your life on my good night. But on my bad night your don't feed me and I run a real risk of starving to death.
Defect	Very good **TEMPTATION** You save my life on my poor night. But then I get the added benefit of not having to pay the slight cost of feeding you on my good night.	Fairly bad **PUNISHMENT** I don't have to pay the slight costs of feeding you on my good nights. But I run a real risk of starving on my poor nights.

Figure D. Vampire bat blood-donor scheme: pay-offs to me from various outcomes

words, although the act of donating blood would increase the chances of the donor dying, this increase was small compared with the increase in the recipient's chances of surviving. Economically speaking, then, it seems plausible that vampire economics conform to the rules of a Prisoner's Dilemma. The blood that the donor gives up is less precious to her (social groups in vampires are female groups) than the same quantity of blood is to the recipient. On her unlucky nights she really would benefit enormously from a gift of blood. But on her lucky nights she would benefit slightly, if she could get away with it, from defecting—refusing to donate blood. 'Getting away with it', of course, means something only if the bats are adopting some kind of Tit for Tat strategy. So, are the other conditions for the evolution of Tit for Tat reciprocation met?

In particular, can these bats recognize one another as individuals? Wilkinson did an experiment with captive bats, proving that they can. The basic idea was to take one bat away for a night and starve it while the others were all fed. The unfortunate starved bat was then returned to the roost, and Wilkinson watched to see who, if anyone, gave it food. The experiment was repeated many times, with the bats taking turns to be the starved victim. The key point was that this population of captive bats was a mixture of two separate groups, taken from caves many miles apart. If vampires are capable of recognizing their friends, the experimentally starved bat should turn out to be fed only by those from its own original cave.

That is pretty much what happened. Thirteen cases of donation were observed. In twelve out of these thirteen, the donor bat was an 'old friend' of the starved victim, taken from the same cave; in only one out of the thirteen cases was the starved victim fed by a 'new friend', not taken from the same cave. Of course this could be a coincidence but we can calculate the odds against this. They come to less than 1 in 500. It is pretty safe to conclude that the bats really were biased in favour of feeding old friends rather than strangers from a different cave.

Vampires are great mythmakers. To devotees of Victorian Gothic they are dark forces that terrorize by night, sapping vital fluids, sacrificing an innocent life merely to gratify a thirst. Combine this with that other Victorian myth, nature red in tooth and claw, and aren't vampires the very incarnation of deepest fears about the world of the selfish gene? As for me, I am sceptical of all myths. If we want to know where the truth lies in particular cases, we have to look. What the Darwinian corpus gives us is not detailed expectations about particular organisms. It gives us something subtler and more valuable: understanding of principle. But if we must have myths, the real facts about vampires could tell a different moral tale. To the bats themselves, not only is blood thicker than water. They rise above the bonds of kinship, forming their own lasting ties of loyal blood-brotherhood. Vampires could form the vanguard of a comfortable new myth, a myth of sharing, mutualistic cooperation. They could herald the benignant idea that, even with selfish genes at the helm, nice guys can finish first.

THE LONG REACH OF THE GENE

An uneasy tension disturbs the heart of the selfish gene theory. It is the tension between gene and individual body as fundamental agent of life. On the one hand we have the beguiling image of independent DNA replicators, skipping like chamois, free and untrammelled down the generations, temporarily brought together in throwaway survival machines, immortal coils shuffling off an endless succession of mortal ones as they forge towards their separate eternities. On the other hand we look at the individual bodies themselves and each one is obviously a coherent, integrated, immensely complicated machine, with a conspicuous unity of purpose. A body doesn't *look* like the product of a loose and temporary federation of warring genetic agents who hardly have time to get acquainted before embarking in sperm or egg for the next leg of the great genetic diaspora. It has one single-minded brain which coordinates a cooperative of limbs and sense organs to achieve one end. The body looks and behaves like a pretty impressive agent in its own right.

In some chapters of this book we have indeed thought of the individual organism as an agent, striving to maximize its success in passing on all its genes. We imagined individual animals making complicated economic 'as if' calculations about the genetic benefits of various courses of action. Yet in other chapters the fundamental rationale was presented from the point of view of genes. Without the gene's eye view of life there is no particular reason why an organism should 'care' about its reproductive success and that of its relatives, rather than, for instance, its own longevity.

How shall we resolve this paradox of the two ways of looking at life? My own attempt to do so is spelled out in *The Extended Phenotype*, the book that, more than anything else I have achieved in my professional life, is my pride and joy. This chapter is a brief distillation of a few of the themes in that book, but really I'd almost rather you stopped reading now and switched to *The Extended Phenotype!*

On any sensible view of the matter Darwinian selection does not work on genes directly. DNA is cocooned in protein, swaddled in membranes, shielded from the world, and invisible to natural selection. If selection tried to choose DNA molecules directly it would hardly find any criterion by which to do so. All genes look alike, just as all recording tapes look alike. The important differences between genes emerge only in their *effects*. This usually means effects on the processes of embryonic development and hence on bodily form and behaviour. Successful genes are genes that, in the environment influenced by all the other genes in a shared embryo, have beneficial effects on that embryo. Beneficial means that they make the embryo likely to develop into a successful adult, an adult likely to reproduce and pass those very same genes on to future generations. The technical word *phenotype* is used for the bodily manifestation of a gene, the effect that a gene, in comparison with its alleles, has on the body, via development. The phenotypic effect of some particular gene might be, say, green eye colour. In practice most genes have more than one phenotypic effect, say green eye colour and curly hair. Natural selection favours some genes rather than others not because of the nature of the genes themselves, but because of their consequences—their phenotypic effects.

Darwinians have usually chosen to discuss genes whose phenotypic effects benefit, or penalize, the survival and reproduction of whole bodies. They have tended not to consider benefits to the gene itself. This is partly why the paradox at the heart of the theory doesn't normally make itself felt. For instance a gene may be successful through improving the running speed of a predator. The whole predator's body, including all its genes, is more successful because it runs faster. Its speed helps it survive to have children; and therfore more copies of all its genes, including the gene for fast running, are passed on. Here the paradox conveniently disappears because what is good for one gene is good for all.

But what if a gene exerted a phenotypic effect that was good for itself but bad for the rest of the genes in the body? This is not a flight of fancy. Cases of it are known, for instance the intriguing phenomenon called meiotic drive. Meiosis, you will remember, is the special kind of cell division that halves the number of chromosomes and gives rise to sperm cells or egg cells. Normal meiosis is a completely fair lottery. Of each pair of alleles, only one can be the lucky one that enters any given sperm or egg. But it is equally likely to be either one of the pair, and if you average over lots of sperms (or eggs) it turns out that half of them contain one allele, half the other. Meiosis is fair, like tossing a penny. But, though we proverbially think of tossing a penny as random, even that is a physical process influenced by a multitude of circumstances—the wind, precisely how hard the penny is flicked, and so on. Meiosis, too, is a physical process, and it can be influenced by genes. What if a mutant gene arose that just happened to have an effect, not upon something obvious like eye colour or curliness of hair, but upon meiosis itself? Suppose it happened to bias meiosis in such a way that it, the mutant gene itself, was more likely than its allelic partner to end up in the egg. There are such genes and they are called segregation distorters. They have a diabolical simplicity. When a segregation distorter arises by mutation, it will spread inexorably through the population at the expense of its allele. It is this that is known as meiotic drive. It will happen even if the effects on bodily welfare, and on the welfare of all the other genes in the body, are disastrous.

Throughout this book we have been alert to the possibility of individual organisms 'cheating' in subtle ways against their social companions. Here we are talking about single genes cheating against the other genes with which they share a body. The geneticist James Crow has called them 'genes that beat the system'. One of the best-known segregation distorters is the so-called *t* gene in mice. When a mouse has two *t* genes it either dies young or is sterile. *t* is therefore said to be 'lethal' in the homozygous state. If a male mouse has only one *t* gene it will be a normal, healthy mouse except in one remarkable respect. If you examine such a male's sperms you will find that up to 95 per cent of them contain the *t* gene, only 5 per cent the normal allele. This is obviously a gross distortion of the 50 per cent ratio that we expect. Whenever, in a wild population, a *t* allele happens to arise by mutation, it immediately spreads like a brushfire. How could it not, when it has such a huge unfair advantage in the meiotic lottery? It spreads so fast that, pretty soon, large numbers of individuals in the population inherit the *t* gene in double dose (that is, from both their parents). These individuals die or are sterile, and before long the whole local population is likely to be driven extinct. There is some evidence that wild populations of mice have, in the past, gone extinct through epidemics of *t* genes.

Not all segregation distorters have such destructive side-effects as *t*. Nevertheless, most of them have at least some adverse consequences. (Almost all genetic side-effects are bad, and a new mutation will normally spread only if its bad effects are outweighed by its good effects. If both good and bad effects apply to the whole body, the net effect can still be good for the body. But if the bad effects are on the body, and the good effects are on the gene alone, from the body's point of view the net effect is all bad.) In spite of its deleterious side-effects, if a segregation distorter arises by mutation it will surely tend to spread through the population. Natural selection (which, after all, works at the genic level) favours the segregation distorter, even though its effects at the level of the individual organism are likely to be bad.

Although segregation distorters exist they aren't very common. We could go on to ask why they aren't common, which is another way of asking why the process of meiosis is normally fair, as scrupulously impartial as tossing a good penny. We'll find that the answer drops out once we have understood why organisms exist anyway.

The individual organism is something whose existence most biologists take for granted, probably because its parts do pull together in such a united and integrated way. Questions about life are conventionally questions about organisms. Biologists ask why organisms do this, why organisms do that. They frequently ask why organisms group themselves into societies. They don't ask—though they should—why living matter groups itself into organisms in the first place. Why isn't the sea still a primordial battleground of free and independent replicators? Why did the ancient replicators club together to make, and reside in, lumbering robots, and why are those robots—individual bodies, you and me—so large and so complicated?

It is hard for many biologists even to see that there is a question here at all. This is because it is second nature for them to pose their questions at the level of the individual organism. Some biologists go so far as to see DNA as a device used by organisms to reproduce themselves, just as an eye is a device used by organisms to see! Readers of this book will recognize that this attitude is an error of great profundity. It is the truth turned crashingly on its head. They will also recognize that the alternative attitude, the selfish gene view of life, has a deep problem of its own. That problem—almost the reverse one—is why individual organisms exist at all, especially in a form so large and coherently purposeful as to mislead biologists into turning the truth upside down. To solve our problem, we have to begin by purging our minds of old attitudes that covertly take the individual organism for granted; otherwise we shall be begging the question. The instrument with which we shall purge our minds is the idea that I call the extended phenotype. It is to this, and what it means, that I now turn.

The phenotypic effects of a gene are normally seen as all the effects that it has on the body in which it sits. This is the conventional definition. But we shall now see that the phenotypic effects of a gene need to be thought of as *all the effects that it has on the world*. It may be that a gene's effects, as a matter of fact, turn out to be confined to the succession of bodies in which the gene sits. But, if so, it will be just as a matter of fact. It will not be something that ought to be part of our very definition. In all this, remember that the phenotypic effects of a gene are the tools by which it levers itself into the next generation. All that I am going to add is that the tools may reach outside the individual body wall. What might it mean in practice to speak of a gene as having an extended phenotypic effect on the world outside the body in which it sits? Examples that spring to mind are artefacts like beaver dams, bird nests, and caddis houses.

Caddis flies are rather nondescript, drab brown insects, which most of us fail to notice as they fly rather clumsily over rivers.

That is when they are adults. But before they emerge as adults they have a rather longer incarnation as larvae walking about the river bottom. And caddis larvae are anything but nondescript. They are among the most remarkable creatures on earth. Using cement of their own manufacture, they skilfully build tubular houses for themselves out of materials that they pick up from the bed of the stream. The house is a mobile home, carried about as the caddis walks, like the shell of a snail or hermit crab except that the animal builds it instead of growing it or finding it. Some species of caddis use sticks as building materials, others fragments of dead leaves, others small snail shells. But perhaps the most impressive caddis houses are the ones built in local stone. The caddis chooses its stones carefully, rejecting those that are too large or too small for the current gap in the wall, even rotating each stone until it achieves the snuggest fit.

Incidentally, why does this impress us so? If we forced ourselves to think in a detached way we surely ought to be more impressed by the architecture of the caddis's eye, or of its elbow joint, than by the comparatively modest architecture of its stone house. After all, the eye and the elbow joint are far more complicated and 'designed' than the house. Yet, perhaps because the eye and elbow joint develop in the same kind of way as our own eyes and elbows develop, a building process for which we, inside our mothers, claim no credit, we are illogically more impressed by the house.

Having digressed so far, I cannot resist going a little further. Impressed as we may be by the caddis house, we are nevertheless, paradoxically, less impressed than we would be by equivalent achievements in animals closer to ourselves. Just imagine the banner headlines if a marine biologist were to discover a species of dolphin that wove large, intricately meshed fishing nets, twenty dolphin-lengths in diameter! Yet we take a spider web for granted, as a nuisance in the house rather than as one of the wonders of the world. And think of the furore if Jane Goodall returned from Gombe stream with photographs of wild chimpanzees building their own houses, well roofed and insulated, of painstakingly selected stones neatly bonded and mortared! Yet caddis larvae, who do precisely that, command only passing interest. It is sometimes said, as though in defence of this double standard, that spiders and caddises achieve their feats of architecture by 'instinct'. But so what? In a way this makes them all the more impressive.

Let us get back to the main argument. The caddis house, nobody could doubt, is an adaptation, evolved by Darwinian selection. It must have been favoured by selection, in very much the same way as, say, the hard shell of lobsters was favoured. It is a protective covering for the body. As such it is of benefit to the whole organism and all its genes. But we have now taught ourselves to see benefits to the organism as incidental, as far as natural selection is concerned. The benefits that actually count are the benefits to those genes that give the shell its protective properties. In the case of the lobster this is the usual story. The lobster's shell is obviously a part of its body. But what about the caddis house?

Natural selection favoured those ancestral caddis genes that caused their possessors to build effective houses. The genes worked on behaviour, presumably by influencing the embryonic development of the nervous system. But what a genticist would actually see is the effect of genes on the shape and other properties of houses. The geneticist should recognize genes 'for' house shape in precisely the same sense as there are genes for, say, leg shape. Admittedly, nobody has actually studied the genetics of caddis houses. To do so you would have to keep careful pedigree records of caddises bred in captivity, and breeding them is difficult. But you don't have to study genetics to be sure that there are, or at least once were, genes influencing differences between caddis houses. All you need is good reason to believe that the caddis house is a Darwinian adaptation. In that case there must have been genes controlling variation in caddis houses, for selection cannot produce adaptations unless there are hereditary differences among which to select.

Although geneticists may think it an odd idea, it is therefore sensible for us to speak of genes 'for' stone shape, stone size, stone hardness, and so on. Any geneticist who objects to this language must, to be consistent, object to speaking of genes for eye colour, genes for wrinkling in peas, and so on. One reason the idea might seem odd in the case of stones is that stones are not living material. Moreover, the influence of genes upon stone properties seems especially indirect. A geneticist might wish to claim that the direct influence of the genes is upon the nervous system that mediates the stone-choosing behaviour, not upon the stones themselves. But I invite such a geneticist to look carefully at what it can ever mean to speak of genes exerting an influence on a nervous system. All that genes can really influence directly is protein synthesis. A gene's influence upon a nervous system, or, for that matter, upon the colour of an eye or the wrinkliness of a pea, is *always* indirect. The gene determines a protein sequence that influences X that influences Y that influences Z that eventually influences the wrinkliness of the seed or the cellular wiring up of the nervous system. The caddis house is only a further extension of this kind of sequence. Stone hardness is an *extended* phenotypic effect of the caddis's genes. If it is legitimate to speak of a gene as affecting the wrinkliness of a pea or the nervous system of an animal (all geneticists think it is) then it must also be legitimate to speak of a gene as affecting the hardness of the stones in a caddis house. Startling thought, isn't it? Yet the reasoning is inescapable.

We are ready for the next step in the argument: genes in one organism can have extended phenotypic effects on the body of another organism. Caddis houses helped us take the previous step; snail shells will help us take this one. The shell plays the same role for a snail as the stone house does for a caddis larva. It is secreted by the snail's own cells, so a conventional geneticist would be happy to speak of genes 'for' shell qualities such as shell thickness. But it turns out that snails parasitized by certain kinds of fluke (flatworm) have extra-thick shells. What can this thickening mean? If the parasitized snails had had extra-thin shells, we'd happily explain this as an obvious debilitating effect on the snail's constitution. But a *thicker* shell? A thicker shell presumably protects the snail better. It looks as though the parasites are actually helping their host by improving its shell. But are they?

We have to think more carefully. If thicker shells are really better for the snail, why don't they have them anyway? The answer probably lies in economics. Making a shell is costly for a snail. It requires energy. It requires calcium and other chemicals that have to be extracted from hard-won food. All these resources, if they were not spent on making shell substance, could be spent on something else such as making more offspring. A snail that spends lots of resources on making an extra-thick shell has bought safety for its own body. But at what cost? It may live longer, but it will be less successful at reproducing and may fail to pass on its genes. Among the genes that fail to be passed on will be the genes for making extra-thick shells. In other words, it is possible for a shell to be too thick as well as (more obviously) too thin. So, when a fluke makes a snail secrete an extra-thick shell, the fluke is not doing the snail a good turn unless the fluke is bearing the economic cost of thickening the shell. And we can safely bet that it isn't being so generous. The fluke is exerting some hidden chemical influence on the snail that forces the snail to shift away from its own 'preferred' thickness of shell. It may be prolonging the snail's life. But it is not helping the snail's genes.

What is in it for the fluke? Why does it do it? My conjecture is the following. Both snail genes and fluke genes stand to gain from the snail's bodily survival, all other things being equal. But survival is not the same thing as reproduction and there is likely to be a trade-off. Whereas snail genes stand to gain from the snail's reproduction, fluke genes don't. This is because any given fluke has no particular expectation that its genes will be housed in its present host's offspring. They might be, but so might those of any of its fluke rivals. Given that snail longevity has to be bought at the cost of some loss in the snail's reproductive success, the fluke genes are 'happy' to make the snail pay that cost, since they have no interest in the snail's reproducing itself. The snail genes are not happy to pay that cost, since their long-term future depends upon the snail reproducing. So, I suggest that fluke genes exert an influence on the shell-secreting cells of the snail, an influence that benefits themselves but is costly to the snail's genes. This theory is testable, though it hasn't been tested yet.

We are now in a position to generalize the lesson of the caddises. If I am right about what the fluke genes are doing, it follows that we can legitimately speak of fluke genes as influencing snail bodies, in just the same sense as snail genes influence snail bodies. It is as if the genes reached outside their 'own' body and manipulated the world outside. As in the case of the caddises, this language might make geneticists uneasy. They are accustomed to the effects of a gene being limited to the body in which it sits. But, again as in the case of the caddises, a close look at what geneticists ever mean by a gene having 'effects' shows that such uneasiness is misplaced. We need to accept only that the change in snail shell is a fluke adaptation. If it is, it has to have come about by Darwinian selection of fluke genes. We have demonstrated that the phenotypic effects of a gene can extend, not only to inanimate objects like stones, but to 'other' living bodies too.

The story of the snails and flukes is only the beginning. Parasites of all types have long been known to exert fascinatingly insidious influences on their hosts. A species of microscopic protozoan parasite called *Nosema*, which infests the larvae of flour beetles, has 'discovered' how to manufacture a chemical that is very special for the beetles. Like other insects, these beetles have a hormone called the juvenile hormone which keeps larvae as larvae. The normal change from larva to adult is triggered by the larva ceasing production of juvenile hormone. The parasite *Nosema* has succeeded in synthesizing (a close chemical analogue of) this hormone. Millions of *Nosema* club together to mass-produce juvenile hormone in the beetle larva's body, thereby preventing it from turning into an adult. Instead it goes on growing, ending up as a giant larva more than twice the weight of a normal adult. No good for propagating beetle genes, but a cornucopia for *Nosema* parasites. Giantism in beetle larvae is an extended phenotypic effect of protozoan genes.

And here is a case history to provoke even more Freudian anxiety than the Peter Pan beetles—parasitic castration! Crabs are parasitized by a creature called *Sacculina*. *Sacculina* is related to barnacles, though you would think, to look at it, that it was a parasitic plant. It drives an elaborate root system deep into the tissues of the unfortunate crab, and sucks nourishment from its body. It is probably no accident that among the first organs that it attacks are the crab's testicles or ovaries; it spares the organs that the crab needs to survive—as opposed to reproduce—till later. The crab is effectively castrated by the parasite. Like a fattened bullock, the castrated crab diverts energy and resources away from reproduction and into its own body—rich pickings for the parasite at the expense of the crab's reproduction. Very much the same story as I conjectured for *Nosema* in the flour beetle and for the fluke in the snail. In all three cases the changes in the host, if we accept that they are Darwinian adaptations for the benefit of the parasite, must be seen as extended phenotypic effects of parasite genes. Genes, then, reach outside their 'own' body to influence phenotypes in other bodies.

To quite a large extent the interests of parasite genes and host genes may coincide. From the selfish gene point of view we can think of *both* fluke genes and snail genes as 'parasites' in the snail body. Both gain from being surrounded by the same protective shell, though they diverge from one another in the precise thickness of shell that they 'prefer'. This divergence arises, fundamentally, from the fact that their method of leaving this snail's body and entering another one is different. For the snail genes the method of leaving is via snail sperms or eggs. For the fluke's genes it is very different. Without going into the details (they are distractingly complicated) what matters is that their genes do not leave the snail's body in the snail's sperms or eggs.

I suggest that the most important question to ask about any parasite is this. Are its genes transmitted to future generations via the same vehicles as the host's genes? If they are not, I would expect it to damage the host, in one way or another. But if they are, the parasite will do all that it can to help the host, not only to survive but to reproduce. Over evolutionary time it will cease to be a parasite, will cooperate with the host, and may eventually merge into the host's tissues and become unrecognizable as a parasite at all. Maybe, as I suggested on page 237, our cells have come far across this evolutionary spectrum: we are all relics of ancient parasitic mergers.

Look at what can happen when parasite genes and host genes do share a common exit. Wood-boring ambrosia beetles (of the species *Xyleborus ferrugineus*) are parasitized by bacteria that not only live in their host's body but also use the host's eggs as their transport into a new host. The genes of such parasites therefore stand to gain from almost exactly the same future circumstances as the genes of their host. The two sets of genes can be expected to 'pull together' for just the same reasons as all the genes of one individual organism normally pull together. It is irrelevant that some of them happen to be 'beetle genes', while others happen to be 'bacterial genes'. Both sets of genes are 'interested' in beetle survival and the propagation of beetle eggs, because both 'see' beetle eggs as their passport to the future. So the bacterial genes share a common destiny with their host's genes, and in my interpretation we should expect the bacteria to cooperate with their beetles in all aspects of life.

It turns out that 'cooperate' is putting it mildly. The service they perform for the beetles could hardly be more intimate. These beetles happen to be haplodiploid, like bees and ants (see Chapter 10). If an egg is fertilized by a male, it always develops into a female. An unfertilized egg develops into a male. Males, in other words, have no father. The eggs that give rise to them develop spontaneously, without being penetrated by a sperm. But, unlike the eggs of bees and ants, ambrosia beetle eggs do need to be penetrated by *something*. This is where the bacteria come in. They prick the unfertilized eggs into action, provoking them to develop into male beetles. These bacteria are, of course, just the kind of parasites that, I argued, should cease to be parasitic and become mutualistic, precisely because they are transmitted in the eggs of the host, together with the host's 'own' genes. Ultimately, their 'own' bodies are likely to disappear, merging into the 'host' body completely.

A revealing spectrum can still be found today among species of hydra—small, sedentary, tentacled animals, like freshwater sea anemones. Their tissues tend to be parasitized by algae. (The 'g' should be pronounced hard. For unknown reasons some biologists, not least in America, have recently taken to saying Algy as in Algernon, not only for the plural 'algae', which is—just—forgivable, but also for the singular 'alga', which is not.) In the species *Hydra vulgaris* and *Hydra attenuata*, the algae are real parasites of the hydras, making them ill. In *Chlorohydra viridissima*, on the other hand, the algae are never absent from the tissues of the hydras, and make a useful contribution to their well-being, providing them with oxygen. Now here is the interesting point. Just as we should expect, in *Chlorohydra* the algae transmit themselves to the next generation by means of the hydra's egg. In the other two species they do not. The interests of alga genes and *Chlorohydra* genes coincide. Both are interested in doing everything in their power to increase production of *Chlorohydra* eggs. But the genes of the other two species of hydra do not 'agree' with the genes of their algae. Not to the same extent, anyway. Both sets of genes may have an interest in the survival of hydra bodies. But only hydra genes care about hydra reproduction. So the algae hang on as debilitating parasites rather than evolving towards benign cooperation. The key point, to repeat it, is that a parasite whose genes aspire to the same destiny as the genes of its host shares all the interests of its host and will eventually cease to act parasitically.

Destiny, in this case, means future generations. *Chlorohydra* genes and alga genes, beetle genes and bacteria genes, can get into the future only via the host's eggs. Therefore, whatever 'calculations' the parasite genes make about optimal policy, in any department of life, will converge on exactly, or nearly exactly, the same optimal policy as similar 'calculations' made by host genes. In the case of the snail and its fluke parasites, we decided that their preferred shell thicknesses were divergent. In the case of the ambrosia beetle and its bacteria, host and parasite will agree in preferring the same wing length, and every other feature of the beetle's body. We can predict this without knowing any details of exactly what the beetles might use their wings, or anything else, for. We can predict it simply from our reasoning that both the beetle genes and the bacterial genes will take whatever steps lie in their power to engineer the same future events—events favourable to the propagation of beetle eggs.

We can take this argument to its logical conclusion and apply it to normal, 'own' genes. Our own genes cooperate with one another, not because they *are* our own but because they share the same outlet—sperm or egg—into the future. If any genes of an organism, such as a human, could discover a way of spreading themselves that did not depend on the conventional sperm or egg route, they would take it and be less cooperative. This is because they would stand to gain by a different set of future outcomes from the other genes in the body. We've already seen examples of genes that bias meiosis in their own favour. Perhaps there are also genes that have broken out of the sperm/egg 'proper channels' altogether and pioneered a sideways route.

There are fragments of DNA that are not incorporated in chromosomes but float freely and multiply in the fluid contents of cells, especially bacterial cells. They go under various names such as viroids or plasmids. A plasmid is even smaller than a virus, and it normally consists of only a few genes. Some plasmids are capable of splicing themselves seamlessly into a chromosome. So smooth is the splice that you can't see the join: the plasmid is indistinguishable from any other part of the chromosome. The same plasmids can also cut themselves out again. This ability of DNA to cut and splice, to jump in and out of chromosomes at the drop of a hat, is one of the more exciting facts that have come to light since the first edition of this book was published. Indeed the recent evidence on plasmids can be seen as beautiful supporting evidence for the conjectures near bottom of page 237 (which seemed a bit wild at the time). From some points of view it does not really matter whether these fragments originated as invading parasites or breakaway rebels. Their likely behaviour will be the same. I shall talk about a breakaway fragment in order to emphasize my point.

Consider a rebel stretch of human DNA that is capable of snipping itself out of its chromosome, floating freely in the cell, perhaps multiplying itself up into many copies, and then splicing itself into another chromosome. What unorthodox alternative routes into the future could such a rebel replicator exploit? We are losing cells continually from our skin; much of the dust in our houses consists of our sloughed-off cells. We must be breathing in one another's cells all the time. If you draw your fingernail across the inside of your mouth it will come away with hundreds of living cells. The kisses and caresses of lovers must transfer multitudes of cells both ways. A stretch of rebel DNA could hitch a ride in any of these cells. If genes could discover a chink of an unorthodox route through to another body (alongside, or instead of, the orthodox sperm or egg route), we must expect natural selection to favour their opportunism and improve it. As for the precise methods that they use, there is no reason why these should be any different from the machinations—all too predictable to a selfish gene/extended phenotype theorist—of viruses.

When we have a cold or a cough, we normally think of the symptoms as annoying byproducts of the virus's activities. But in some cases it seems more probable that they are deliberately engineered by the virus to help it to travel from one host to another. Not content with simply being breathed into the atmosphere, the virus makes us sneeze or cough explosively. The rabies virus is transmitted in saliva when one animal bites another.

In dogs, one of the symptoms of the disease is that normally peaceful and friendly animals become ferocious biters, foaming at the mouth. Ominously too, instead of staying within a mile or so of home like normal dogs, they turn into restless wanderers, propagating the virus far afield. It has even been suggested that the well-known hydrophobic symptom encourages the dog to shake the wet foam from its mouth—and with it the virus. I do not know of any direct evidence that sexually transmitted diseases increase the libido of sufferers, but I conjecture that it would be worth looking into. Certainly at least one alleged aphrodisiac, Spanish Fly, is said to work by inducing an itch... and making people itch is just the kind of thing viruses are good at.

The point of comparing rebel human DNA with invading parasitic viruses is that there really isn't any important difference between them. Viruses may well, indeed, have originated as collections of breakaway genes. If we want to erect any distinction, it should be between genes that pass from body to body via the orthodox route of sperms or eggs, and genes that pass from body to body via unorthodox, 'sideways' routes. Both classes may include genes that originated as 'own' chromosomal genes. And both classes may include genes that originated as external, invading parasites. Or perhaps, as I speculated on page 237, all 'own' chromosomal genes should be regarded as mutually parasitic on one another. The important difference between my two classes of genes lies in the divergent circumstances from which they stand to benefit in the future. A cold virus gene and a breakaway human chromosomal gene agree with one another in 'wanting' their host to sneeze. An orthodox chromosomal gene and a venereally transmitted virus agree with one another in wanting their host to copulate. It is an intriguing thought that both would want the host to be sexually attractive. More, an orthodox chromosomal gene and a virus that is transmitted inside the host's egg would agree in wanting the host to succeed not just in its courtship but in every detailed aspect of its life, down to being a loyal, doting parent and even grandparent.

The caddis lives inside its house, and the parasites that I have so far discussed have lived inside their hosts. The genes, then, are physically close to their extended phenotypic effects, as close as genes ordinarily are to their conventional phenotypes. But genes can act at a distance; extended phenotypes can extend a long way. One of the longest that I can think of spans a lake. Like a spider web or a caddis house, a beaver dam is among the true wonders of the world. It is not entirely clear what its Darwinian purpose is, but it certainly must have one, for the beavers expend so much time and energy to build it. The lake that it creates probably serves to protect the beaver's lodge from predators. It also provides a convenient waterway for travelling and for transporting logs. Beavers use flotation for the same reason as Canadian lumber companies use rivers and eighteenth-century coal merchants used canals. Whatever its benefits, a beaver lake is a conspicuous and characteristic feature of the landscape. It is a phenotype, no less than the beaver's teeth and tail, and it has evolved under the influence of Darwinian selection. Darwinian selection has to have genetic variation to work on. Here the choice must have been between good lakes and less good lakes. Selection favoured beaver genes that made good lakes for transporting trees, just as it favoured genes that made good teeth for felling them. Beaver lakes are extended phenotypic effects of beaver genes, and they can extend over several hundreds of yards. A long reach indeed!

Parasites, too, don't have to live inside their hosts; their genes can express themselves in hosts at a distance. Cuckoo nestlings don't live inside robins or reed-warblers; they don't suck their blood or devour their tissues, yet we have no hesitation in labelling them as parasites. Cuckoo adaptations to manipulate the behaviour of foster parents can be looked upon as extended phenotypic action at a distance by cuckoo genes.

It is easy to empathize with foster parents duped into incubating the cuckoo's eggs. Human egg collectors, too, have been fooled by the uncanny resemblance of cuckoo eggs to, say, meadow-pipit eggs or reed-warbler eggs (different races of female cuckoos specialize in different host species). What is harder to understand is the behaviour of foster parents later in the season, towards young cuckoos that are almost fledged. The cuckoo is usually much larger, in some cases grotesquely larger, than its 'parent'. I am looking at a photograph of an adult dunnock, so small in comparison to its monstrous foster child that it has to perch on its back in order to feed it. Here we feel less sympathy for the host. We marvel at its stupidity, its gullibility. Surely any fool should be able to see that there is something wrong with a child like that.

I think that cuckoo nestlings must be doing rather more than just 'fooling' their hosts, more than just pretending to be something that they aren't. They seem to act on the host's nervous system in rather the same way as an addictive drug. This is not so hard to sympathize with, even for those with no experience of addictive drugs. A man can be aroused, even to erection, by a printed photograph of a woman's body. He is not 'fooled' into thinking that the pattern of printing ink really is a woman. He knows that he is only looking at ink on paper, yet his nervous system responds to it in the same kind of way as it might respond to a real woman. We may find the attractions of a particular member of the opposite sex irresistible, even though the better judgment of our better self tells us that a liaison with that person is not in anyone's long-term interests. The same can be true of the irresistible attractions of unhealthy food. The dunnock probably has no conscious awareness of its long-term best interests, so it is even easier to understand that its nervous system might find certain kinds of stimulation irresistible.

So enticing is the red gape of a cuckoo nestling that it is not uncommon for ornithologists to see a bird dropping food into the mouth of a baby cuckoo sitting in some other bird's nest! A bird may be flying home, carrying food for its own young. Suddenly, out of the corner of its eye, it sees the red super-gape of a young cuckoo, in the nest of a bird of some quite different species. It is diverted to the alien nest where it drops into the cuckoo's mouth the food that had been destined for its own young. The 'irresistibility theory' fits with the views of early German ornithologists who referred to foster parents as behaving like 'addicts' and to the cuckoo nestling as their 'vice'. It is only fair to add that this kind of language finds less favour with some modern experimenters. But there's no doubt that if we do assume that the cuckoo's gape is a powerful drug-like superstimulus, it becomes very much easier to explain what is going on. It becomes easier to sympathize with the behaviour of the diminutive parent standing on the back of its monstrous child. It is not being stupid. 'Fooled' is the wrong word to use. Its nervous system is being controlled, as irresistibly as if it were a helpless drug addict, or as if the cuckoo were a scientist plugging electrodes into its brain.

But even if we now feel more personal sympathy for the manipulated foster parent, we can still ask why natural selection has allowed the cuckoos to get away with it. Why haven't host nervous systems evolved resistance to the red gape drug? Maybe selection hasn't yet had time to do its work. Perhaps cuckoos have only in recent centuries started parasitizing their present hosts, and will in a few centuries be forced to give them up and victimize other species. There is some evidence to support this theory. But I can't help feeling that there must be more to it than that.

In the evolutionary 'arms race' between cuckoos and any host species, there is sort of built-in unfairness, resulting from unequal costs of failure. Each individual cuckoo nestling is descended from a long line of ancestral cuckoo nestlings, every single one of whom must have succeeded in manipulating its foster parent. Any cuckoo nestling that lost its hold, even momentarily, over its host would have died as a result. But each individual foster parent is descended from a long line of ancestors many of whom never encountered a cuckoo in their lives. And those that did have a cuckoo in their nest could have succumbed to it and still lived to rear another brood next season. The point is that there is an asymmetry in the cost of failure. Genes for failure to resist enslavement by cuckoos can easily be passed down the generations of robins or dunnocks. Genes for failure to enslave foster parents cannot be passed down the generations of cuckoos. This is what I meant by 'built-in unfairness', and by 'asymmetry in the cost of failure'. The point is summed up in one of Aesop's fables: 'The rabbit runs faster than the fox, because the rabbit is running for his life while the fox is only running for his dinner.' My colleague John Krebs and I have dubbed this the 'life/dinner principle'.

Because of the life/dinner principle, animals might at times behave in ways that are not in their own best interests, manipulated by some other animal. Actually, in a sense they are acting in their own best interests: the whole point of the life/dinner principle is that they theoretically could resist manipulation but it would be too costly to do so. Perhaps to resist manipulation by a cuckoo you need bigger eyes or a bigger brain, which would have overhead costs. Rivals with a genetic tendency to resist manipulation would actually be less successful in passing on genes, because of the economic costs of resisting.

But we have once again slipped back into looking at life from the point of view of the individual organism rather than its genes.

When we talked about flukes and snails we accustomed ourselves to the idea that a parasite's genes could have phenotypic
effects on the host's body, in exactly the same way as any animal's genes have phenotypic effects on its 'own' body. We showed
that the very idea of an 'own' body was a loaded assumption.
In one sense, all the genes in a body are 'parasitic' genes, whether
we like to call them the body's 'own' genes or not. Cuckoos came
into the discussion as an example of parasites not living inside
the bodies of their hosts. They manipulate their hosts in much
the same way as internal parasites do, and the manipulation, as
we have now seen, can be as powerful and irresistible as any
internal drug or hormone. As in the case of internal parasites, we
should now rephrase the whole matter in terms of genes and
extended phenotypes.

In the evolutionary arms race between cuckoos and hosts,
advances on each side took the form of genetic mutations arising
and being favoured by natural selection. Whatever it is about the
cuckoo's gape that acts like a drug on the host's nervous system,
it must have originated as a genetic mutation. This mutation
worked via its effect on, say, the colour and shape of the young
cuckoo's gape. But even this was not its most immediate effect.
Its most immediate effect was upon unseen chemical happen-
ings inside cells. The effect of genes on colour and shape of gape
is itself indirect. And now here is the point. Only a little more
indirect is the effect of the same cuckoo genes on the behaviour
of the besotted host. In exactly the same sense as we may speak
of cuckoo genes having (phenotypic) effects on the colour and
shape of cuckoo gapes, so we may speak of cuckoo genes having
(extended phenotypic) effects on host behaviour. Parasite genes
can have effects on host bodies, not just when the parasite lives
inside the host where it can manipulate by direct chemical
means, but when the parasite is quite separate from the host and

manipulates it from a distance. Indeed, as we are about to see, even chemical influences can act outside the body.

Cuckoos are remarkable and instructive creatures. But almost any wonder among the vertebrates can be surpassed by the insects. They have the advantage that there are just so many of them; my colleague Robert May has aptly observed that 'to a good approximation, all species are insects.' Insect 'cuckoos' defy listing; they are so numerous and their habit has been reinvented so often. Some examples that we'll look at have gone beyond familiar cuckooism to fulfil the wildest fantasies that *The Extended Phenotype* might have inspired.

A bird cuckoo deposits her egg and disappears. Some ant cuckoo females make their presence felt in more dramatic fashion. I don't often give Latin names, but *Bothriomyrmex regicidus* and *B. decapitans* tell a story. These two species are both parasites on other species of ants. Among all ants, of course, the young are normally fed not by parents but by workers, so it is workers that any would-be cuckoo must fool or manipulate. A useful first step is to dispose of the workers' own mother with her propensity to produce competing brood. In these two species the parasite queen, all alone, steals into the nest of another ant species. She seeks out the host queen, and rides about on her back while she quietly performs, to quote Edward Wilson's artfully macabre understatement, 'the one act for which she is uniquely specialized: slowly cutting off the head of her victim'. The murderess is then adopted by the orphaned workers, who unsuspectingly tend her eggs and larvae. Some are nurtured into workers themselves, who gradually replace the original species in the nest. Others become queens who fly out to seek pastures new and royal heads yet unsevered.

But sawing off heads is a bit of a chore. Parasites are not accustomed to exerting themselves if they can coerce a stand-in.

My favourite character in Wilson's *The Insect Societies* is *Monomorium santschii*. This species, over evolutionary time, has lost its worker caste altogether. The host workers do everything for their parasites, even the most terrible task of all. At the behest of the invading parasite queen, they actually perform the deed of murdering their own mother. The usurper doesn't need to use her jaws. She uses mind-control. How she does it is a mystery; she probably employs a chemical, for ant nervous systems are generally highly attuned to them. If her weapon is indeed chemical, then it is as insidious a drug as any known to science. For think what it accomplishes. It floods the brain of the worker ant, grabs the reins of her muscles, woos her from deeply ingrained duties and turns her against her own mother. For ants, matricide is an act of special genetic madness and formidable indeed must be the drug that drives them to it. In the world of the extended phenotype, ask not how an animal's behaviour benefits its genes; ask instead whose genes it is benefiting.

It is hardly surprising that ants are exploited by parasites, not just other ants but an astonishing menagerie of specialist hangers-on. Worker ants sweep a rich flow of food from a wide catchment area into a central hoard which is a sitting target for freeloaders. Ants are also good agents of protection: they are well-armed and numerous. The aphids of Chapter 10 could be seen as paying out nectar to hire professional bodyguards. Several butterfly species live out their caterpillar stage inside an ants' nest. Some are straightforward pillagers. Others offer something to the ants in return for protection. Often they bristle, literally, with equipment for manipulating their protectors. The caterpillar of a butterfly called *Thisbe irenea* has a sound-producing organ in its head for summoning ants, and a pair of telescopic spouts near its rear end which exude seductive nectar. On its shoulders stands another pair of nozzles, which cast an altogether more subtle spell. Their secretion seems to be not food but a volatile potion that has a dramatic impact upon the ants' behaviour. An ant coming under the influence leaps clear into the air. Its jaws open wide and it turns aggressive, far more eager than usual to attack, bite, and sting any moving object. Except, significantly, the caterpillar responsible for drugging it. Moreover, an ant under the sway of a dope-peddling caterpillar eventually enters a state called 'binding', in which it becomes inseparable from its caterpillar for a period of many days. Like an aphid, then, the caterpillar employs ants as bodyguards, but it goes one better. Whereas aphids rely on the ants' normal aggression against predators, the caterpillar administers an aggression-arousing drug and it seems to slip them something addictively binding as well.

I have chosen extreme examples. But, in more modest ways, nature teems with animals and plants that manipulate others of the same or of different species. In all cases in which natural selection has favoured genes for manipulation, it is legitimate to speak of those same genes as having (extended phenotypic) effects on the body of the manipulated organism. It doesn't matter in which body a gene physically sits. The target of its manipulation may be the same body or a different one. Natural selection favours those genes that manipulate the world to ensure their own propagation. This leads to what I have called the central theorem of the extended phenotype: *An animal's behaviour tends to maximize the survival of the genes 'for' that behaviour, whether or not those genes happen to be in the body of the particular animal performing it.* I was writing in the context of animal behaviour, but the theorem could apply, of course, to colour, size, shape—to anything.

It is finally time to return to the problem with which we started, to the tension between individual organism and gene as rival candidates for the central role in natural selection. In earlier chapters I made the assumption that there was no problem, because individual reproduction was equivalent to gene survival. I assumed there that you can say either 'The organism works to propagate all its genes' or 'The genes work to force a succession of organisms to propagate them.' They seemed like two equivalent ways of saying the same thing, and which form of words you chose seemed a matter of taste. But somehow the tension remained.

One way of sorting this whole matter out is to use the terms 'replicator' and 'vehicle'. The fundamental units of natural selection, the basic things that survive or fail to survive, that form lineages of identical copies with occasional random mutations, are called replicators. DNA molecules are replicators. They generally, for reasons that we shall come to, gang together into large communal survival machines or 'vehicles'. The vehicles that we know best are individual bodies like our own. A body, then, is not a replicator; it is a vehicle. I must emphasize this, since the point has been misunderstood. Vehicles don't replicate themselves; they work to propagate their replicators. Replicators don't behave, don't perceive the world, don't catch prey or run away from predators; they make vehicles that do all those things. For many purposes it is convenient for biologists to focus their attention at the level of the vehicle. For other purposes it is convenient for them to focus their attention at the level of the replicator. Gene and individual organism are not rivals for the same starring role in the Darwinian drama. They are cast in different, complementary, and in many respects equally important roles, the role of replicator and the role of vehicle.

The replicator/vehicle terminology is helpful in various ways. For instance it clears up a tiresome controversy over the level at which natural selection acts. Superficially it might seem logical to place 'individual selection' on a sort of ladder of levels of selection, halfway between the 'gene selection' advocated in Chapter 3 and the 'group selection' criticized in Chapter 7. 'Individual selection' seems vaguely to be a middle way between two extremes, and many biologists and philosophers have been seduced into this facile path and treated it as such. But we can now see that it isn't like that at all. We can now see that the organism and the group of organisms are true rivals for the vehicle role in the story, but neither of them is even a *candidate* for the replicator role. The controversy between 'individual selection' and 'group selection' is a real controversy between alternative vehicles. The controversy between individual selection and gene selection isn't a controversy at all, for gene and organism are candidates for different, and complementary, roles in the story, the replicator and the vehicle.

The rivalry between individual organism and group of organisms for the vehicle role, being a real rivalry, can be settled. As it happens the outcome, in my view, is a decisive victory for the individual organism. The group is too wishy-washy an entity. A herd of deer, a pride of lions, or a pack of wolves has a certain rudimentary coherence and unity of purpose. But this is paltry in comparison to the coherence and unity of purpose of the body of an individual lion, wolf, or deer. That this is true is now widely accepted, but *why* is it true? Extended phenotypes and parasites can again help us.

We saw that when the genes of a parasite work together with each other, but in opposition to the genes of the host (which all work together with *each* other), it is because the two sets of genes have different methods of leaving the shared vehicle, the host's body. Snail genes leave the shared vehicle via snail sperm and eggs. Because all snail genes have an equal stake in every sperm and every egg, because they all participate in the same unpartisan meiosis, they work together for the common good, and therefore tend to make the snail body a coherent, purposeful vehicle. The real reason why a fluke is recognizably separate from its host, the reason why it doesn't merge its purposes and its identity with the purposes and identity of the host, is that the fluke genes don't share the snail genes' method of leaving the shared vehicle, and don't share in the snail's meiotic lottery—they have a lottery of their own. Therefore, to that extent and that extent only, the two vehicles remain separated as a snail and a recognizably distinct fluke inside it. If fluke genes were passed on in snail eggs and sperms, the two bodies would evolve to become as one flesh. We mightn't even be able to tell that there ever had been two vehicles.

'Single' individual organisms such as ourselves are the ultimate embodiment of many such mergers. The group of organisms—the flock of birds, the pack of wolves—does not merge into a single vehicle, precisely because the genes in the flock or the pack do not share a common method of leaving the present vehicle. To be sure, packs may bud off daughter packs. But the genes in the parent pack don't pass to the daughter pack in a single vessel in which all have an equal share. The genes in a pack of wolves don't all stand to gain from the same set of events in the future. A gene can foster its own future welfare by favouring its own individual wolf, at the expense of other individual wolves. An individual wolf, therefore, is a vehicle worthy of the name. A pack of wolves is not. Genetically speaking, the reason for this is that all the cells except the sex cells in a wolf's body have the same genes, while, as for the sex cells, all the genes have an equal chance of being in each one of them. But the cells in a *pack* of wolves do not have the same genes, nor do they have the same chance of being in the cells of sub-packs that are budded off. They have everything to gain by struggling against rivals in other wolf bodies (although the fact that a wolf pack is likely to be a kin group will mitigate the struggle).

The essential quality that an entity needs, if it is to become an effective gene vehicle, is this. It must have an impartial exit channel into the future, for all the genes inside it. This is true of an individual wolf. The channel is the thin stream of sperms, or eggs, which it manufactures by meiosis. It is not true of the pack of wolves. Genes have something to gain from selfishly promoting the welfare of their own individual bodies, at the expense of other genes in the wolf pack. A beehive, when it swarms, appears to reproduce by broad-fronted budding, like a wolf pack. But if we look more carefully we find that, as far as the genes are concerned, their destiny is largely shared. The future of the genes in the swarm is, at least to a large extent, lodged in the ovaries of one queen. This is why—it is just another way of expressing the message of earlier chapters—the bee colony looks and behaves like a truly integrated single vehicle.

Everywhere we find that life, as a matter of fact, is bundled into discrete, individually purposeful vehicles like wolves and beehives. But the doctrine of the extended phenotype has taught us that it needn't have been so. Fundamentally, all that we have a right to expect from our theory is a battleground of replicators, jostling, jockeying, fighting for a future in the genetic hereafter. The weapons in the fight are phenotypic effects, initially direct chemical effects in cells but eventually feathers and fangs and even more remote effects. It undeniably happens to be the case that these phenotypic effects have largely become bundled up into discrete vehicles, each with its genes disciplined and ordered by the prospect of a shared bottleneck of sperms or eggs funnelling them into the future. But this is not a fact to be taken for granted. It is a fact to be questioned and wondered at in its own right. Why did genes come together into large vehicles, each with a single genetic exit route? Why did genes choose to gang up and make large bodies for themselves to live in? In *The Extended*

Phenotype I attempt to work out an answer to this difficult problem. Here I can sketch only a part of that answer—although, as might be expected after seven years, I can also now take it a little further.

I shall divide the question up into three. Why did genes gang up in cells? Why did cells gang up in many-celled bodies? And why did bodies adopt what I shall call a 'bottlenecked' life cycle?

First then, why did genes gang up in cells? Why did those ancient replicators give up the cavalier freedom of the primeval soup and take to swarming in huge colonies? Why do they cooperate? We can see part of the answer by looking at how modern DNA molecules cooperate in the chemical factories that are living cells. DNA molecules make proteins. Proteins work as enzymes, catalysing particular chemical reactions. Often a single chemical reaction is not sufficient to synthesize a useful end-product. In a human pharmaceutical factory the synthesis of a useful chemical needs a production line. The starting chemical cannot be transformed directly into the desired end-product. A series of intermediates must be synthesized in strict sequence. Much of a research chemist's ingenuity goes into devising pathways of feasible intermediates between starting chemicals and desired end-products. In the same way single enzymes in a living cell usually cannot, on their own, achieve the synthesis of a useful end-product from a given starting chemical. A whole set of enzymes is necessary, one to catalyse the transformation of the raw material into the first intermediate, another to catalyse the transformation of the first intermediate into the second, and so on.

Each of these enzymes is made by one gene. If a sequence of six enzymes is needed for a particular synthetic pathway, all six genes for making them must be present. Now it is quite likely that there are two alternative pathways for arriving at that same end-product, each needing six different enzymes, and with nothing to choose between the two of them. This kind of thing happens in chemical factories. Which pathway is chosen may be historical accident, or it may be a matter of more deliberate planning by chemists. In nature's chemistry the choice will never, of course, be a deliberate one. Instead it will come about through natural selection. But how can natural selection see to it that the two pathways are not mixed, and that cooperating groups of compatible genes emerge? In very much the same way as I suggested with my analogy of the German and English rowers (Chapter 5). The important thing is that a gene for a stage in pathway 1 will flourish in the presence of genes for other stages in pathway 1, but not in the presence of pathway 2 genes. If the population already happens to be dominated by genes for pathway 1, selection will favour other genes for pathway 1, and penalize genes for pathway 2. And vice versa. Tempting as it is, it is positively wrong to speak of the genes for the six enzymes of pathway 2 being selected 'as a group'. Each one is selected as a separate selfish gene, but it flourishes only in the presence of the right set of other genes.

Nowadays this cooperation between genes goes on within cells. It must have started as rudimentary cooperation between self-replicating molecules in the primeval soup (or whatever primeval medium there was). Cell walls perhaps arose as a device to keep useful chemicals together and stop them leaking away. Many of the chemical reactions in the cell actually go on in the fabric of membranes; a membrane acts as a combined conveyor belt and test-tube rack. But cooperation between genes did not stay limited to cellular biochemistry. Cells came together (or failed to separate after cell division) to form many-celled bodies.

This brings us to the second of my three questions. Why did cells gang together; why the lumbering robots? This is another question about cooperation. But the domain has shifted from the world of molecules to a larger scale. Many-celled bodies out-grow the microscope. They can even become elephants or whales. Being big is not necessarily a good thing: most organisms are bacteria and very few are elephants. But when the ways of mak-ing a living that are open to small organisms have all been filled, there are still prosperous livings to be made by larger organisms. Large organisms can eat smaller ones, for instance, and can avoid being eaten by them.

The advantages of being in a club of cells don't stop with size. The cells in the club can specialize, each thereby becoming more efficient at performing its particular task. Specialist cells serve other cells in the club and they also benefit from the efficiency of other specialists. If there are many cells, some can specialize as sensors to detect prey, others as nerves to pass on the message, others as stinging cells to paralyse the prey, muscle cells to move tentacles and catch the prey, secretory cells to dissolve it and yet others to absorb the juices. We must not forget that, at least in modern bodies like our own, the cells are a clone. All contain the same genes, although different genes will be turned on in the different specialist cells. Genes in each cell type are directly bene-fiting their own copies in the minority of cells specialized for reproduction, the cells of the immortal germ line.

So, to the third question. Why do bodies participate in a 'bottlenecked' life cycle?

To begin with, what do I mean by bottlenecked? No matter how many cells there may be in the body of an elephant, the ele-phant began life as a single cell, a fertilized egg. The fertilized egg is a narrow bottleneck which, during embryonic development, widens out into the trillions of cells of an adult elephant. And no matter how many cells, of no matter how many specialized types, cooperate to perform the unimaginably complicated task of run-ning an adult elephant, the efforts of all those cells converge on the final goal of producing single cells again—sperms or eggs. The elephant not only has its beginning in a single cell, a fertilized egg. Its end, meaning its goal or end-product, is the production of single cells, fertilized eggs of the next generation. The life cycle of the broad and bulky elephant both begins and ends with a narrow bottleneck. This bottlenecking is characteristic of the life cycles of all many-celled animals and most plants. Why? What is its significance? We cannot answer this without considering what life might look like without it.

It will be helpful to imagine two hypothetical species of seaweed called bottle-wrack and splurge-weed. Splurge-weed grows as a set of straggling, amorphous branches in the sea. Every now and then branches break off and drift away. These breakages can occur anywhere in the plants, and the fragments can be large or small. As with cuttings in a garden, they are capable of growing just like the original plant. This shedding of parts is the species's method of reproducing. As you will notice, it isn't really different from its method of growing, except that the growing parts become physically detached from one another.

Bottle-wrack looks the same and grows in the same straggly way. There is one crucial difference, however. It reproduces by releasing single-celled spores which drift off in the sea and grow into new plants. These spores are just cells of the plant like any others. As in the case of splurge-weed, no sex is involved. The daughters of a plant consist of cells that are clone-mates of the cells of the parent plant. The only difference between the two species is that splurge-weed reproduces by hiving off chunks of itself consisting of indeterminate numbers of cells, while bottle-wrack reproduces by hiving off chunks of itself always consisting of single cells.

By imagining these two kinds of plant, we have zeroed in on the crucial difference between a bottlenecked and an unbottlenecked life cycle. Bottle-wrack reproduces by squeezing itself, every generation, through a single-celled bottleneck. Splurge-weed just grows and breaks in two. It hardly can be said to possess discrete 'generations', or to consist of discrete 'organisms', at all. What about bottle-wrack? I'll spell it out soon, but we can already see an inkling of the answer. Doesn't bottle-wrack already seem to have a more discrete, 'organismy' feel to it?

Splurge-weed, as we have seen, reproduces by the same process as it grows. Indeed it scarcely reproduces at all. Bottle-wrack, on the other hand, makes a clear separation between growth and reproduction. We may have zeroed in on the difference, but so what? What is the significance of it? Why does it matter? I have thought a long time about this and I think I know the answer. (Incidentally, it was harder to work out that there was a question than to think of the answer!) The answer can be divided into three parts, the first two of which have to do with the relationship between evolution and embryonic development.

First, think about the problem of evolving a complex organ from a simpler one. We don't have to stay with plants, and for this stage of the argument it might be better to switch to animals because they have more obviously complicated organs. Again there is no need to think in terms of sex; sexual versus asexual reproduction is a red herring here. We can imagine our animals reproducing by sending off nonsexual spores, single cells that, mutations aside, are genetically identical to one another and to all the other cells in the body.

The complicated organs of an advanced animal like a human or a woodlouse have evolved by gradual degrees from the simpler organs of ancestors. But the ancestral organs did not literally change themselves into the descendant organs, like swords being beaten into ploughshares. Not only *did* they not. The point I want to make is that in most cases they *could* not. There is only a

limited amount of change that can be achieved by direct transformation in the 'swords to ploughshares' manner. Really radical
change can be achieved only by going 'back to the drawing
board', throwing away the previous design and starting afresh.
When engineers go back to the drawing board and create a new
design, they do not necessarily throw away the ideas from the
old design. But they don't literally try to deform the old physical
object into the new one. The old object is too weighed down with
the clutter of history. Maybe you can beat a sword into a plough-
share, but try 'beating' a propellor engine into a jet engine! You
can't do it. You have to discard the propellor engine and go back
to the drawing board.

Living things, of course, were never designed on drawing
boards. But they do go back to fresh beginnings. They make a
clean start in every generation. Every new organism begins as a
single cell and grows anew. It inherits the *ideas* of ancestral design,
in the form of the DNA program, but it does not inherit the physical organs of its ancestors. It does not inherit its parent's heart
and *remould* it into a new (and possibly improved) heart. It starts
from scratch, as a single cell, and grows a new heart, using the
same design program as its parent's heart, to which improvements may be added. You see the conclusion I am leading up to.
One important thing about a 'bottlenecked' life cycle is that it
makes possible the equivalent of going back to the drawing
board.

Bottlenecking of the life cycle has a second, related consequence. It provides a 'calendar' that can be used to regulate the
processes of embryology. In a bottlenecked life cycle, every fresh
generation marches through approximately the same parade of
events. The organism begins as a single cell. It grows by cell division. And it reproduces by sending out daughter cells. Presumably
it eventually dies, but that is less important than it seems to us

mortals; as far as this discussion is concerned the end of the cycle is reached when the present organism reproduces and a new generation's cycle begins. Although in theory the organism could reproduce at any time during its growth phase, we can expect that eventually an optimum time for reproduction would emerge. Organisms that released spores when they were too young or too old would end up with fewer descendants than rivals that built up their strength and then released a massive number of spores when in the prime of life.

The argument is moving towards the idea of a stereotyped, regularly repeating life cycle. Not only does each generation begin with a single-celled bottleneck. It also has a growth phase—'childhood'—of rather fixed duration. The fixed duration, the stereotypy, of the growth phase makes it possible for particular things to happen at particular times during embryonic development, as if governed by a strictly observed calendar. To varying extents in different kinds of creature, cell divisions during development occur in rigid sequence, a sequence that recurs in each repetition of the life cycle. Each cell has its own location and time of appearance in the roster of cell divisions. In some cases, incidentally, this is so precise that embryologists can give a name to each cell, and a given cell in one individual organism can be said to have an exact counterpart in another organism.

So, the stereotyped growth cycle provides a clock, or calendar, by means of which embryological events may be triggered. Think of how readily we ourselves use the cycles of the earth's daily rotation, and its yearly circumnavigation of the sun, to structure and order our lives. In the same way, the endlessly repeated growth rhythms imposed by a bottlenecked life cycle will—it seems almost inevitable—be used to order and structure embryology. Particular genes can be switched on and off at particular times because the bottleneck/growth-cycle calendar ensures that there *is* such a thing as a particular time. Such well-tempered regulations of gene activity are a prerequisite for the evolution of embryologies capable of crafting complex tissues and organs. The precision and complexity of an eagle's eye or a swallow's wing couldn't emerge without clockwork rules for what is laid down when.

The third consequence of a bottlenecked life history is a genetic one. Here, the example of bottle-wrack and splurge-weed serves us again. Assuming, again for simplicity, that both species reproduce asexually, think about how they might evolve. Evolution requires genetic change, mutation. Mutation can happen during any cell division. In splurge-weed, cell lineages are broad-fronted, the opposite of bottlenecked. Each branch that breaks apart and drifts away is many-celled. It is therefore quite possible that two cells in a daughter will be more distant relatives of one another than either is to cells in the parent plant. (By 'relatives', I literally mean cousins, grandchildren and so on. Cells have definite lines of descent and these lines are branching, so words like second cousin can be used of cells in a body without apology.) Bottle-wrack differs sharply from splurge-weed here. All cells in a daughter plant are descended from a single spore cell, so all cells in a given plant are closer cousins (or whatever) of one another than of any cell in another plant.

This difference between the two species has important genetic consequences. Think of the fate of a newly mutated gene, first in splurge-weed, then in bottle-wrack. In splurge-weed, the new mutation can arise in any cell, in any branch of the plant. Since daughter plants are produced by broad-fronted budding, lineal descendants of the mutant cell can find themselves sharing daughter plants and grand-daughter plants with unmutated cells which are relatively distant cousins of themselves. In bottle-wrack, on the other hand, the most recent common ancestor of all the cells in a plant is no older than the spore that provided the plant's bottlenecked beginning. If that spore contained the mutant gene, all the cells of the new plant will contain the mutant gene. If the spore did not, they will not. Cells in bottle-wrack will be more genetically uniform within plants than cells in splurge-weed (give or take an occasional reverse-mutation). In bottle-wrack, the individual plant will be a unit with a genetic identity, will deserve the name individual. Plants of splurge-weed will have less genetic identity, will be less entitled to the name 'individual' than their opposite numbers in bottle-wrack.

This is not just a matter of terminology. With mutations around, the cells within a plant of splurge-weed will not have all the same genetic interests at heart. A gene in a splurge-weed cell stands to gain by promoting the reproduction of its cell. It does not necessarily stand to gain by promoting the reproduction of its 'individual' plant. Mutation will make it unlikely that the cells within a plant are genetically identical, so they won't collaborate wholeheartedly with one another in the manufacture of organs and new plants. Natural selection will choose among cells rather than 'plants'. In bottle-wrack, on the other hand, all the cells within a plant are likely to have the same genes, because only very recent mutations could divide them. Therefore they will happily collaborate in manufacturing efficient survival machines. Cells in different plants are more likely to have different genes. After all, cells that have passed through different bottlenecks may be distinguished by all but the most recent mutations—and this means the majority. Selection will therefore judge rival plants, not rival cells as in splurge-weed. So we can expect to see the evolution of organs and contrivances that serve the whole plant.

By the way, strictly for those with a professional interest, there is an analogy here with the argument over group selection. We can think of an individual organism as a 'group' of cells. A form of group selection can be made to work, provided some means can be found for increasing the ratio of between-group variation to within-group variation. Bottle-wrack's reproductive habit has exactly the effect of increasing this ratio; splurge-weed's habit has just the opposite effect. There are also similarities, which may be revealing but which I shall not explore, between 'bottle-necking' and two other ideas that have dominated this chapter. Firstly the idea that parasites will cooperate with hosts to the extent that their genes pass to the next generation in the same reproductive cells as the genes of the hosts—squeezing through the same bottleneck. And secondly the idea that the cells of a sex-ually reproducing body cooperate with each other only because meiosis is scrupulously fair.

To sum up, we have seen three reasons why a bottlenecked life history tends to foster the evolution of the organism as a discrete and unitary vehicle. The three may be labelled, respectively, 'back to the drawing board', 'orderly timing-cycle', and 'cellular uni-formity'. Which came first, the bottlenecking of the life cycle, or the discrete organism? I should like to think that they evolved together. Indeed I suspect that the essential, defining feature of an individual organism *is* that it is a unit that begins and ends with a single-celled bottleneck. If life cycles become bottle-necked, living material seems bound to become boxed into dis-crete, unitary organisms. And the more that living material is boxed into discrete survival machines, the more will the cells of those survival machines concentrate their efforts on that special class of cells that are destined to ferry their shared genes through the bottleneck into the next generation. The two phenomena, bottlenecked life cycles and discrete organisms, go hand in hand. As each evolves, it reinforces the other. The two are mutually enhancing, like the spiralling feelings of a woman and a man during the progress of a love affair.

The Extended Phenotype is a long book and its argument cannot easily be crammed into one chapter. I have been obliged to adopt here a condensed, rather intuitive, even impressionistic style. I hope, nevertheless, that I have succeeded in conveying the flavour of the argument.

Let me end with a brief manifesto, a summary of the entire selfish gene/extended phenotype view of life. It is a view, I maintain, that applies to living things everywhere in the universe. The fundamental unit, the prime mover of all life, is the replicator. A replicator is anything in the universe of which copies are made. Replicators come into existence, in the first place, by chance, by the random jostling of smaller particles. Once a replicator has come into existence it is capable of generating an indefinitely large set of copies of itself. No copying process is perfect, however, and the population of replicators comes to include varieties that differ from one another. Some of these varieties turn out to have lost the power of self-replication, and their kind ceases to exist when they themselves cease to exist. Others can still replicate, but less effectively. Yet other varieties happen to find themselves in possession of new tricks: they turn out to be even better self-replicators than their predecessors and contemporaries. It is their descendants that come to dominate the population. As time goes by, the world becomes filled with the most powerful and ingenious replicators.

Gradually, more and more elaborate ways of being a good replicator are discovered. Replicators survive, not only by virtue of their own intrinsic properties, but by virtue of their consequences on the world. These consequences can be quite indirect. All that is necessary is that eventually the consequences, however tortuous and indirect, feed back and affect the success of the replicator at getting itself copied.

The success that a replicator has in the world will depend on what kind of a world it is—the pre-existing conditions. Among the most important of these conditions will be other replicators and their consequences. Like the English and German rowers, replicators that are mutually beneficial will come to predominate in each other's presence. At some point in the evolution of life on our earth, this ganging up of mutually compatible replicators began to be formalized in the creation of discrete vehicles—cells and, later, many-celled bodies. Vehicles that evolved a bottlenecked life cycle prospered, and became more discrete and vehicle-like.

This packaging of living material into discrete vehicles became such a salient and dominant feature that, when biologists arrived on the scene and started asking questions about life, their questions were mostly about vehicles—individual organisms. The individual organism came first in the biologist's consciousness, while the replicators—now known as genes—were seen as part of the machinery used by individual organisms. It requires a deliberate mental effort to turn biology the right way up again, and remind ourselves that the replicators come first, in importance as well as in history.

One way to remind ourselves is to reflect that, even today, not all the phenotypic effects of a gene are bound up in the individual body in which it sits. Certainly in principle, and also in fact, the gene reaches out through the individual body wall and manipulates objects in the world outside, some of them inanimate, some of them other living beings, some of them a long way away. With only a little imagination we can see the gene as sitting at the centre of a radiating web of extended phenotypic power. And an object in the world is the centre of a converging web of influences from many genes sitting in many organisms. The long reach of the gene knows no obvious boundaries. The whole world is criss-crossed with causal arrows joining genes to phenotypic effects, far and near.

It is an additional fact, too important in practice to be called incidental but not necessary enough in theory to be called inevitable, that these causal arrows have become bundled up. Replicators are no longer peppered freely through the sea; they are packaged in huge colonies—individual bodies. And phenotypic consequences, instead of being evenly distributed throughout the world, have in many cases congealed into those same bodies. But the individual body, so familiar to us on our planet, did not have to exist. The only kind of entity that has to exist in order for life to arise, anywhere in the universe, is the immortal replicator.

EPILOGUE TO 40TH ANNIVERSARY EDITION

S cientists, unlike politicians, can take pleasure in being wrong. A politician who changes his mind is accused of 'flip-flopping'. Tony Blair boasted that he had 'not got a reverse gear'. Scientists on the whole prefer to see their ideas vindicated, but an occasional reversal gains respect, especially when graciously acknowledged. I have never heard of a scientist being maligned as a flip-flopper.

In some ways I would quite like to find ways to recant the central message of *The Selfish Gene*. So many exciting things are fast happening in the world of genomics, it would seem almost inevitable—even tantalizing—that a book with the word 'gene' in the title would, forty years on, need drastic revision if not outright discarding. This might indeed be so, were it not that 'gene' in this book is used in a special sense, tailored to evolution rather than embryology. My definition is the population geneticists' definition adopted by George C. Williams, one of the acknowledged heroes of the book, now lost to us along with John Maynard Smith and Bill Hamilton: 'A gene is defined as any portion of chromosomal material that potentially lasts for enough generations to serve as a unit of natural selection.' I pushed it to a somewhat facetious conclusion: 'To be strict, this book should be called... *The slightly selfish big bit of chromosome and the even more selfish little bit of chromosome.'* As opposed to the embryologist's concern with how genes affect phenotypes, we have here the neo-Darwinist's concern with changes in frequencies of entities in populations. Those entities are genes in the Williams sense (Williams later called that sense the 'codex'). Genes can be counted and their frequency is the measure of their success. One of the central messages of this book is that the individual organism doesn't have this property. An organism has a frequency of one, and therefore cannot 'serve as a unit of natural selection'. Not in the same sense of *replicator* anyway. If the organism is a unit of natural selection, it is in the quite different sense of gene 'vehicle'. The measure of its success is the frequency of its genes in future generations, and the quantity it strives to maximize is what Hamilton defined as 'inclusive fitness'.

A gene achieves its numerical success in the population by virtue of its (phenotypic) effects on individual bodies. A successful gene is represented in many bodies over a long period of time. It helps those bodies to survive long enough to reproduce in the environment. But the environment means not just the external environment of the body—trees, water, predators, etc.—but also the internal environment, and especially the other genes with which the selfish gene shares a succession of bodies through the population and down the generations. It follows that natural selection favours genes that flourish in the company of other genes in the breeding population. Genes are indeed 'selfish' in the sense promoted in this book. They are also *cooperative* with other genes with which they share, not just the present particular body, but bodies in general, generated by the species' gene pool. A sexually reproducing population is a cartel of mutually compatible, cooperating genes: cooperating today because they have flourished by cooperating through many generations of similar bodies in the ancestral past. The important point to understand (it is much misunderstood) is that the cooperativeness is favoured, not because a group of genes is naturally selected as a whole, but because individual genes are separately selected against the background of the other genes likely to be met in a body, and this means the other genes in the species' gene pool. The pool, that is, from which every individual of a sexually reproducing species draws its genes as a sample. The genes of the species (but not other species) are continually meeting each other—and cooperating with each other—in a succession of bodies.

We still don't really understand what drove the origin of sexual reproduction. But a consequence of sexual reproduction was the invention of *the species* as the habitat of cooperating cartels of mutually compatible genes. As explained in the chapter called 'The long reach of the gene', the key to the cooperation is that, in every generation, all the genes in a body share the same 'bottlenecked' exit route to the future: the sperms or eggs in which they aspire to sail into the next generation. *The Cooperative Gene* would have been an equally appropriate title for this book, and the book itself would not have changed at all. I suspect that a whole lot of mistaken criticisms could have been avoided.

Another good title would have been *The Immortal Gene*. As well as being more poetic than 'selfish', 'immortal' captures a key part of the book's argument. The high fidelity of DNA copying—mutations are rare—is essential to evolution by natural selection. High fidelity means that genes, in the form of exact informational copies, can survive for millions of years. Successful ones, that is. Unsuccessful ones, by definition, don't. The difference wouldn't be significant if the potential lifespan of a piece of genetic information was short anyway. To look at it another way, every living individual has been built, during its embryonic development, by genes which can trace their ancestry through a very large number of generations, in a very large number of individuals. Living animals have inherited the genes that helped huge numbers of ancestors to survive. That is why living animals have what it takes to survive—and reproduce. The details of what it takes vary from species to species—predator or prey, parasite or host, adapted to water or land, underground or forest canopy—but the general rule remains.

A central point of the book is the one developed by my friend the great Bill Hamilton, whose death I still mourn. Animals are expected to look after not only their own children but other genetic relatives. The simple way to express it, and the one that I favour, is 'Hamilton's Rule': a gene for altruism will spread if the cost to the altruist, C, is less than the value, B, to the beneficiary, devalued by the coefficient of relatedness, r, between them. r is a proportion between 0 and 1. It has the value 1 for identical twins; 0.5 for offspring and full siblings; 0.25 for grandchildren, half-siblings, and nieces; 0.125 for first cousins. But when is it zero? What is the meaning of zero on this scale? This is harder to say, but it is important and it was not fully spelled out in the first edition of *The Selfish Gene*. Zero does *not* mean that the two individuals share no genes in common. All humans share more than 99 percent of our genes, more than 90 percent with a mouse, and three-quarters of our genes with a fish. These high percentages have confused many people into misunderstanding kin selection, including some distinguished scientists. But those figures are not what is meant by r. Where r is 0.5 for my brother (say), it is zero for a *random member of the background population with whom I might be competing*. For purposes of theorizing about the evolution of altruism, r between first cousins is 0.125 only when compared to the reference background population ($r = 0$), which is the rest of the population to whom altruism potentially might have been shown: competitors for food and space, fellow travellers through time in the environment of the species. The 0.5 (0.125, etc.) refers to the *additional* relatedness *over and above* the background population, whose relatedness approaches zero.

Genes in the Williams sense are things you can count as the generations go by, and it doesn't matter what their molecular nature is; it doesn't matter, for instance, that they are split up into a series of 'exons' (expressed) separated by mostly inert 'introns' (ignored by the translation machinery). Molecular genomics is a fascinating subject but it doesn't heavily impinge on the 'gene's eye view' of evolution which is the central theme of the book. To put the point another way, *The Selfish Gene* is quite likely a valid account of life on other planets even if the genes on those other planets have no connection with DNA. Nevertheless, there are ways in which the details of modern molecular genetics, the detailed study of DNA, can be gathered into the gene's eye fold and it turns out that they vindicate my view of life rather than casting doubt on it. I'll come on to this after what may seem like a radical change of subject, beginning with a specific question which obviously stands for any number of similar questions.

How closely related are you to Queen Elizabeth II? As it happens, I know I'm her 15th cousin twice removed. Our common ancestor is Richard Plantagenet, 3rd Duke of York (1412–1460). One of Richard's sons was King Edward IV, from whom Queen Elizabeth is descended. Another son was George, Duke of Clarence, from whom I am descended (allegedly drowned in a butt of Malmsey wine). You may not know it but you are very probably closer to the Queen than 15th cousin and so am I and so is the postman. There are so many different ways of being somebody's distant cousin, and we are all related to each other in many of those ways. I know that I am my wife's twelfth cousin twice removed (common ancestor George Hastings, 1st Earl of Huntingdon, 1488–1544). But it is highly probable that I am a closer cousin to her in various unknown ways (various pathways through our respective ancestries) and it is absolutely certain I am also her more distant cousin in many more ways. We all are. You and the Queen might simultaneously be ninth cousins six times removed, and twentieth cousins 4 times removed, and thirtieth cousins 8 times removed. All of us, regardless of where in the world we live, are not only cousins of each other. We are cousins in hundreds of different ways. This is just another way of saying we are all members of the background population among whom r, the coefficient of relatedness, approaches zero. I could calculate r between me and the Queen using the one pathway for which records exist, but it would, as the definition requires, be so close to zero as to make no difference.

The reason for all that bewildering multiplicity of cousinship is sex. We have two parents, four grandparents, eight great grandparents, and so on, up to astronomical numbers. If you go on multiplying by two back to the time of William the Conqueror, the number of your ancestors (and mine, and the Queen's and the postman's) would be at least a billion, which is more than the world population at the time. That calculation alone proves that, wherever you come from, we share many of our ancestors (ultimately all if you go sufficiently far back) and are cousins of each other many times over.

All that complexity disappears if you look at cousinship from the gene's point of view (the point of view advocated, in different ways, throughout this book) as opposed to the individual organism's point of view (as has been conventional among biologists). Stop asking: What kind of cousin am I to my wife (the postman, the Queen)? Instead, ask the question from the point of view of a single gene, say my gene for blue eyes: What relation is my blue eye gene to the postman's blue eye gene? Polymorphisms such as the ABO blood groups go way back in history, and are shared by other apes and even monkeys. The A gene in a human sees the equivalent gene in a chimp as a closer cousin than the B gene in a human. As for the SRY gene on the Y chromosome, which determines maleness, my SRY gene 'looks upon' the SRY gene of a kangaroo as its kissing cousin.

Or we can look at relatedness from a mitochondrion's point of view. Mitochondria are tiny bodies teeming in all our cells, vital to our survival. They reproduce asexually and retain the remnants of their own genomes (they are remotely descended from free-living bacteria). By the Williams definition, a mitochondrial genome can be thought of as a single 'gene'. We get our mitochondria from our mothers only. So if we were now to ask how close is the cousinship of your mitochondria to the Queen's mitochondria there is a single answer. We may not know what that answer is, but we do know that her mitochondria and yours are cousins in only one way, not hundreds of ways as is the case from the point of view of the body as a whole. Trace your ancestry back through the generations, but always only through the maternal line and you follow a single narrow (mitochondrial) thread, as opposed to the ever branching thread of 'whole organism pedigrees'. Do the same for the Queen, following her narrow maternal thread back through the generations. Sooner or later the two threads will meet and now, by simply counting generations along the two threads, you can easily calculate your mitochondrial cousinship to the Queen.

What you can do for mitochondria, you can in principle do for any particular gene, and this illustrates the difference between a gene's point of view and an organism's point of view. From a whole organism point of view you have two parents, four grandparents, eight great grandparents, etc. But, like a mitochondrion, each gene has only one parent, one grandparent, one great grandparent, etc. I have one gene for blue eyes and the Queen has two. In principle we could trace the generations back and discover the cousinship between my blue eye gene and each of the Queen's two. The common ancestor of our two genes is called the 'coalescence point'. Coalescence analysis has become a flourishing branch of genetics and very fascinating it is. Can you see how congenial it is to the 'gene's eye view' that this whole book espouses? We are not talking about altruism any more. The gene's eye view is flexing its muscles in other domains, in this case looking back at ancestry.

You can even investigate the coalescence point between two alleles in one individual body. Prince Charles has blue eyes and we can assume that he has a pair of blue eye alleles opposite each other on Chromosome 15. How closely related to each other are Prince Charles' two blue eye genes, one from his father, one from his mother? In this case we know one possible answer, only because royal pedigrees are documented in ways that most of our pedigrees are not. Queen Victoria had blue eyes and Prince Charles is descended from Victoria in two ways: via King Edward VII on his mother's side; and via Princess Alice of Hesse on his father's side. There's a 50 percent probability that one of Victoria's blue eye genes peeled off two copies of itself, one of which went to her son, Edward VII, and the other to her daughter Princess Alice. Further copies of these two sibling genes could easily have passed down the generations to Queen Elizabeth II on one side and Prince Philip on the other, whence they were reunited in Prince Charles. This would mean the 'coalescence' point of Charles's two genes was Victoria. We do not—cannot—know whether this actually is true for Charles' blue eye genes. But statistically it has to be true that many of his pairs of genes coalesce back in Victoria. And the same kind of reasoning applies to pairs of your genes, and pairs of my genes. Even though we may not have Prince Charles' well-documented pedigree to consult, any pair of your genes could, in principle, look back at their common ancestor, the coalescence point at which they were 'peeled off' from a single ancestral gene.

Now, here's something interesting. Although I can't establish the exact coalescence point of any particular allelic pair of my genes, geneticists can in principle take all the pairs of genes from any one individual and, by considering all possible pathways back through the past (actually not all possible pathways because there are too many, but a statistical sample of them), derive a pattern of coalescences across the whole genome. Heng Li and Richard Durbin of the Sanger Institute in Cambridge realized a remarkable thing: the pattern of coalescences among pairs of genes in the genome of a single individual gives us enough information to reconstruct demographic details about datable moments in the prehistory of an entire species.

In our discussion of coalescence between pairs of genes, one from the father and one from the mother, the word 'gene' means something a bit more fluid than the normal usage of molecular biologists. Indeed, you could say that the coalescence geneticists have reverted to something a bit like my 'slightly selfish big bit of chromosome and even more selfish little bit of chromosome.' Coalescence analysis is studying chunks of DNA which might be larger or even smaller than a molecular biologist's understanding of a single gene but which can still be seen as cousins of each other, having been 'peeled off' from a shared ancestor some definite number of generations ago.

When a gene (in that sense) 'peels off' two copies of itself and gives one to each of two offspring, the descendants of those two copies may, over time, accumulate differences due to mutation. These may be 'under the radar' in the sense that they don't show up as phenotypic differences. The mutated differences between them are proportional to the time that has elapsed since the split, a fact that biologists make good use of, over much greater time spans, in the so-called 'molecular clock'. Moreover, the pairs of genes whose cousinship we are calculating needn't have the same phenotypic effects as each other. I have one blue eye gene from my father paired with one brown eye gene from my mother. Although these genes are different, even they must have a coalescence somewhere in the past: the moment when a particular gene in a shared ancestor of my two parents peeled off one copy for one child and another copy to its sibling. That coalescence (unlike the two copies of Victoria's blue eye gene) was a long time ago, and the pair of genes has had a long time to accumulate differences, not least the difference in the eye colours that they mediate.

Now, I said that the coalescence pattern within one individual's genome can be used to reconstruct details of demographic prehistory. Any individual's genome can do this. As it happens, I am one of the people in the world who has had his complete genome sequenced. This was for a television programme called *Sex, Death and the Meaning of Life* which I presented on Channel Four in 2012. Yan Wong, my co-author of *The Ancestor's Tale*, from whom I learned everything I know about coalescence theory and much else besides, seized upon this and did the necessary Li/ Durbin style calculations using my genome, and my genome alone, to make inferences about human history. He found a large number of coalescences around 60,000 years ago. This suggests that the breeding population in which my ancestors were embedded was small 60,000 years ago. There were few people around, so the chance of a pair of modern genes coalescing in the same ancestor back at that time was high. There were fewer coalescences 300,000 years ago, suggesting that the effective population size was larger. These figures can be plotted as a graph of effective population size against time. Here's the pattern he found, and it is the same pattern as the originators of the technique would expect to find from any European genome.

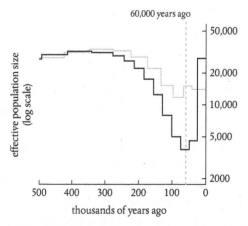

60,000 years ago

From R. Dawkins and Y. Wong (2006) *The Ancestor's Tale*, 2nd edition
Image courtesy of Y. Wong

The black line shows the estimates of effective population size at various times in history based upon my genome (coalescences between genes from my father and my mother). It shows that the effective population size among my ancestral population plummeted around 60,000 years ago. The grey line shows the equivalent pattern derived from the genome of a Nigerian man. It also shows a drop in population around the same time, but a less dramatic one. Perhaps whatever calamity caused the drop was less severe in Africa than in Eurasia.

Incidentally Yan was my undergraduate pupil in New College, Oxford, before I started learning more from him than he learns from me. He then became a graduate student of Alan Grafen, whom I had also tutored as an undergraduate, who subsequently became my graduate student and whom I have described as being now my intellectual mentor. So Yan is both my student and my grandstudent—a neat memetic analogue to the point I was making earlier about how we are related in multiple ways—although the direction of cultural inheritance is more complicated than this simple formulation implies.

To summarize, the gene's eye view of life, the central theme of this book, illuminates not just the evolution of altruism and selfishness, as expounded in previous editions. It also illuminates the deep past, in ways of which I had no inkling when I first wrote *The Selfish Gene* and which are expounded more fully in relevant passages (largely written by Yan, my co-author) of the second edition (2016) of *The Ancestor's Tale*. So powerful is the gene's eye view, the genome of a single individual is sufficient to make quantitatively detailed inferences about historical demography. What else might it be capable of? As foreshadowed by the Nigerian comparison, future analyses of individuals from different parts of the world could give a geographic dimension to these demographic signals from the past.

Might the gene's eye view penetrate the remote past in yet other ways? Several of my books have developed an idea which I called 'The Genetic Book of the Dead'. The gene pool of a species is a mutually supportive cartel of genes that have survived in particular environments of the past, both distant and recent. This makes it a kind of negative imprint of those environments. A sufficiently knowledgeable geneticist should be able to read out, from the genome of an animal, the environments in which its ancestors survived. In principle, the DNA in a mole *Talpa europaea* should be eloquent of an underground world, a world of damp, subterranean darkness, smelling of worms, leaf decay, and beetle larvae. The DNA of a dromedary, *Camelus dromedarius*, if we but knew how to read it, would spell out a coded description of ancient ancestral deserts, dust storms, dunes, and thirst. The DNA of *Tursiops truncatus*, the common bottlenose dolphin, spells, in a language that we may one day decipher, 'open sea, pursue fish fast, avoid killer whales'. But the same dolphin DNA also contains paragraphs about earlier worlds in which the genes survived: on land when the ancestors escaped the attentions of tyrannosaurs and allosaurs long enough to breed. Then, before that, parts of the DNA surely spell out descriptions of even older feats of survival, back in the sea, when the ancestors were fish, pursued by sharks and even eurypterids (giant sea scorpions). Active research on 'The Genetic Book of the Dead' lies in the future. Will it colour the epilogue of the fiftieth anniversary edition of *The Selfish Gene?*

ENDNOTES

The following notes refer to the original eleven chapters only. Although the text of these chapters is almost identical to the first edition, the page numbers are different as the type has been completely reset. Each note is referenced by an asterisk in the main text.

CHAPTER 1

Why are people?

p. 1 *...all attempts to answer that question before 1859 are worthless...*

Some people, even non-religious people, have taken offence at the quotation from Simpson. I agree that, when you first read it, it sounds terribly philistine and gauche and intolerant, a bit like Henry Ford's 'History is more or less bunk'. But, religious answers apart (I am familiar with them; save your stamp), when you are actually challenged to think of pre-Darwinian answers to the questions 'What is man?' 'Is there a meaning to life?' 'What are we for?', can you, as a matter of fact, think of any that are not now worthless except for their (considerable) historic interest? There is such a thing as being just plain wrong, and that is what, before 1859, all answers to those questions were.

p. 3 *I am not advocating a morality based on evolution.*

Critics have occasionally misunderstood The Selfish Gene to be advocating selfishness as a principle by which we should live! Others, perhaps because they read the book by title only or never made it past the first two pages, have thought that I was saying that, whether we like it or not, selfishness and other nasty ways are an inescapable part of our nature. This error is easy to fall into if you think, as many people unaccountably seem to, that genetic 'determination' is for keeps—absolute and irreversible. In fact genes 'determine' behaviour only in a statistical sense (see also pp. 46–50). A good analogy is the widely conceded generalization that 'A red sky at night is the shepherd's delight'. It may be a statistical fact that a good red sunset portends a fine day on the morrow, but we would not bet a large sum on it. We know perfectly well that the weather

is influenced in very complex ways by many factors. Any weather fore-
cast is subject to error. It is a statistical forecast only. We don't see red
sunsets as irrevocably determining fine weather the next day, and no
more should we think of genes as irrevocably determining anything.
There is no reason why the influence of genes cannot easily be reversed
by other influences. For a full discussion of 'genetic determinism', and
why misunderstandings have arisen, see chapter 2 of *The Extended
Phenotype*, and my paper 'Sociobiology: The New Storm in a Teacup'. I've
even been accused of claiming that human beings are fundamentally all
Chicago gangsters! But the essential point of my Chicago gangster analogy (p. 2) was, of course, that:

> knowledge about the kind of world in which a man has prospered tells
> you something about that man. It had nothing to do with the particular
> qualities of Chicago gangsters. I could just as well have used the ana-
> logy of a man who had risen to the top of the Church of England, or
> been elected to the Athenaeum. In any case it was not people but genes
> that were the subject of my analogy.

I have discussed this, and other over-literal misunderstandings, in my
paper 'In Defence of Selfish Genes', from which the above quotation is
taken.

I must add that the occasional political asides in this chapter make
uncomfortable rereading for me in 1989. 'How many times must this
[the need to restrain selfish greed to prevent the destruction of the
whole group] have been said in recent years to the working people of
Britain?' (p. 10) makes me sound like a Tory! In 1975, when it was written, a socialist government which I had helped to vote in was battling
desperately against 23 per cent inflation, and was obviously concerned about high wage claims. My remark could have been taken
from a speech by any Labour minister of the time. Now that Britain
has a government of the new right, which has elevated meanness and
selfishness to the status of ideology, my words seem to have acquired
a kind of nastiness by association, which I regret. It is not that I take
back what I said. Selfish short-sightedness still has the undesirable
consequences that I mentioned. But nowadays, if one were seeking
examples of selfish short-sightedness in Britain, one would not look
first at the working class. Actually, it is probably best not to burden a

scientific work with political asides at all, since it is remarkable how quickly these date. The writings of politically aware scientists of the 1930s—J. B. S. Haldane and Lancelot Hogben, for instance—are today significantly marred by their anachronistic barbs.

p. 7 ...it is possible that the female improves the male's sexual performance by eating his head.

I first learned this odd fact about male insects during a research lecture by a colleague on caddis flies. He said that he wished he could breed caddises in captivity but, try as he would, he could not persuade them to mate. At this the Professor of Entomology growled from the front row, as if it were the most obvious thing to have overlooked: 'Haven't you tried cutting their heads off?'

p. 14 ...the fundamental unit of selection is not the species, nor the group, nor even, strictly, the individual. It is the gene...

Since writing my manifesto of genic selection, I have had second thoughts about whether there may not also be a *kind* of higher-level selection occasionally operating during the long haul of evolution. I hasten to add that, when I say 'higher-level', I do not mean anything to do with 'group selection'. I am talking about something much more subtle and much more interesting. My feeling now is that not only are some individual organisms better at surviving than others; whole classes of organisms may be better at *evolving* than others. Of course, the evolving that we are talking about here is still the same old evolution, mediated via selection on genes. Mutations are still favoured because of their impact on the survival and reproductive success of individuals. But a major new mutation in basic embryological plan can also open up new floodgates of radiating evolution for millions of years to come. There can be a kind of higher-level selection for embryologies that lend themselves to evolution: a selection in favour of evolvability. This kind of selection may even be cumulative and therefore progressive, in ways that group selection is not. These ideas are spelt out in my paper 'The Evolution of Evolvability', which was largely inspired by playing with Blind Watchmaker, a computer program simulating aspects of evolution.

CHAPTER 2

The replicators

p. 18 The simplified account I shall give [of the origin of life] is probably not too far from the truth.

There are many theories of the origin of life. Rather than labour through them, in *The Selfish Gene* I chose just one to illustrate the main idea. But I wouldn't wish to give the impression that this was the only serious candidate, or even the best one. Indeed, in *The Blind Watchmaker*, I deliberately chose a different one for the same purpose, A. G. Cairns-Smith's clay theory. In neither book did I commit myself to the particular hypothesis chosen. If I wrote another book I should probably take the opportunity to try to explain yet another viewpoint, that of the German mathematical chemist Manfred Eigen and his colleagues. What I am always trying to get over is something about the fundamental properties that must lie at the heart of any good theory of the origin of life on any planet, notably the idea of self-replicating genetic entities.

p. 21 'Behold a virgin shall conceive...'

Several distressed correspondents have queried the mistranslation of 'young woman' into 'virgin' in the biblical prophecy, and have demanded a reply from me. Hurting religious sensibilities is a perilous business these days, so I had better oblige. Actually it is a pleasure, for scientists can't often get satisfyingly dusty in the library indulging in a real academic footnote. The point is in fact well known to biblical scholars, and not disputed by them. The Hebrew word in Isaiah is עלמה (*almah*), which undisputedly means 'young woman', with no implication of virginity. If 'virgin' had been intended, בתולה (*bethulah*) could have been used instead (the ambiguous English word 'maiden' illustrates how easy it can be to slide between the two meanings). The 'mutation' occurred when the pre-Christian Greek translation known as the Septuagint rendered *almah* into παθθένος (*parthenos*), which really does usually mean virgin. Matthew (not, of course, the Apostle and contemporary of Jesus, but the gospel-maker writing long afterwards) quoted Isaiah in what seems to be a derivative of the Septuagint version (all but two of the fifteen Greek words are identical) when he said, 'Now all this was done, that it might be fulfilled which was spoken of the Lord by the prophet, saying, Behold, a virgin shall be with child, and shall bring forth a son, and they shall call his name Emmanuel' (Authorized English translation). It is widely accepted among Christian scholars that the story of the virgin birth of Jesus was a late interpolation, put in presumably by Greek-speaking disciples in order that the (mistranslated) prophecy should be seen to be fulfilled. Modern versions such as the *New English Bible* correctly give 'young woman' in Isaiah. They equally correctly leave 'virgin' in Matthew, since there they are translating from his Greek.

p. 25 *Now they swarm in huge colonies, safe inside gigantic lumbering robots...*

This purple passage (a rare—well, fairly rare—indulgence) has been quoted and requoted in gleeful evidence of my rabid 'genetic determinism'. Part of the problem lies with the popular, but erroneous, associations of the word 'robot'. We are in the golden age of electronics, and robots are no longer rigidly inflexible morons but are capable of learning, intelligence, and creativity. Ironically, even as long ago as 1920 when Karel Capek coined the word, 'robots' were mechanical beings that ended up with human feelings, like falling in love. People who think that robots are by definition more 'deterministic' than human beings are muddled (unless they are religious, in which case they might consistently hold that humans have some divine gift of free will denied to mere machines). If, like most of the critics of my 'lumbering robot' passage, you are not religious, then face up to the following question: What on earth do you think you are, if not a robot, albeit a very complicated one? I have discussed all this in *The Extended Phenotype*, pp. 15–17.

The error has been compounded by yet another telling 'mutation'. Just as it seemed theologically necessary that Jesus should have been born of a virgin, so it seems demonologically necessary that any 'genetic determinist' worth his salt must believe that genes 'control' every aspect of our behaviour. I wrote of the genetic replicators: 'they created us, body and mind' (p. 25). This has been duly misquoted (e.g. in *Not in Our Genes* by Rose, Kamin, and Lewontin (p. 287), and previously in a scholarly paper by Lewontin) as '[they] *control* us, body and mind' (emphasis mine). In the context of my chapter, I think it is obvious what I meant by 'created', and it is very different from 'control'. Anybody can see that, as a matter of fact, genes do not con-

trol their creations in the strong sense criticized as 'determinism'. We effortlessly (well, fairly effortlessly) defy them every time we use contraception.

CHAPTER 3

Immortal coils

p. 30 ...*impossible to disentangle the contribution of one gene from that of another.*

Here, and also on pages 110–13, is my answer to critics of genetic 'atomism'. Strictly it is an anticipation, not an answer, since it predates the
criticism! I am sorry that it will be necessary to quote myself so fully, but
the relevant passages of *The Selfish Gene* seem to be disquietingly easy to
miss! For example, in 'Caring Groups and Selfish Genes' (in *The Panda's
Thumb*), S. J. Gould stated:

> There is no gene 'for' such unambiguous bits of morphology as your left
> kneecap or your fingernail. Bodies cannot be atomized into parts, each con-
> structed by an individual gene. Hundreds of genes contribute to the building
> of most body parts...

Gould wrote that in a criticism of *The Selfish Gene*. But now look at my
actual words (p. 30):

> The manufacture of a body is a cooperative venture of such intricacy
> that it is almost impossible to disentangle the contribution of one gene
> from that of another. A given gene will have many different effects on
> quite different parts of the body. A given part of the body will be influ-
> enced by many genes, and the effect of any one gene depends on inter-
> action with many others.

And again (p. 46):

> However independent and free genes may be in their journey through the
> generations, they are very much *not* free and independent agents in their
> control of embryonic development. They collaborate and interact in inex-
> tricably complex ways, both with each other, and with their external
> environment. Expressions like 'gene for long legs' or 'gene for altruistic
> behaviour' are convenient figures of speech, but it is important to under-
> stand what they mean. There is no gene which single-handedly builds a
> leg, long or short. Building a leg is a multigene cooperative enterprise.

Influences from the external environment too are indispensable; after all, legs are actually made of food! But there may well be a single gene which, *other things being equal*, tends to make legs longer than they would have been under the influence of the gene's allele.

I amplified the point in my next paragraph by an analogy with the effects of fertilizer on the growth of wheat. It is almost as though Gould was so sure, in advance, that I must be a naïve atomist, that he overlooked the extensive passages in which I made exactly the same interactionist point as he was later to insist upon.

Gould goes on:

Dawkins will need another metaphor: genes caucusing, forming alliances, showing deference for a chance to join a pact, gauging probable environments.

In my rowing analogy (pp. 110–12), I had already done precisely what Gould later recommended. Look at this rowing passage also to see why Gould, though we agree over so much, is wrong to assert that natural selection 'accepts or rejects entire organisms because suites of parts, interacting in complex ways, confer advantages'. The true explanation of the 'cooperativeness' of genes is that:

Genes are selected, not as 'good' in isolation, but as good at working against the background of other genes in the gene pool. A good gene must be compatible with, and complementary to, the other genes with whom it has to share a long succession of bodies. (p. 110)

I have written a fuller reply to criticisms of genetic atomism in *The Extended Phenotype*, especially on pp. 116–17 and 239–47.

p. 36 *The definition I want to use comes from G. C. Williams.*

Williams's exact words, in *Adaptation and Natural Selection*, are:

I use the term gene to mean 'that which segregates and recombines with appreciable frequency.'…A gene could be defined as any hereditary information for which there is a favorable or unfavorable selection bias equal to several or many times its rate of endogenous change.

Williams's book has now become widely, and rightly, regarded as a classic, respected by 'sociobiologists' and critics of sociobiology alike. I think it is clear that Williams never thought of himself as advocating anything new or revolutionary in his 'genic selectionism', and no more did I in 1976. We both thought that we were simply reasserting a fundamental principle of Fisher, Haldane, and Wright, the founding fathers of 'neo-Darwinism' in the 1930s. Nevertheless, perhaps because of our uncompromising language, some people, including Sewall Wright himself, apparently take exception to our view that 'the gene is the unit of selection'. Their basic reason is that natural selection sees organisms, not the genes inside them. My reply to views such as Wright's is in *The Extended Phenotype*, especially pp. 238–47. Williams's most recent thoughts on the question of the gene as the unit of selection, in his 'Defense of Reductionism in Evolutionary Biology', are as penetrating as ever. Some philosophers, for example, D. L. Hull, K. Sterelny and P. Kitcher, and M. Hampe and S. R. Morgan, have also recently made useful contributions to clarifying the issue of the 'units of selection'. Unfortunately there are other philosophers who have confused it.

p. 43 ...the individual is too large and too temporary a genetic unit...

Following Williams, I made much of the fragmenting effects of meiosis in my argument that the individual organism cannot play the role of replicator in natural selection. I now see that this was only half the story. The other half is spelled out in *The Extended Phenotype* (pp. 97–9) and in my paper 'Replicators and Vehicles'. If the fragmenting effects of meiosis were the whole story, an asexually reproducing organism like a female stick-insect would be a true replicator, a sort of giant gene. But if a stick insect is changed—say it loses a leg—the change is not passed on to future generations. Genes alone pass down the generations, whether reproduction is sexual or asexual. Genes, therefore, really are replicators. In the case of an asexual stick-insect, the entire genome (the set of all its genes) is a replicator. But the stick-insect itself is not. A stick-insect body is not moulded as a replica of the body of the previous generation. The body in any one generation grows afresh from an egg, under the direction of its genome, which is a replica of the genome of the previous generation.

All printed copies of this book will be the same as one another. They will be replicas but not replicators. They will be replicas not because they have copied one another, but because all have copied the same printing plates. They do not form a lineage of copies, with some books being ancestral to others. A lineage of copies would exist if we xeroxed a page of a book, then xeroxed the xerox, then xeroxed the xerox of the xerox, and so on. In this lineage of pages, there really would be an ancestor/descendant relationship. A new blemish that showed up anywhere along the series would be shared by descendants but not by ancestors. An ancestor/descendant series of this kind has the potential to evolve.

Superficially, successive generations of stick-insect bodies appear to constitute a lineage of replicas. But if you experimentally change one member of the lineage (for instance by removing a leg), the change is not passed on down the lineage. By contrast, if you experimentally change one member of the lineage of genomes (for instance by X-rays), the change will be passed on down the lineage. This, rather than the fragmenting effect of meiosis, is the fundamental reason for saying that the individual organism is not the 'unit of selection'—not a true replicator. It is one of the most important consequences of the universally admitted fact that the 'Lamarckian' theory of inheritance is false.

p. 51 *Another theory, due to Sir Peter Medawar...*

I have been taken to task (not, of course, by Williams himself or even with his knowledge) for attributing this theory of ageing to P. B. Medawar, rather than to G. C. Williams. It is true that many biologists, especially in America, know the theory mainly through Williams's 1957 paper, 'Pleiotropy, Natural Selection, and the Evolution of Senescence'. It is also true that Williams elaborated the theory beyond Medawar's treatment. Nevertheless my own judgement is that Medawar spelled out the essential core of the idea in 1952 in *An Unsolved Problem in Biology* and in 1957 in *The Uniqueness of the Individual*. I should add that I find Williams's development of the theory very helpful, since it makes clear a necessary step in the argument (the importance of 'pleiotropy' or multiple gene effects) which Medawar did not explicitly emphasize. W. D. Hamilton has more recently taken this kind of theory even further in his paper, 'The Moulding of Senescence by Natural Selection'. Incidentally, I have had many interesting letters from doctors but none, I think, commented on my speculations about 'fooling' genes as to the age of the body they are in (pp. 53–4). The idea still doesn't strike me as obviously silly, and if it were right wouldn't it be rather important, medically?

p. 55 What is the good of sex?

The problem of what sex is good for is still as tantalizing as ever, despite some thought-provoking books, notably those by M. T. Ghiselin, G. C. Williams, J. Maynard Smith, and G. Bell, and a volume edited by R. Michod and B. Levin. To me, the most exciting new idea is W. D. Hamilton's parasite theory, which has been explained in non-technical language by Jeremy Cherfas and John Gribbin in *The Redundant Male*.

p. 57 ...the surplus DNA is...a parasite, or at best a harmless but useless passenger...(see also p. 237)

My suggestion that surplus, untranslated DNA might be a self-interested parasite has been taken up and developed by molecular biologists (see papers by Orgel and Crick, and Doolittle and Sapienza) under the catchphrase 'Selfish DNA'. S. J. Gould, in *Hen's Teeth and Horse's Toes*, has made the provocative (to me!) claim that, despite the historical origins of the idea of selfish DNA, 'The theories of selfish genes and selfish DNA could not be more different in the structures of explanation that nurture them.' I find his reasoning wrong but interesting, which, incidentally, he has been kind enough to tell me, is how he usually finds mine. After a preamble on 'reductionism' and 'hierarchy' (which, as usual, I find neither wrong nor interesting), he goes on:

> Dawkins's selfish genes increase in frequency because they have effects on bodies, aiding them in their struggle for existence. Selfish DNA increases in frequency for precisely the opposite reason— because it has no effect on bodies...

I see the distinction that Gould is making, but I cannot see it as a fundamental one. On the contrary, I still see selfish DNA as a special case of the whole theory of the selfish gene, which is precisely how the idea of selfish DNA originally arose. (This point, that selfish DNA is a special case, is perhaps even clearer on page 237 of this book than in the passage from page 57 cited by Doolittle and Sapienza, and Orgel and Crick. Doolittle and Sapienza, by the way, use the phrase 'selfish genes', rather than 'selfish DNA', in their title.) Let me reply to Gould with the following analogy. Genes that give wasps their yellow and black stripes increase in frequency because this ('warning') colour pattern powerfully stimulates the brains of other animals. Genes that give tigers their yellow and black stripes increase in frequency 'for precisely the opposite reason'—because ideally this (cryptic) colour pattern does not stimulate the brains of other animals at all. There is indeed a distinction here, closely analogous (at a different hierarchical level!) to Gould's distinction, but it is a subtle distinction of detail. We should hardly wish to claim that the two cases 'could not be more different in the structures of explanation that nurture them'. Orgel and Crick hit the nail on the head when they make the analogy between selfish DNA and cuckoo eggs: cuckoo eggs, after all, escape detection by looking exactly like host eggs.

Incidentally, the latest edition of the *Oxford English Dictionary* lists a new meaning of 'selfish' as 'Of a gene or genetic material: tending to be perpetuated or to spread although of no effect on the phenotype.' This is an admirably concise definition of 'selfish DNA', and the second supporting quotation actually concerns selfish DNA. In my opinion, however, the final phrase, 'although of no effect on the phenotype', is unfortunate. Selfish genes *may* not have effects on the phenotype, but many of them do. It would be open to the lexicographers to claim that they intended to confine the meaning to 'selfish DNA', which really does have no phenotypic effects. But their first supporting quotation, which is from *The Selfish Gene*, includes selfish genes that do have phenotypic effects. Far it be from me, however, to cavil at the honour of being quoted in the *Oxford English Dictionary*!

I have discussed selfish DNA further in *The Extended Phenotype* (pp. 156–64).

CHAPTER 4
The gene machine

p. 63 Brains may be regarded as analogous in function to computers.
Statements like this worry literal-minded critics. They are right, of course, that brains differ in many respects from computers. Their internal methods of working, for instance, happen to be very different from the particular kind of computers that our technology has developed. This is no way reduces the truth of my statement about their being analogous in function. Functionally, the brain plays precisely the role of

on-board computer—data processing, pattern recognition, short-term and long-term data storage, operation coordination, and so on.

Whilst we are on computers, my remarks about them have become gratifyingly—or frighteningly, depending on your view—dated. I wrote (p. 62) that 'you could pack only a few hundred transistors into a skull.' Transistors today are combined in integrated circuits. The number of transistor-equivalents that you could pack into a skull today must be up in the billions. I also stated (p. 66) that computers, playing chess, had reached the standard of a good amateur. Today, chess programs that beat all but very serious players are commonplace on cheap home computers, and the best programs in the world now present a serious challenge to grand masters. Here, for instance, is the *Spectator*'s chess correspondent Raymond Keene, in the issue of 7 October 1988:

> It is still something of a sensation when a titled player is beaten by a computer, but not, perhaps, for much longer. The most dangerous metal monster so far to challenge the human brain is the quaintly named 'Deep Thought', no doubt in homage to Douglas Adams. Deep Thought's latest exploit has been to terrorise human opponents in the US Open Championship, held in August in Boston. I still do not have DT's overall rating performance to hand, which will be the acid test of its achievement in an open Swiss system competition, but I have seen a remarkably impressive win against the strong Canadian Igor Ivanov, a man who once defeated Karpov! Watch closely; this may be the future of chess.

There follows a move-by-move account of the game. This is Keene's reaction to Deep Thought's Move 22:

> A wonderful move... The idea is to centralise the queen... and this concept leads to remarkably speedy success... The startling outcome... Black's queen's wing is now utterly demolished by the queen penetration.

Ivanov's reply to this is described as:

> A desperate fling, which the computer contemptuously brushes aside... The ultimate humiliation. DT ignores the queen recapture, steering instead for a snap checkmate... Black resigns.

Not only is Deep Thought one of the world's top chess players. What I find almost more striking is the language of human consciousness that the commentator feels obliged to use. Deep Thought 'contemptuously brushes aside' Ivanov's 'desperate fling'. Deep Thought is described as 'aggressive'. Keene speaks of Ivanov as 'hoping' for some outcome, but his language shows that he would be equally happy using a word like 'hope' for Deep Thought. Personally I rather look forward to a computer program winning the world championship. Humanity needs a lesson in humility.

p. 68 There is a civilization 200 light-years away, in the constellation of Andromeda.

A for Andromeda and its sequel, Andromeda Breakthrough, are inconsistent about whether the alien civilization hails from the enormously distant Andromeda galaxy, or a nearer star in the constellation of Andromeda as I said. In the first novel the planet is placed 200 light-years away, well within our own galaxy. In the sequel, however, the same aliens are located in the Andromeda galaxy, which is about 2 million light-years away. Readers of my page 68 may replace '200' with '2 million' according to taste. For my purpose the relevance of the story remains undiminished.

Fred Hoyle, the senior author of both these novels, is an eminent astronomer and the author of my favourite of all science fiction stories, The Black Cloud. The superb scientific insight deployed in his novels makes a poignant contrast to his spate of more recent books written jointly with C. Wickramasinghe. Their misrepresenting of Darwinism (as a theory of pure chance) and their waspish attacks on Darwin himself do nothing to assist their otherwise intriguing (though implausible) speculations on interstellar origins of life. Publishers should correct the misapprehension that a scholar's distinction in one field implies authority in another. And as long as that misapprehension exists, distinguished scholars should resist the temptation to abuse it.

p. 71 ...strategies and tricks of the living trade...

This strategic way of talking about an animal or plant, or a gene, as if it were consciously working out how best to increase its success—for instance picturing 'males as high-stake high-risk gamblers, and females as safe investors' (p. 73)—has become commonplace among working biologists. It is a language of convenience which is harmless unless it happens to fall into the hands of those ill-equipped to understand it. Or over-equipped to misunderstand it? I can, for example, find no other way to make sense of an article criticizing The Selfish Gene in the journal

Philosophy, by someone called Mary Midgley, which is typified by its first sentence: 'Genes cannot be selfish or unselfish, any more than atoms can be jealous, elephants abstract or biscuits teleological.' My own 'In Defence of Selfish Genes', in a subsequent issue of the same journal, is a full reply to this incidentally highly intemperate and vicious paper. It seems that some people, educationally over-endowed with the tools of philosophy, cannot resist poking in their scholarly apparatus where it isn't helpful. I am reminded of P. B. Medawar's remark about the attractions of 'philosophy-fiction' to 'a large population of people, often with well-developed literary and scholarly tastes, who have been educated far beyond their capacity to undertake analytical thought'.

p. 76 *Perhaps consciousness arises when the brain's simulation of the world becomes so complete that it must include a model of itself.*

I discuss the idea of brains simulating worlds in my 1988 Gifford Lecture, 'Worlds in Microcosm'. I am still unclear whether it really can help us much with the deep problem of consciousness itself, but I confess to being pleased that it caught the attention of Sir Karl Popper in his Darwin Lecture. The philosopher Daniel Dennett has offered a theory of consciousness that takes the metaphor of computer simulation further. To understand his theory, we have to grasp two technical ideas from the world of computers: the idea of a virtual machine, and the distinction between serial and parallel processors. I'll have to get the explanation of these out of the way first.

A computer is a real machine, hardware in a box. But at any particular time it is running a program that makes it look like another machine, a virtual machine. This has long been true of all computers, but modern 'user-friendly' computers bring home the point especially vividly. At the time of writing, the market leader in user-friendliness is widely agreed to be the Apple Macintosh. Its success is due to a wired-in suite of programs that make the real hardware machine—whose mechanisms are, as with any computer, forbiddingly complicated and not very compatible with human intuition—*look* like a different kind of machine: a virtual machine, specifically designed to mesh with the human brain and the human hand. The virtual machine known as the Macintosh User Interface is recognizably a machine. It has buttons to press, and slide controls like a hi-fi set. But it is a *virtual* machine. The buttons and sliders are not made of metal or plastic. They are pictures on the screen, and you press them or slide them by moving a virtual finger about the screen. As a human you feel in control, because you are accustomed to moving things around with your finger. I have been an intensive programmer and user of a wide variety of digital computers for twenty-five years, and I can testify that using the Macintosh (or its imitators) is a qualitatively different experience from using any earlier type of computer. There is an effortless, natural feel to it, almost as if the virtual machine were an extension of one's own body. To a remarkable extent the virtual machine allows you to use intuition instead of looking up the manual.

I now turn to the other background idea that we need to import from computer science, the idea of serial and parallel processors. Today's digital computers are mostly serial processors. They have one central calculating mill, a single electronic bottleneck through which all data have to pass when being manipulated. They can create an illusion of doing many things simultaneously because they are so fast. A serial computer is like a chess master 'simultaneously' playing twenty opponents but actually rotating around them. Unlike the chess master, the computer rotates so swiftly and quietly around its tasks that each human user has the illusion of enjoying the computer's exclusive attention. Fundamentally, however, the computer is attending to its users serially.

Recently, as part of the quest for ever more dizzying speeds of performance, engineers have made genuinely parallel-processing machines. One such is the Edinburgh Supercomputer, which I was recently privileged to visit. It consists of a parallel array of some hundreds of 'transputers', each one equivalent in power to a contemporary desk-top computer. The supercomputer works by taking the problem it has been set, subdividing it into smaller tasks that can be tackled independently, and farming out the tasks to gangs of transputers. The transputers take the sub-problem away, solve it, hand in the answer and report for a new task. Meanwhile other gangs of transputers are reporting in with their solutions, so the whole supercomputer gets to the final answer orders of magnitude faster than a normal serial computer could.

I said that an ordinary serial computer can create an illusion of being a parallel processor, by rotating its 'attention' sufficiently fast around a number of tasks. We could say that there is a *virtual* parallel processor sitting atop serial hardware. Dennett's idea is that the human brain has done exactly the reverse. The hardware of the brain is fundamentally parallel, like that of the Edinburgh machine. And it runs software designed to create an illusion of serial processing: a serially processing virtual machine riding on top of parallel architecture. The salient feature of the subjective experience of thinking, Dennett thinks, is the serial 'one-thing-after-another', 'Joycean', stream of consciousness. He believes that most animals lack this serial experience, and use brains directly in their native, parallel-processing mode. Doubtless the human brain, too, uses its parallel architecture directly for many of the routine tasks of keeping a complicated survival machine ticking over. But, in addition, the human brain evolved a software virtual machine to simulate the illusion of a serial processor. The mind, with its serial stream of consciousness, is a virtual machine, a 'user-friendly' way of experiencing the brain, just as the 'Macintosh User Interface' is a user-friendly way of experiencing the physical computer inside its grey box.

It is not obvious why we humans needed a serial virtual machine, when other species seem quite happy with their unadorned parallel machines. Perhaps there is something fundamentally serial about the more difficult tasks that a wild human is called upon to do, or perhaps Dennett is wrong to single us out. He further believes that the development of the serial software has been a largely cultural phenomenon, and again it is not obvious to me why this should be particularly likely. But I should add that, at the time of my writing, Dennett's paper is unpublished and my account is based on recollections of his 1988 Jacobsen Lecture in London. The reader is advised to consult Dennett's own account when it is published, rather than rely on my doubtless imperfect and impressionistic—maybe even embellished—one.

The psychologist Nicholas Humphrey, too, has developed a tempting hypothesis of how the evolution of a capacity to simulate might have led to consciousness. In his book, *The Inner Eye*, Humphrey makes a convincing case that highly social animals like us and chimpanzees have to become expert psychologists. Brains have to juggle with, and simulate, many aspects of the world. But most aspects of the world are pretty simple in comparison to brains themselves. A social animal lives in a world of others, a world of potential mates, rivals, partners, and enemies. To survive and prosper in such a world, you have to become good at predicting what these other individuals are going to do next. Predicting what is going to happen in the inanimate world is a piece of cake compared with predicting what is going to happen in the social world. Academic psychologists, working scientifically, aren't really very good at predicting human behaviour. Social companions, using minute movements of the facial muscles and other subtle cues, are often astonishingly good at reading minds and second-guessing behaviour. Humphrey believes that this 'natural psychological' skill has become highly evolved in social animals, almost like an extra eye or other complicated organ. The 'inner eye' is the evolved social-psychological organ, just as the outer eye is the visual organ.

So far, I find Humphrey's reasoning convincing. He goes on to argue that the inner eye works by self-inspection. Each animal looks inwards to its own feelings and emotions, as a means of understanding the feelings and emotions of others. The psychological organ works by self-inspection. I am not so sure whether I agree that this helps us to understand consciousness, but Humphrey is a graceful writer and his book is persuasive.

p. 78 A gene for altruistic behaviour...

People sometimes get all upset about genes 'for' altruism, or other apparently complicated behaviour. They think (wrongly) that in some sense the complexity of the behaviour must be contained within the gene. How can there be a single gene for altruism, they ask, when all that a gene does is encode one protein chain? But to speak of a gene 'for' something only ever means that a *change* in the gene causes a *change* in the something. A single genetic *difference*, by changing some detail of the molecules in cells, causes a *difference* in the already complex embryonic processes, and hence in, say, behaviour.

For instance, a mutant gene in birds 'for' brotherly altruism will certainly not be solely responsible for an entirely new complicated behaviour pattern. Instead, it will alter some already existing, and probably already complicated, behaviour pattern. The most likely precursor in this case is parental behaviour. Birds routinely have the complicated nervous apparatus needed to feed and care for their own offspring. This has, in turn, been built up over many generations of slow, step-by-step evolution, from antecedents of its own. (Incidentally, sceptics about genes for fraternal care are often inconsistent: why aren't they just as sceptical about genes for equally complicated parental care?) The pre-existing behaviour pattern—parental care in this case—will be mediated by a convenient rule of thumb, such as 'Feed all squawking, gaping things in the nest.' The gene 'for feeding younger brothers and sisters' could work, then, by accelerating the age at which this rule of thumb matures in development. A fledgling bearing the fraternal gene as a new mutation will simply activate its 'parental' rule of thumb a little earlier than a normal bird. It will treat the squawking, gaping things in its parents' nest—its younger brothers and sisters—as if they were squawking, gaping things in its own nest—its children. Far from being a brand new, complicated behavioural innovation, 'fraternal behaviour' would originally arise as a slight variant in the developmental timing of already-existing behaviour. As so often, fallacies arise when we forget the essential gradualism of evolution, the fact that adaptive evolution proceeds by small, step-by-step alterations of pre-existing structures or behaviour.

p. 79 Hygienic bees

If the original book had had footnotes, one of them would have been devoted to explaining—as Rothenbuhler himself scrupulously did—that the bee results were not quite so neat and tidy. Out of the many colonies that, according to theory, should not have shown hygienic behaviour, one nevertheless did. In Rothenbuhler's own words, 'We cannot disregard this result, regardless of how much we would like to, but we are basing the genetic hypothesis on the other data.' A mutation in the anomalous colony is a possible explanation, though it is not very likely.

p. 81 This is the behaviour that can be broadly labelled communication.

I now find myself dissatisfied with this treatment of animal communication. John Krebs and I have argued in two articles that most animal signals are best seen as neither informative nor deceptive, but rather as *manipulative*. A signal is a means by which one animal makes use of another animal's muscle power. A nightingale's song is not information, not even deceitful information. It is persuasive, hypnotic, spellbinding oratory. This kind of argument is taken to its logical conclusion in *The Extended Phenotype*, part of which I have abridged in Chapter 13 of this book. Krebs and I argue that signals evolve from an interplay of what we call mind-reading and manipulation. A startlingly different approach to the whole matter of animal signals is that of Amotz Zahavi. In a note to Chapter 9, I discuss Zahavi's views far more sympathetically than in the first edition of this book.

CHAPTER 5

Aggression: stability and the selfish machine

p. 90 ...evolutionarily stable strategy...

I now like to express the essential idea of an ESS in the following more economical way. An ESS is a strategy that does well against copies of itself. The rationale for this is as follows. A successful strategy is one that dominates the population. Therefore it will tend to encounter copies of itself. Therefore it won't stay successful unless it does well against copies of itself. This definition is not so mathematically precise as Maynard Smith's, and it cannot replace his definition because it is actually incomplete. But it does have the virtue of encapsulating, intuitively, the basic ESS idea.

The ESS way of thinking has become more widespread among biologists now than when this chapter was written. Maynard Smith himself has summarized developments up to 1982 in his book *Evolution and the Theory of Games*. Geoffrey Parker, another of the leading contributors to the field, has written a slightly more recent account. Robert Axelrod's *The Evolution of Cooperation* makes use of ESS theory, but I won't discuss it here, since one of my two new chapters, 'Nice guys finish first', is largely devoted to explaining Axelrod's work. My own writings on the subject of ESS theory since the first edition of this book are an article called 'Good Strategy or Evolutionarily Stable Strategy?', and the joint papers on digger wasps discussed in the note to p. 98.

p. 97 ...retaliator emerges as evolutionarily stable.

This statement, unfortunately, was wrong. There was an error in the original Maynard Smith and Price paper, and I repeated it in this chapter, even exacerbating it by making the rather foolish statement that prober–retaliator is 'almost' an ESS (if a strategy is 'almost' an ESS, then it is not an ESS and will be invaded). Retaliator looks superficially like an ESS because, in a population of retaliators, no other strategy does better. But dove does equally well since, in a population of retaliators, it is indistinguishable in its behaviour from retaliator. Dove therefore can drift into the population. It is what happens next that is the problem. J. S. Gale and the Revd L. J. Eaves did a dynamic computer simulation in which they took a population of model animals through a large number of generations of evolution. They showed that the true ESS in this game is in fact a stable mixture of hawks and bullies. This is not the only error in the early ESS literature that has been exposed by dynamic treatment of this type. Another nice example is an error of my own, discussed in my notes to Chapter 9.

p. 98 *Unfortunately, we know too little at present to assign realistic numbers to the costs and benefits of various outcomes in nature.*

We now have some good field measurements of costs and benefits in nature, which have been plugged into particular ESS models. One of the best examples comes from great golden digger wasps in North America. Digger wasps are not the familiar social wasps of our autumn jam-pots, which are neuter females working for a colony. Each female digger wasp is on her own, and she devotes her life to providing shelter and food for a succession of her larvae. Typically, a female begins by digging a long bore-hole into the earth, at the bottom of which is a hollowed-out chamber. She then sets off to hunt prey (katydids or longhorned grass-hoppers in the case of the great golden digger wasp). When she finds one she stings it to paralyse it, and drags it back into her burrow. Having accumulated four or five katydids she lays an egg on the top of the pile and seals up the burrow. The egg hatches into a larva, which feeds on the katydids. The point about the prey being paralysed rather than killed, by the way, is that they don't decay but are eaten alive and are therefore fresh. It was this macabre habit, in the related Ichneumon wasps, that provoked Darwin to write: 'I cannot persuade myself that a beneficent and omnipotent God would have designedly created the Ich-neumonidae with the express intention of their feeding within the living bodies of Caterpillars...' He might as well have used the example of a French chef boiling lobsters alive to preserve the flavour. Returning to the life of the female digger wasp, it is a solitary one except that other females are working independently in the same area, and sometimes they occupy one another's burrows rather than go to the trouble of digging a new one.

Dr Jane Brockmann is a sort of wasp equivalent of Jane Goodall. She came from America to work with me at Oxford, bringing her copious records of almost every event in the life of two entire populations of individually identified female wasps. These records were so complete that individual wasp time-budgets could be drawn up. Time is an economic commodity: the more time spent on one part of life, the less is available for other parts. Alan Grafen joined the two of us and taught us how to think correctly about time-costs and reproductive benefits. We found evidence for a true mixed ESS in a game played between female wasps in a population in New Hampshire, though we failed to find such evidence in another population in Michigan. Briefly, the New Hampshire wasps either dig their own nests or enter a nest that another wasp has dug. According to our interpretation, wasps can benefit by entering because some nests are abandoned by their original diggers and are reusable. It does not pay to enter a nest that is occupied, but an enterer has no way of knowing which nests are occupied and which abandoned. She runs the risk of going for days in double-occupation, at the end of which she may come home to find the burrow sealed up, and all her efforts in vain—the other occupant has laid her egg and will reap the benefits. If too much entering is going on in a population, available burrows become scarce, the chance of double-occupation goes up, and it therefore pays to dig. Conversely, if plenty of wasps are digging, the high availability of burrows favours entering. There is a critical frequency of entering in the population at which digging and entering are equally profitable. If the actual frequency is below the critical frequency, natural selection favours entering, because there is a good supply of available abandoned burrows. If the actual frequency is higher than the critical frequency, there is a shortage of available burrows and natural selection favours digging. So a balance is maintained in the population. The detailed, quantitative evidence suggests that this is a true mixed ESS, each individual wasp having a probability of digging or entering, rather than the population containing a mixture of digging and entering specialists.

p. 104 *The neatest demonstration I know of this form of behavioural asymmetry...*

An even clearer demonstration than Tinbergen's of the 'resident always wins' phenomenon comes from N. B. Davies's research on speckled wood butterflies. Tinbergen's work was done before ESS theory was invented, and my ESS interpretation in the first edition of this book was made with hindsight. Davies conceived his butterfly study in the light of ESS theory. He noticed that individual male butterflies in Wytham Wood, near Oxford, defended patches of sunlight. Females were attracted to sun patches, so a sun patch was a valuable resource, something worth fighting over. There were more males than sun patches and the surplus waited their chance in the leafy canopy. By catching males and releasing them one after the other, Davies showed that whichever of two individuals was released first into a sun patch was treated, by both individuals, as the 'owner'. Whichever male arrived second in the sun patch was treated as the 'intruder'. The intruder always, without exception, promptly conceded defeat, leaving the owner in sole control. In a final *coup de grâce* experiment, Davies managed to 'fool' both butterflies into 'thinking' that they were the owner and the other was the intruder. Only under these conditions did a really serious, prolonged fight break out. By the way, in all those cases where, for simplicity, I have spoken as though there was a single pair of butterflies there was really, of course, a statistical sample of pairs.

p. 106 *Paradoxical ESS*

Another incident that conceivably might represent a paradoxical ESS was recorded in a letter to *The Times* (of London, 7 December 1977) from a Mr James Dawson: 'For some years I have noticed that a gull using a flag pole as a vantage point invariably makes way for another gull wishing to alight on the post and this irrespectively of the size of the two birds.'

The most satisfying example of a paradoxical strategy known to me involves domestic pigs in a Skinner box. The strategy is stable in the same sense as an ESS, but it is better called a DSS ('developmentally stable strategy') because it arises during the animals' own lifetimes rather than over evolutionary time. A Skinner box is an apparatus in which an animal learns to feed itself by pressing a lever, food then being automatically delivered down a chute. Experimental psychologists are accustomed to putting pigeons or rats in small Skinner boxes, where they soon learn to press delicate little levers for a food reward. Pigs can learn the same thing in a scaled-up Skinner box with a very undelicate snout-lever (I saw a research film of this many years ago and I well recall almost dying of laughter). B. A. Baldwin and G. B. Meese trained pigs in a Skinner sty, but there is an added twist to the tale. The snout-lever was at one end of the sty; the food dispenser at the other. So the pig had to press the lever, then race up to the other end of the sty to get the food, then rush back to the lever, and so on. This sounds all very well, but Baldwin and Meese put *pairs* of pigs into the apparatus. It now became possible for one pig to exploit the other. The 'slave' pig rushed back and forth pressing the bar. The 'master' pig sat by the food chute and ate the food as it was dispensed. Pairs of pigs did indeed settle down into a stable 'master/slave' pattern of this kind, one working and running, the other doing most of the eating.

Now for the paradox. The labels 'master' and 'slave' turned out to be all topsy-turvy. Whenever a pair of pigs settled down to a stable pattern, the pig that ended up playing the 'master' or 'exploiting' role was the pig that, in all other ways, was subordinate. The so-called 'slave' pig, the one that did all the work, was the pig that was usually dominant. Anybody knowing the pigs would have predicted that, on the contrary, the dominant pig would have been the master, doing most of the eating; the subordinate pig should have been the hard-working and scarcely-eating slave.

How could this paradoxical reversal arise? It is easy to understand, once you start thinking in terms of stable strategies. All that we have to do is scale the idea down from evolutionary time to developmental time, the time-scale on which a relationship between two individuals develops. The strategy 'If dominant, sit by the food trough; if subordinate, work the lever' sounds sensible, but would not be stable. The subordinate pig, having pressed the lever, would come sprinting over, only to find the dominant pig with its front feet firmly in the trough and impossible to dislodge. The subordinate pig would soon give up pressing the lever, for the habit would never be rewarded. But now consider the reverse strategy: 'If dominant, work the lever; if subordinate, sit by the food trough.' This would be stable, even though it has the paradoxical result that the subordinate pig gets most of the food. All that is necessary is that there should be *some* food left for the dominant pig when he charges up from the other end of the sty. As soon as he arrives, he has no difficulty in tossing the subordinate pig out of the trough. As long as there is a crumb left to reward him, his habit of working the lever, and thereby inadvertently stuffing the subordinate pig, will persist. And the subordinate pig's habit of reclining idly by the trough is rewarded too. So the whole 'strategy', 'If dominant behave as a "slave", if subordinate behave as a "master"', is rewarded and therefore stable.

p. 106 ...a kind of dominance hierarchy [in crickets]...

Ted Burk, then my graduate student, found further evidence for this kind of pseudo-dominance hierarchy in crickets. He also showed that a male cricket is more likely to court females if he has recently won a fight against another male. This should be called the 'Duke of Marlborough Effect', after the following entry in the diary of the first Duchess of Marlborough: 'His Grace returned from the wars today and pleasured me twice in his top-boots.' An alternative name might be suggested by the following report from the magazine *New Scientist* about changes in levels of the masculine hormone testosterone: 'Levels doubled in tennis players during the 24 hours before a big match. Afterwards, the levels in winners stayed up, but in losers they dropped.'

p. 109 ...the ESS concept as one of the most important advances in evolutionary theory since Darwin.

This sentence is a bit over the top. I was probably over-reacting to the then prevalent neglect of the ESS idea in the contemporary biological literature, especially in America. The term does not occur anywhere in E. O. Wilson's massive *Sociobiology*, for instance. It is neglected no longer, and I can now take a more judicious and less evangelical view. You don't actually *have* to use ESS language, provided you think clearly enough. But it is a great aid to thinking clearly, especially in those cases—which in practice is most cases—where detailed genetic knowledge is not available. It is sometimes said that ESS models assume that reproduction is asexual, but this statement is misleading if taken to mean a positive assumption of asexual as opposed to sexual reproduction. The truth is rather that ESS models don't bother to commit themselves about the details of the genetic system. Instead they assume that, in some vague sense, like begets like. For many purposes this assumption is adequate. Indeed its vagueness can even be beneficial, since it concentrates the mind on essentials and away from details, such as genetic dominance, which are usually unknown in particular cases. ESS thinking is most useful in a negative role; it helps us to avoid theoretical errors that might otherwise tempt us.

p. 113 *Progressive evolution may be not so much a steady upward climb as a series of discrete steps from stable plateau to stable plateau.*

This paragraph is a fair summary of one way of expressing the now well-known theory of punctuated equilibrium. I am ashamed to say that, when I wrote my conjecture, I, like many biologists in England at the time, was totally ignorant of that theory, although it had been published three years earlier. I have since, for instance in *The Blind Watchmaker*, become somewhat petulant—perhaps too much so—over the way the theory of punctuated equilibrium has been oversold. If this has hurt anybody's feelings, I regret it. They may like to note that, at least in 1976, my heart was in the right place.

CHAPTER 6
Genesmanship

p. 117 *…I have never been able to understand why they have been so neglected…*

Hamilton's 1964 papers are neglected no longer. The history of their earlier neglect and subsequent recognition makes an interesting quantitative study in its own right, a case study in the incorporation of a 'meme' into the meme pool. I trace the progress of this meme in the notes to Chapter 11.

p. 117 *…I shall assume that we are talking about genes that are rare…*

The device of assuming that we are talking about a gene that is rare in the population as a whole was a bit of a cheat, to make the measuring of relatedness easy to explain. One of Hamilton's main achievements was to show that his conclusions follow *regardless* of whether the gene concerned is rare or common. This turns out to be an aspect of the theory that people find difficult to understand.

The problem of measuring relatedness trips many of us up in the following way. Any two members of a species, whether they belong to the same family or not, usually share more than 90 per cent of their genes.

What, then, are we talking about when we speak of the relatedness between brothers as, or between first cousins as? The answer is that brothers share of their genes *over and above* the 90 per cent (or whatever it is) that all individuals share in any case. There is a kind of baseline relatedness, shared by all members of a species; indeed, to a lesser extent, shared by members of other species. Altruism is expected towards individuals whose relatedness is higher than the baseline, whatever the baseline happens to be.

In the first edition, I evaded the problem by using the trick of talking about rare genes. This is correct as far as it goes, but it doesn't go far enough. Hamilton himself wrote of genes being 'identical by descent', but this presents difficulties of its own, as Alan Grafen has shown. Other writers did not even acknowledge that there was a problem, and simply spoke of absolute percentages of shared genes, which is a definite and positive error. Such careless talk did lead to serious misunderstandings. For instance a distinguished anthropologist, in the course of a bitter attack on 'sociobiology' published in 1978, tried to argue that if we took kin selection seriously we should expect all humans to be altruistic to one another, since all humans share more than 99 per cent of their genes. I have been given a brief reply to this error in my 'Twelve Misunderstandings of Kin Selection' (it rates as Misunderstanding Number 5). The other eleven misunderstandings are worth a look, too.

Alan Grafen gives what may be the definitive solution to the problem of measuring relatedness in his 'Geometric View of Relatedness', which I shall not attempt to expound here. And in another paper, 'Natural Selection, Kin Selection and Group Selection', Grafen clears up a further common and important problem, namely the widespread misuse of Hamilton's concept of 'inclusive fitness'. He also tells us the right and wrong way to calculate costs and benefits to genetic relatives.

p. 121 ... *armadillos ... it would be well worth somebody's while going out to South America to have a look.*

No developments are reported on the armadillo front, but some spectacular new facts have come to light for another group of 'cloning' animals—aphids. It has long been known that aphids (e.g. greenfly) reproduce asexually as well as sexually. If you see a crowd of aphids on a plant, the chances are that they are all members of an identical female clone, while those on the next-door plant will be members of a different clone. Theoretically these conditions are ideal for the evolution of kin-selected altruism. No actual instances of aphid altruism were known, however, until sterile 'soldiers' were discovered in a Japanese species of aphids by Shigeyuki Aoki, in 1977, just too late to appear in the first edition of this book. Aoki has since found the phenomenon in a number of different species, and has good evidence that it has evolved at least four times independently in different groups of aphids.

Briefly, Aoki's story is this. Aphid 'soldiers' are an anatomically distinct caste, just as distinct as the castes of traditional social insects like ants. They are larvae that do not mature to full adulthood, and they are therefore sterile. They neither look nor behave like their non-soldier larval contemporaries, to whom they are, however, *genetically* identical. Soldiers are typically larger than non-soldiers; they have extra-big front legs which make them look almost scorpion-like; and they have sharp horns pointing forward from the head. They use these weapons to fight and kill would-be predators. They often die in the process, but even if they don't it is still correct to think of them as genetically 'altruistic' because they are sterile.

In terms of selfish genes, what is going on here? Aoki does not mention precisely what determines which individuals become sterile soldiers and which become normal reproductive adults, but we can safely say that it must be an environmental, not a genetic difference—obviously, since the sterile soldiers and the normal aphids on any one plant are genetically identical. However, there must be genes for the capacity to be environmentally switched into either of the two developmental pathways. Why has natural selection favoured these genes, even though some of them end up in the bodies of sterile soldiers and are therefore not passed on? Because, thanks to the soldiers, copies of those very same genes have been saved in the bodies of the reproductive non-soldiers. The rationale is just the same as for all social insects (see Chapter 10), except that in other social insects, such as ants or termites, the genes in the sterile 'altruists' have only a *statistical* chance of helping copies of themselves in non-sterile reproductives. Aphid altruistic genes enjoy certainty rather than statistical likelihood since aphid soldiers are clone-mates of the reproductive sisters whom they benefit. In some respects Aoki's aphids provide the neatest real-life illustration of the power of Hamilton's ideas.

Should aphids, then, be admitted to the exclusive club of truly social insects, traditionally the bastion of ants, bees, wasps, and termites? Entomological conservatives could blackball them on various grounds. They lack a long-lived old queen, for instance. Moreover, being a true clone, the aphids are no more 'social' than the cells of your body. There is a single animal feeding on the plant. It just happens to have its body divided up into physically separate aphids, some of which play a specialized defensive role just like white blood corpuscles in the human body. 'True' social insects, the argument goes, cooperate in spite of not being part of the same organism, whereas Aoki's aphids cooperate because they do belong to the same 'organism'. I cannot get worked up about this semantic issue. It seems to me that, so long as you understand what is going on among ants, aphids, and human cells, you should be at liberty to call them social or not, as you please. As for my own preference, I have reasons for calling Aoki's aphids social organisms, rather than parts of a single organism. There are crucial properties of a single organism which a single aphid possesses, but which a clone of aphids does not possess. The argument is spelled out in *The Extended Phenotype*, in the chapter called 'Rediscovering the Organism', and also in the new chapter of the present book called 'The long reach of the gene'.

p. 122 *Kin selection is emphatically not a special case of group selection.*

The confusion over the difference between group selection and kin selection has not disappeared. It may even have got worse. My remarks stand with redoubled emphasis except that, by a thoughtless choice of words, I introduced a quite separate fallacy of my own at the top of page 102 of the first edition of this book. I said in the original (it is one of the few things I have altered in the text of this edition): 'We simply expect that second cousins should tend to receive as much altruism as offspring or siblings.' As S. Altmann has pointed out, this is obviously wrong. It is wrong for a reason that has nothing to do with the point I was trying to argue at the time. If an altruistic animal has a cake to give to relatives, there is no reason at all for it to give every relative a slice, the size of the slices being determined by the closeness of relatedness. Indeed, this would lead to absurdity since all members of the species, not to mention other species, are at least distant relatives who could therefore each claim a carefully measured crumb! On the contrary, if there is a close relative in the vicinity, there is no reason to give a distant relative any cake at all. Subject to other complications like laws of diminishing returns, the whole cake should be given to the closest relative available. What I of course meant to say was 'We simply expect that second cousins should be as likely to receive altruism as offspring or siblings' (p. 122), and this is what now stands.

p. 122 *He deliberately excludes offspring: they don't count as kin!*

I expressed the hope that E. O. Wilson would change his definition of kin selection in future writings, so as to include offspring as 'kin'. I am happy to report that, in his *On Human Nature*, the offending phrase, 'other than offspring', has indeed—I am not claiming any credit for this!—been omitted. He adds, 'Although kin are defined so as to include offspring, the term kin selection is ordinarily used only if at least some other relatives, such as brothers, sisters, or parents, are also affected.' This is unfortunately an accurate statement about ordinary usage by biologists, which simply reflects the fact that many biologists still lack a gut understanding of what kin selection is fundamentally all about. They *still* wrongly think of it as something extra and esoteric, over and above ordinary 'individual selection'. It isn't. Kin selection follows from the fundamental assumptions of neo-Darwinism as night follows day.

p. 124 *But what a complicated calculation...*

The fallacy that the theory of kin selection demands unrealistic feats of calculation by animals is revived without abatement by successive generations of students. Not just young students, either. *The Use and Abuse of Biology*, by the distinguished social anthropologist Marshall Sahlins, could be left in decent obscurity had it not been hailed as a 'withering attack' on 'sociobiology'. The following quotation, in the context of whether kin selection could work in humans, is almost too good to be true:

> In passing it needs to be remarked that the epistemological problems presented by a lack of linguistic support for calculating r, coefficients of relationship, amount to a serious defect in the theory of kin selection. Fractions are of very rare occurrence in the world's languages, appearing in Indo-European and in the archaic civilizations of the Near and Far East, but they are generally lacking among the so-called primitive peoples. Hunters and gatherers generally do not have counting systems

beyond one, two, and three. I refrain from comment on the even greater problem of how animals are supposed to figure out how that r [ego, first cousins] = ⅛.

This is not the first time I have quoted this highly revealing passage, and I may as well quote my own rather uncharitable reply to it, from 'Twelve Misunderstandings of Kin Selection':

> A pity, for Sahlins, that he succumbed to the temptation to 'refrain from comment' on how animals are supposed to 'figure out' r. The very absurdity of the idea he tried to ridicule should have set mental alarm bells ringing. A snail shell is an exquisite logarithmic spiral, but where does the snail keep its log tables; how indeed does it read them, since the lens in its eye lacks 'linguistic support' for calculating m, the coefficient of refraction? How do green plants 'figure out' the formula of chlorophyll?

The fact is that if you thought about anatomy, physiology, or almost any aspect of biology, not just behaviour, in Sahlins's way you would arrive at his same non-existent problem. The embryological development of any bit of an animal's or plant's body requires complicated mathematics for its complete description, but this does not mean that the animal or plant must itself be a clever mathematician! Very tall trees usually have huge buttresses flaring out like wings from the base of their trunks. Within any one species, the taller the tree, the relatively larger the buttresses. It is widely accepted that the shape and size of these buttresses are close to the economic optimum for keeping the tree erect, although an engineer would require quite sophisticated mathematics to demonstrate this. It would never occur to Sahlins or anyone else to doubt the theory underlying buttresses simply on the grounds that trees lack the mathematical expertise to do the calculations. Why, then, raise the problem for the special case of kin-selected behaviour? It can't be because it is behaviour as opposed to anatomy, because there are plenty of other examples of behaviour (other than kin-selected behaviour, I mean) that Sahlins would cheerfully accept without raising his 'epistemological' objection; think, for instance, of my own illustration (pp. 124–5) of the complicated calculations that in some sense we all must do whenever we catch a ball. One cannot help wondering: are there social scientists who are quite happy with the theory of natural selection generally but who, for quite extraneous reasons that may have roots in the history of their subject, desperately want to find something—*anything*—wrong with the theory of *kin selection specifically*?

p. 128 …we have to think how animals might actually go about estimating who their close relations are… We know who our relations are because we are told…

The whole subject of kin recognition has taken off in a big way since this book was written. Animals, including ourselves, seem to show remarkably subtle abilities to discriminate relatives from nonrelatives, often by smell. A recent book, *Kin Recognition in Animals*, summarizes what is now known. The chapter on humans by Pamela Wells shows that the statement above ('We know who our relations are because we are told') needs to be supplemented: there is at least circumstantial evidence that we are capable of using various nonverbal cues, including the smell of our relatives' sweat. The whole subject is, for me, epitomized by the quotation with which she begins:

> all good kumrads you can tell
> by their altruistic smell
> e. e. cummings

Relatives might need to recognize one another for reasons other than altruism. They might also want to strike a balance between outbreeding and inbreeding, as we shall see in the next note.

p. 129 …the injurious effects of recessive genes which appear with inbreeding. (For some reason many anthropologists do not like this explanation.)

A lethal gene is one that kills its possessor. A recessive lethal, like any recessive gene, doesn't exert its effect unless it is in double dose. Recessive lethals get by in the gene pool, because most individuals possessing them have only one copy and therefore never suffer the effects. Any given lethal is rare, because if it ever gets common it meets copies of itself and kills off its carriers. There could nevertheless be lots of different types of lethal, so we could still all be riddled with them. Estimates vary as to how many different ones there are lurking in the human gene pool. Some books reckon as many as two lethals, on average, per person. If a random male mates with a random female, the chances are that his lethals will not match hers and their children will not suffer. But if a brother mates with a sister, or a father with a daughter, things are ominously different. However rare my lethal recessives may be in the population at large, and however rare my sister's lethal recessives may be in the population at large, there is a disquietingly high chance that hers and mine are the same. If you do the sums, it turns out that, for every lethal recessive that I possess, if I mate with my sister one in eight of our offspring will be born dead or will die young. Incidentally, dying in adolescence is even more 'lethal', genetically speaking, than dying at birth: a stillborn child doesn't waste so much of the parents' vital time and energy. But, whichever way you look at it, close incest is not just mildly deleterious. It is potentially catastrophic. Selection for active incest-avoidance could be as strong as any selection pressure that has been measured in nature.

Anthropologists who object to Darwinian explanations of incest-avoidance perhaps do not realize what a strong Darwinian case they are opposing. Their arguments are sometimes so weak as to suggest desperate special pleading. They commonly say, for instance: 'If Darwinian selection had really built into us an instinctive revulsion against incest, we wouldn't need to forbid it. The taboo only grows up because people have incestuous lusts. So the rule against incest cannot have a "biological" function, it must be purely "social".' This objection is rather like the following: 'Cars don't need locks on the ignition switch because they have locks on the doors. Therefore ignition locks cannot be anti-theft devices; they must have some purely ritual significance!' Anthropologists are also fond of stressing that different cultures have different taboos, indeed different definitions of kinship. They seem to think that this, too, undermines Darwinian aspirations to explain incest-avoidance. But one might as well say that sexual desire cannot be a Darwinian adaptation because different cultures prefer to copulate in different positions. It seems to me highly plausible that incest-avoidance in humans, no less than in other animals, is the consequence of strong Darwinian selection.

Not only is it a bad thing to mate with those genetically too close to you. Too-distant outbreeding can also be bad because of genetic incompatibilities between different strains. Exactly where the ideal intermediate falls is not easy to predict. Should you mate with your first cousin? With your second or third cousin? Patrick Bateson has tried to ask Japanese quail where their own preferences lie along the spectrum. In an experimental setup called the Amsterdam Apparatus, birds were invited to choose among members of the opposite sex arrayed behind miniature shop-windows. They preferred first cousins over both full siblings and unrelated birds. Further experiments suggested that young quail learn the attributes of their clutch-companions, and then, later in life, tend to choose sexual partners that are quite like their clutch-mates but not too like them.

Quail, then, seem to avoid incest by their own internal lack of desire for those with whom they have grown up. Other animals do it by observing social laws, socially imposed rules of dispersal. Adolescent male lions, for instance, are driven out of the parental pride where female relatives remain to tempt them, and breed only if they manage to usurp another pride. In chimpanzee and gorilla societies it tends to be the young females who leave to seek mates in other bands. Both dispersal patterns, as well as the quail's system, are to be found among the various cultures of our own species.

p. 134 *Since [cuckoo hosts] are not in danger of being parasitized by members of their own species...*

This is probably true of most species of birds. Nevertheless, we should not be surprised to find some birds parasitizing nests of their own species. And the phenomenon is, indeed, being found in an increasing number of species. This is especially so now that new molecular techniques are coming in for establishing who is related to whom. Actually, the selfish gene theory might expect it to happen even more often than we so far know.

p. 136 *Kin selection in lions*

Bertram's emphasis on kin selection as the prime mover of cooperation in lions has been challenged by C. Packer and A. Pusey. They claim that in many prides the two male lions are not related. Packer and Pusey suggest that reciprocal altruism is at least as likely as kin selection as an explanation for cooperation in lions. Probably both sides are right. Chapter 12 emphasizes that reciprocation ('Tit for Tat') can evolve only if a critical quorum of reciprocators can initially be mustered. This ensures that a would-be partner has a decent chance of being a reciprocator. Kinship is perhaps the most obvious way for this to happen. Relatives naturally tend to resemble one another, so even if the critical frequency is not met in the population at large it may be met within the family. Perhaps cooperation in lions got its start through the kin-effects suggested by Bertram, and this provided the necessary conditions for reciprocation to be favoured. The disagreement over lions can be settled only by facts, and facts, as ever, tell us only about the particular case, not the general theoretical argument.

p. 137 If C is my identical twin…

It is now widely understood that an identical twin is theoretically as valuable to you as you are to yourself—as long as the twin really is guaranteed identical. What is not so widely understood is that the same is true of a guaranteed monogamous mother. If you know for certain that your mother will continue to produce your father's children and only your father's children, your mother is as genetically valuable to you as an identical twin, or as yourself. Think of yourself as an offspring-producing machine. Then your monogamous mother is a (full) sibling-producing machine, and full siblings are as genetically valuable to you as your own offspring. Of course this neglects all kinds of practical considerations. For instance, your mother is older than you, though whether this makes her a better or worse bet for future reproduction than you yourself depends on particular circumstances—we can't give a general rule.

That argument assumes that your mother can be relied upon to continue producing your father's children, as opposed to some other male's children. The extent to which she can be relied upon depends upon the mating system of the species. If you are a member of a habitually promiscuous species, you obviously cannot count on your mother's offspring being your full siblings. Even under ideally monogamous conditions, there is one apparently inescapable consideration that tends to make your mother a worse bet than you are yourself. Your father may die. With the best will in the world, if your father is dead your mother can hardly be expected to go on producing his children, can she?

Well, as a matter of fact she can. The circumstances under which this can happen are obviously of great interest for the theory of kin selection. As mammals we are used to the idea that birth follows copulation after a fixed and rather short interval. A human male can father young posthumously, but not after he has been dead longer than nine months (except with the aid of deep-freezing in a sperm-bank). But there are several groups of insects in which a female stores sperm inside herself for the whole of her life, eking it out to fertilize eggs as the years go by, often long years after the death of her mate. If you are a member of a species that does this, you can potentially be really very sure that your mother will continue to be a good 'genetic bet'. A female ant mates only in a single mating flight, early in her life. The female then loses her wings and never mates again. Admittedly, in many ant species the female mates with several males on her mating flight. But if you happen to belong to one of those species whose females are always monogamous, you really can regard your mother as at least as good a genetic bet as you are yourself. The great point about being a young ant, as opposed to a young mammal, is that it doesn't matter whether your father is dead (indeed, he almost certainly is dead!). You can be pretty sure that your father's sperm are living on after him, and that your mother can continue to make full siblings for you.

It follows that, if we are interested in the evolutionary origins of sibling care and of phenomena like the insect soldiers, we should look with special attention towards those species in which females store sperm for life. In the case of ants, bees, and wasps there is, as Chapter 10 discusses, a special genetic peculiarity—haplodiploidy—that may have predisposed them to become highly social. What I am arguing here is that haplodiploidy is not the only predisposing factor. The habit of lifetime sperm-storage may have been at least as important. Under ideal conditions it can make a mother as genetically valuable, and as worthy of 'altruistic' help, as an identical twin.

p. 138 ...social anthropologists might have interesting things to say.

This remark now makes me blush with embarrassment. I have since learned that social anthropologists not only have things to say about the 'mother's brother effect': many of them have for years spoken of little else! The effect that I 'predicted' is an empirical fact in a large number of cultures that has been well known to anthropologists for decades. Moreover, when I suggested the specific hypothesis that 'in a society with a high degree of marital infidelity, maternal uncles should be more altruistic than "fathers", since they have more grounds for confidence in their relatedness to the child' (p. 138), I regrettably overlooked the fact that Richard Alexander had already made the same suggestion (a footnote acknowledging this was inserted in later printings of the first edition of this book). The hypothesis has been tested, by Alexander himself among others, using quantitative counts from the anthropological literature, with favourable results.

CHAPTER 7

Family planning

p. 142 *Wynne-Edwards... has been mainly responsible for promulgating the idea of group selection.*

Wynne-Edwards is generally treated more kindly than academic heretics often are. By being wrong in an unequivocal way, he is widely credited (though I personally think this point is rather overdone) with having provoked people into thinking more clearly about selection. He himself made a magnanimous recantation in 1978, when he wrote:

> The general consensus of theoretical biologists at present is that credible models cannot be devised, by which the slow march of group selection could overtake the much faster spread of selfish genes that bring gains in individual fitness. I therefore accept their opinion.

Magnanimous these second thoughts may have been, but unfortunately he has had third ones: his latest book re-recants.

Group selection, in the sense in which we have all long understood it, is even more out of favour among biologists than it was when my first edition was published. You could be forgiven for thinking the opposite: a generation has grown up, especially in America, that scatters the name 'group selection' around like confetti. It is littered over all kinds of cases that used to be (and by the rest of us still are) clearly and straightforwardly understood as something else, say kin selection. I suppose it is futile to become too annoyed by such semantic parvenus. Nevertheless, the whole issue of group selection was very satisfactorily settled a decade ago by John Maynard Smith and others, and it is irritating to find that we are now two generations, as well as two nations, divided only by a common language. It is particularly unfortunate that philosophers, now belatedly entering the field, have started out muddled by this recent caprice of terminology. I recommend Alan Grafen's essay 'Natural

Selection, Kin Selection and Group Selection' as a clear-thinking, and I hope now definitive, sorting out of the neo-group-selection problem.

Battle of the generations

p. 160 R. L. Trivers, in 1972, neatly solved the problem...

Robert Trivers, whose papers of the early 1970s were among the most important inspirations for me in writing the first edition of this book, and whose ideas especially dominate Chapter 8, has finally produced his own book, *Social Evolution*. I recommend it, not only for its content but for its style: clear-thinking, academically correct but with just enough anthropomorphic irresponsibility to tease the pompous, and spiced with personal autobiographical asides. I cannot resist quoting one of these: it is so characteristic. Trivers is describing his excitement on observing the relationship between two rival male baboons in Kenya: 'There was another reason for my excitement and this was an unconscious identification with Arthur. Arthur was a superb young male in his prime...' Trivers's new chapter on parent–offspring conflict brings the subject up to date. There is indeed rather little to add to his paper of 1974, apart from some new factual examples. The theory has stood the test of time. More detailed mathematical and genetic models have confirmed that Trivers's largely verbal arguments do indeed follow from currently accepted Darwinian theory.

p. 175 According to him the parent will always win.

Alexander has generously conceded, in his 1980 book *Darwinism and Human Affairs* (p. 39), that he was wrong to argue that parental victory in parent–offspring conflict follows inevitably from fundamental Darwinian assumptions. It now seems to me that his thesis, that parents enjoy an asymmetrical advantage over their offspring in the battle of the generations, could be bolstered by a different kind of argument, which I learnt from Eric Charnov.

Charnov was writing about social insects and the origins of sterile castes, but his argument applies more generally and I shall put it in general terms. Consider a young female of a monogamous species, not necessarily an insect, on the threshold of adult life. Her dilemma is whether to leave and try to reproduce on her own, or to stay in the parental nest and help rear her younger brothers and sisters. Because of the breeding habits of her species, she can be confident that her mother will go on giving her full brothers and sisters for a long time to come. By Hamilton's logic, these sibs are just as genetically 'valuable' to her as her own off-spring would be. As far as genetic relatedness is concerned, the young female will be indifferent between the two courses of action; she doesn't 'care' whether she goes or stays. But her parents will be far from indifferent to which she does. Looked at from the point of view of her old mother, the choice is between grandchildren or children. New children are twice as valuable, genetically speaking, as new grandchildren. If we speak of conflict between parents and offspring over whether the offspring leaves or stays and helps at the nest, Charnov's point is that the conflict is a walk-over for the parents for the very good reason that only the parents see it as a conflict at all!

It is a bit like a race between two athletes, where one has been offered £1,000 only if he wins, while his opponent has been promised £1,000 whether he wins or loses. We should expect that the first runner will try harder and that, if the two are otherwise evenly matched, he will proba-bly win. Actually Charnov's point is stronger than this analogy suggests, because the costs of running flat out are not so great as to deter many people, whether they are financially rewarded or not. Such Olympic ide-als are too much of a luxury for the Darwinian games: effort in one direction is always paid for as lost effort in another direction. It is as if the more effort you put into any one race, the less likely you are to win future races because of exhaustion.

Conditions will vary from species to species, so we can't always fore-cast the results of Darwinian games. Nevertheless, if we consider only closeness of genetic relatedness and assume a monogamous mating sys-tem (so that the daughter can be sure that her sibs are full sibs), we can expect an old mother to succeed in manipulating her young adult daugh-ter into staying and helping. The mother has everything to gain, while the daughter herself will have no inducement to resist her mother's manipulation because she is genetically indifferent between the avail-able choices.

Once again, it is important to stress that this has been an 'other things being equal' kind of argument. Even though other things will usually not be equal, Charnov's reasoning could still be useful to Alexander or anyone else advocating a parental manipulation theory. In any case, Alexander's practical arguments for expecting parental victory—parents being bigger, stronger, and so on—are well taken.

CHAPTER 9
Battle of the Sexes

p. 182 ... how much more severe must be the conflict between mates, who are not related to each other?

As so often, this opening sentence hides an implicit 'other things being equal'. Obviously mates are likely to have a great deal to gain from cooperation. This emerges again and again throughout the chapter. After all, mates are likely to be engaged in a nonzero sum game, a game in which both can increase their winnings by cooperating, rather than one's gain necessarily being the other's loss (I explain this idea in Chapter 12). This is one of the places in the book where my tone swung too far towards the cynical, selfish view of life. At the time it seemed necessary, since the dominant view of animal courtship had swung far in the other direction. Nearly universally, people had uncritically assumed that mates would cooperate unstintingly with each other. The possibility of exploitation wasn't even considered. In this historical context the apparent cynicism of my opening sentence is understandable, but today I would adopt a softer tone. Similarly, at the end of this chapter my remarks about human sexual roles now seem to me naïvely worded. Two books that go more thoroughly into the evolution of human sex differences are Martin Daly and Margo Wilson's *Sex, Evolution, and Behavior*, and Donald Symons's *The Evolution of Human Sexuality*.

p. 184 ... the number of children a male can have is virtually unlimited. Female exploitation begins here.

It now seems misleading to emphasize the disparity between sperm and egg size as the basis of sex roles. Even if one sperm is small and cheap, it is far from cheap to make millions of sperms and successfully inject them into a female against all the competition. I now prefer the following approach to explaining the fundamental asymmetry between males and females.

Suppose we start with two sexes that have none of the particular attributes of males and females. Call them by the neutral names A and B. All we need specify is that every mating has to be between an A and a B. Now, any animal, whether an A or a B, faces a trade-off. Time and effort devoted to fighting with rivals cannot be spent on rearing existing offspring, and vice versa. Any animal can be expected to balance its effort between these rival claims. The point I am about to come to is that the As may settle at a different balance from the Bs and that, once they do, there is likely to be an escalating disparity between them.

To see this, suppose that the two sexes, the As and the Bs, differ from one another, right from the start, in whether they can most influence their success by investing in children or by investing in fighting (I'll use 'fighting' to stand for all kinds of direct competition within one sex). Initially the difference between the sexes can be very slight, since my point will be that there is an inherent tendency for it to grow. Say the As start out with fighting making a greater contribution to their reproductive success than parental behaviour does; the Bs, on the other hand, start out with parental behaviour contributing slightly more than fighting to variation in *their* reproductive success. This means, for example, that although an A of course benefits from parental care, the difference between a successful carer and an unsuccessful carer among the As is smaller than the difference between a successful fighter and an unsuccessful fighter among the As. Among the Bs, just the reverse is true. So, for a given amount of effort, an A can do itself good by fighting, whereas a B is more likely to do itself good by shifting its effort away from fighting and towards parental care.

In subsequent generations, therefore, the As will fight a bit more than their parents, the Bs will fight a bit less and care a bit more than their parents. Now the difference between the best A and the worst A with respect to fighting will be even greater, the difference between the best A and the worst A with respect to caring will be even less. Therefore an A has even more to gain by putting its effort into fighting, even less to gain by putting its effort into caring. Exactly the opposite will be true of the Bs as the generations go by. The key idea here is that a small initial difference between the sexes can be self-enhancing: selection can start with an initial, slight difference and make it grow larger and larger, until the As become what we now call males, the Bs what we now call females. The initial difference can be small enough to arise at random. After all, the starting conditions of the two sexes are unlikely to be exactly identical.

As you will notice, this is rather like the theory, originating with Parker, Baker, and Smith and discussed on page 185, of the early separation of primitive gametes into sperms and eggs. The argument just given is more general. The separation into sperms and eggs is only one aspect of a more basic separation of sexual roles. Instead of treating the sperm–egg separation as primary, and tracing all the characteristic attributes of males and females back to it, we now have an argument that explains the sperm–egg separation and other aspects all in the same way. We have to assume only that there are two sexes who have to mate with each other; we need know nothing more about those sexes. Starting from this minimal assumption, we positively expect that, however equal the two sexes may be at the start, they will diverge into two sexes specializing in opposite and complementary reproductive techniques. The separation between sperms and eggs is a symptom of this more general separation, not the cause of it.

p. 196 Let us take Maynard Smith's method of analysing aggressive contests, and apply it to sex.

This idea of trying to find an evolutionarily stable mix of strategies within one sex, balanced by an evolutionarily stable mix of strategies in the other sex, has now been taken further by Maynard Smith himself and, independently but in a similar direction, by Alan Grafen and Richard Sibly. Grafen and Sibly's paper is the more technically advanced, Maynard Smith's the easier to explain in words. Briefly, he begins by considering two strategies, guard and desert, which can be adopted by either sex. As in my 'coy/fast and faithful/philanderer' model, the interesting question is: What combinations of strategies among males are stable against what combinations of strategies among females? The answer depends upon our assumption about the particular economic circumstances of the species. Interestingly, though, however much we vary the economic assumptions, we don't have a whole continuum of quantitatively varying stable outcomes. The model tends to home in one of only four stable outcomes. The four outcomes are named after animal species that exemplify them. There is the Duck (male deserts, female guards), the Stickleback (female deserts, male guards), the Fruit-fly (both desert), and the Gibbon (both guard).

And here is something even more interesting. Remember from Chapter 5 that ESS models can settle at either of two outcomes, both equally stable? Well, that is true of this Maynard Smith model, too. What is especially interesting is that particular pairs, as opposed to other pairs, of these outcomes are jointly stable under the same economic circumstances. For instance, under one range of circumstances, both Duck and Stickleback are stable. Which of the two actually arises depends upon luck or, more precisely, upon accidents of evolutionary history—initial conditions. Under another range of circumstances, both Gibbon and Fruit-fly are stable. Again, it is historical accident that determines which of the two occurs in any given species. But there are no circumstances in which Gibbon and Duck are jointly stable, no circumstances in which Duck and Fruit-fly are jointly stable. This 'stablemate' (to coin a double pun) analysis of congenial and uncongenial combinations of ESSs has interesting consequences for our reconstructions of evolutionary history. For instance, it leads us to expect that certain kinds of transitions between mating systems in evolutionary history will be probable, others improbable. Maynard Smith explores these historical networks in a brief survey of mating patterns throughout the animal kingdom, ending with the memorable rhetorical question: 'Why don't male mammals lactate?'

p. 198 ...*it can be shown that really there would be no oscillation. The system would converge to a stable state.*

I am sorry to say that this statement is wrong. It is wrong in an interesting way, however, so I have left the error in and shall now take some time to expose it. It is actually the same kind of error as Gale and Eaves spotted in Maynard Smith and Price's original paper (see note to page 97). My error was pointed out by two mathematical biologists working in Austria, P. Schuster and K. Sigmund.

I had correctly worked out the ratios of faithful to philanderer males, and of coy to fast females, at which the two kinds of males were equally successful, and the two kinds of females were equally successful. This is indeed an equilibrium, but I failed to check whether it was a *stable* equilibrium. It could have been a precarious knife-edge rather than a secure valley. In order to check for stability, we have to see what would happen if we perturb the equilibrium slightly (push a ball off a knife-edge and you lose it; push it away from the centre of a valley and it comes back). In my particular numerical example, the equilibrium ratio for males was faithful and philanderer. Now, what if by chance the proportion of philanderers in the population increases to a value slightly higher than equilibrium? In order for the equilibrium to qualify as stable and self-correcting, it is necessary that philanderers should immediately start doing slightly less well. Unfortunately, as Schuster and Sigmund showed, this is not what happens. On the contrary, philanderers start doing better! Their frequency in the population, then, far from being self-stabilizing, is self-enhancing. It increases—not for ever, but only up to a point. If you simulate the model dynamically on a computer, as I have now done, you get an endlessly repeating cycle. Ironically, this is precisely the cycle that I described hypothetically on pages 197–8, but I thought that I was doing it purely as an explanatory device, just as I had with hawks and doves. By analogy with hawks and doves I assumed, quite wrongly, that the cycle was hypothetical only, and that the system would really settle into a stable equilibrium. Schuster and Sigmund's parting-shot leaves no more to be said:

> Briefly, then, we can draw two conclusions:
>
> (a) that the battle of sexes has much in common with predation; and
> (b) that the behaviour of lovers is oscillating like the moon, and unpredictable as the weather.

Of course, people didn't need differential equations to notice this before.

p. 202 ...cases of paternal devotion...common among fish. Why?

Tamsin Carlisle's undergraduate hypothesis about fish has now been tested comparatively by Mark Ridley, in the course of an exhaustive review of paternal care in the entire animal kingdom. His paper is an astonishing *tour de force* which, like Carlisle's hypothesis itself, also began as an undergraduate essay written for me. Unfortunately, he did not find in favour of the hypothesis.

p. 206 ...a kind of unstable, runaway process.

R. A. Fisher's runaway theory of sexual selection, which he stated extremely briefly, has now been spelt out mathematically by R. Lande and others.

It has become a difficult subject, but it can be explained in nonmathematical terms provided sufficient space is given over to it. It does need a whole chapter, however, and I devoted one to it in *The Blind Watchmaker* (Chapter 8), so I'll say no more about it here.

Instead, I'll deal with one problem about sexual selection that I have never sufficiently emphasized in any of my books. How is the necessary variation maintained? Darwinian selection can function only if there is a good supply of genetic variation to work upon. If you try to breed, say, rabbits for ever longer ears, to begin with you'll succeed. The average rabbit in a wild population will have medium-sized ears (by rabbit standards; by our standards, of course, it will have very long ones). A few rabbits will have shorter than average ears and a few longer than average. By breeding only from those with the longest ears you'll succeed in increasing the average in later generations. For a while. But if you *continue* to breed from those with the longest ears there will come a time when the necessary variation is no longer available. They'll all have the 'longest' ears, and evolution will grind to a halt. In normal evolution this sort of thing is not a problem, because most environments don't carry on exerting consistent and unswerving pressure in one direction. The 'best' length for any particular bit of an animal will normally not be 'a bit longer than the present average, whatever the present average may be'. The best length is more likely to be a fixed quantity, say three inches. But sexual selection really can have the embarrassing property of chasing an ever-driving 'optimum'. Female fashion really could desire ever longer male ears, no matter how long the ears of the current population might already be. So variation really could seriously run out. And yet sexual selection does seem to have worked; we do see absurdly exaggerated male ornaments. We seem to have a paradox, which we may call the paradox of the vanishing variation.

Lande's solution to the paradox is mutation. There will always be enough mutation, he thinks, to fuel sustained selection. The reason people had doubted this before was that they thought in terms of one gene at a time: mutation rates at any one genetic locus are too low to resolve the paradox of vanishing variation. Lande reminded us that 'tails' and other things that sexual selection works on are influenced by an indefinitely large number of different genes—'polygenes'—whose small effects all add up. Moreover, as evolution goes on, it will be a shifting set of polygenes that are relevant: new genes will be recruited into the set that influence variation in 'tail length', and old ones lost. Mutation can affect any of this large and shifting set of genes, so the paradox of vanishing variation itself vanishes.

W. D. Hamilton's answer to the paradox is different. He answers it in the same way he answers most questions nowadays: 'Parasites'. Think back to the rabbit ears. The best length for rabbit ears presumably depends on various acoustic factors. There is no particular reason to expect these factors to change in a consistent and sustained direction as generations go by. The best length for rabbit ears may not be absolutely constant, but still selection is unlikely to push so far in any particular direction that it strays outside the range of variation easily thrown up by the present gene pool. Hence no paradox of vanishing variation.

But now look at the kind of violently fluctuating environment provided by parasites. In a world full of parasites there is strong selection in favour of ability to resist them. Natural selection will favour whichever individual rabbits are least vulnerable to the parasites that happen to be around. The crucial point is that these will not always be the same parasites. Plagues come and go. Today it may be myxomatosis, next year the rabbit equivalent of the Black Death, the year after that rabbit AIDS, and so on. Then, after say a ten-year cycle it may be back to myxomatosis, and so on. Or the myxomatosis virus itself may evolve to counter whatever counteradaptations the rabbits come up with. Hamilton pictures cycles of counteradaptation and counter-counteradaptation endlessly rolling through time and forever perversely updating the definition of the 'best' rabbit.

The upshot of all this is that there is something importantly different about adaptations for disease-resistance as compared with adaptations to the physical environment. Whereas there may be a pretty fixed 'best' length for a rabbit's legs to be, there is no fixed 'best' rabbit as far as disease-resistance is concerned. As the currently most dangerous disease changes, so the currently 'best' rabbit changes. Are parasites the only selective forces that work this way? What about predators and prey, for instance? Hamilton agrees that they are basically like parasites. But they don't evolve so fast as many parasites. And parasites are more likely than predators or prey to evolve detailed gene-for-gene counteradaptations.

Hamilton takes the cyclical challenges offered by parasites and makes them the basis for an altogether grander theory, his theory of why sex exists at all. But here we are concerned with his use of parasites to solve the paradox of vanishing variation in sexual selection. He believes that hereditary disease-resistance among males is the most important criterion by which females choose them. Disease is such a powerful scourge that females will benefit greatly from any ability they may have to diagnose it in potential mates. A female who behaves like a good diagnostic doctor and chooses only the healthiest male for mate will tend to gain healthy genes for her children. Now, because the definition of the 'best rabbit' is always changing, there will always be something important for females to choose between, when they look the males over. There will always be some 'good' males and some 'bad' males. They won't all become 'good' after generations of selection, because by then the parasites will have changed and so the definition of a 'good' rabbit will have changed. Genes for resisting one strain of myxoma virus will not be good at resisting the next strain of myxoma virus that mutates on to the scene. And so on, through indefinite cycles of evolving pestilence. Parasites never let up, so females cannot let up in their relentless search for healthy mates.

How will the males respond to being scrutinized by females acting as doctors? Will genes for faking good health be favoured? To begin with, maybe, but selection will then act on females to sharpen up their diagnostic skills and sort out the fakes from the really healthy. In the end, Hamilton believes, females will become such good doctors that males will be forced, if they advertise at all, to advertise honestly. If any sexual advertisement becomes exaggerated in males it will be because it is a genuine indicator of health. Males will evolve so as to make it easy for females to see that they are healthy—if they are. Genuinely healthy males will be pleased to advertise the fact. Unhealthy ones, of course, will not, but what can they do? If they don't at least *try* to display a health certificate, females will draw the worst conclusions. By the way, all this talk of doctors would be misleading if it suggested that females are interested in curing males. Their only interest is diagnosis, and it is not an altruistic interest. And I'm assuming that it is no longer necessary to apologize for metaphors like 'honesty' and 'drawing conclusions'.

To return to the point about advertising, it is as though males are forced by the females to evolve clinical thermometers permanently sticking out of their mouths, clearly displayed for females to read. What might these 'thermometers' be? Well, think of the spectacularly long tail of a male bird of paradise. We have already seen Fisher's elegant explanation for this elegant adornment. Hamilton's explanation is altogether more down-to-earth. A common symptom of disease in a bird is diarrhoea. If you have a long tail, diarrhoea is likely to mess it up. If you want to conceal the fact that you are suffering from diarrhoea, the best way to do it would be to avoid having a long tail. By the same token, if you want to advertise the fact that you are *not* suffering from diarrhoea, the best way to do so would be to have a very long tail. That way, the fact that your tail is clean will be the more conspicuous. If you don't have much of a tail at all, females cannot see whether it is clean or not, and will conclude the worst. Hamilton would not wish to commit himself to this *particular* explanation for bird of paradise tails, but it is a good example of the *kind* of explanation that he favours.

I used the simile of females acting as diagnostic doctors and males making their task easy by sporting 'thermometers' all over the place. Thinking about other diagnostic standbys of the doctor, the blood pressure meter and the stethoscope, led me to a couple of speculations about human sexual selection. I'll briefly present them, though I admit that I find them less plausible than pleasing. First, a theory about why humans have lost the penis bone. An erect human penis can be so hard and stiff that people jokingly express scepticism that there is no bone inside. As a matter of fact lots of mammals do have a stiffening bone, the baculum or os penis, to help the erection along. What's more, it is common among our relatives the primates; even our closest cousin the chimpanzee has one, although admittedly a very tiny one which may be on its evolutionary way out. There seems to have been a tendency to reduce the os penis in the primates; our species, along with a couple of monkey species, has lost it completely. So, we have got rid of the bone that in our ancestors presumably made it easy to have a nice stiff penis. Instead, we rely entirely on a hydraulic pumping system, which one cannot but feel is a costly and roundabout way of doing things. And, notoriously, erection can fail—unfortunate, to say the least, for the genetic success of a male in the wild. What is the obvious remedy? A bone in the penis, of course. So why don't we evolve one? For once, biologists of the 'genetic constraints' brigade cannot cop out with 'Oh, the necessary variation just couldn't arise.' Until recently our ancestors had precisely such a bone and we have actually gone out of our way to lose it! Why?

Erection in humans is accomplished purely by pressure of blood. It is unfortunately not plausible to suggest that erection hardness is the equivalent of a doctor's blood pressure meter used by females to gauge male health. But we are not tied to the metaphor of the blood pressure meter. If, for *whatever* reason, erection failure is a sensitive early warning of certain kinds of ill health, physical or mental, a version of the theory can work. All that females need is a dependable tool for diagnosis. Doctors don't use an erection test in routine health check-ups—they prefer to ask you to stick out your tongue. But erection failure is a known early warning of diabetes and certain neurological diseases. Far more commonly it results from psychological factors—depression, anxiety, stress, overwork, loss of confidence and all that. (In nature, one might imagine males low in the 'peck order' being afflicted in this way. Some monkeys use the erect penis as a threat signal.) It is not implausible that, with natural selection refining their diagnostic skills, females could glean all sorts of clues about a male's health, and the robustness of his ability to cope with stress, from the tone and bearing of his penis. But a bone would get in the way! Anybody can grow a bone in the penis; you don't have to be particularly healthy or tough. So selection pressure from females forced males to lose the os penis, because then only genuinely healthy or strong males could present a really stiff erection and the females could make an unobstructed diagnosis.

There is a possible zone of contention here. How, it might be said, were the females who imposed the selection supposed to know whether the stiffness that they felt was bone or hydraulic pressure? After all, we began with the observation that a human erection can feel like bone. But I doubt if the females were really that easily fooled. They too were under selection, in their case not to lose bone but to gain judgement. And don't forget, the female is exposed to the very same penis when it is not erect, and the contrast is extremely striking. Bones cannot detumesce (though admittedly they can be retracted). Perhaps it is the impressive double life of the penis that guarantees the authenticity of the hydraulic advertisement.

Now to the 'stethoscope'. Consider another notorious problem of the bedroom, snoring. Today it may be just a social inconvenience. Once upon a time it could have been life or death. In the depths of a quiet night snoring can be remarkably loud. It could summon predators from far and wide to the snorer and the group among whom he is lying. Why, then, do so many people do it? Imagine a sleeping band of our ancestors in some pleistocene cave, males snoring each on a different note, females kept awake with nothing to do but listen (I suppose it is true that males snore the more). Are the males providing the females with deliberately advertised and amplified stethoscopic information? Could the precise quality and timbre of your snore be diagnostic of the health of your respiratory tract? I don't mean to suggest that people snore only when they are ill. Rather, the snore is like a radio carrier-frequency, which drones on regardless; it is a clear signal which is *modulated*, in diagnostically sensitive ways, by the condition of the nose and throat. The idea of females preferring the clear trumpet note of unobstructed bronchi over virus-blown snorts is all very well, but I confess that I find it hard to imagine females positively going for a snorer at all. Still, personal intuition is notoriously unreliable. Perhaps at least this would make a research project for an insomniac doctor. Come to think of it, she might be in a good position to test the other theory as well.

These two speculations should not be taken too seriously. They will have succeeded if they bring home the principle of Hamilton's theory about how females try to choose healthy males. Perhaps the most interesting thing about them is that they point up the link between Hamilton's parasite theory and Amotz Zahavi's 'handicap' theory. If you follow through the logic of my penis hypothesis, males are handicapped by the loss of the bone and the handicap is not just incidental. The hydraulic advertisement gains its effectiveness precisely *because* erection sometimes fails. Darwinian readers will certainly have picked up this 'handicap' implication and it may have aroused in them grave suspicions. I ask them to suspend judgement until they have read the next note, on a new way of looking at the handicap principle itself.

p. 207 ... [Zahavi's]... maddeningly contrary 'handicap principle'

In the first edition I wrote: 'I do not believe this theory, although I am not quite so confident in my scepticism as I was when I first heard it.' I'm glad I added that 'although', because Zahavi's theory is now looking a lot more plausible than when I wrote the passage. Several respected theoreticians have recently started taking it seriously. Most worrying for me,

408 · ENDNOTES TO CHAPTER 9

these include my colleague Alan Grafen who, as has been said in print before, 'has the most annoying habit of always being right'. He has translated Zahavi's verbal ideas into a mathematical model and claims that it works. And that it is not a fancy, esoteric travesty of Zahavi such as others have played with, but a direct mathematical translation of Zahavi's idea itself. I shall discuss Grafen's original ESS version of his model, although he himself is now working on a full genetic version which will in some ways supersede the ESS model. This doesn't mean that the ESS model is actually wrong. It is a good approximation. Indeed, all ESS models, including the ones in this book, are approximations in the same sense.

The handicap principle is potentially relevant to all situations in which individuals try to judge the quality of other individuals, but we shall speak of males advertising to females. This is for the sake of clarity; it is one of those cases where the sexism of pronouns is actually useful. Grafen notes that there are at least four approaches to the handicap principle. These can be called the Qualifying Handicap (any male who has survived in spite of his handicap must be pretty good in other respects, so females choose him); the Revealing Handicap (males perform some onerous task in order to expose their otherwise concealed abilities); the Conditional Handicap (only high-quality males develop a handicap at all); and finally Grafen's preferred interpretation, which he calls the Strategic Choice Handicap (males have private information about their own quality, denied to females, and use this information to 'decide' whether to grow a handicap and how large it should be). Grafen's Strategic Choice Handicap interpretation lends itself to ESS analysis. There is no prior assumption that the advertisements that males adopt will be costly or handicapping. On the contrary, they are free to evolve any kind of advertisement, honest or dishonest, costly or cheap. But Grafen shows that, given this freedom to start with, a handicap system would be likely to emerge as evolutionarily stable.

Grafen's starting assumptions are the following four:

1. Males vary in real quality. Quality is not some vaguely snobbish idea like unthinking pride in one's old college or fraternity (I once received a letter from a reader which concluded: 'I hope you won't think this an arrogant letter, but after all I am a Balliol man'). Quality, for Grafen, means that there are such things as good males and bad males in the sense that females would benefit

genetically if they mated with good males and avoided bad ones. It means something like muscular strength, running speed, ability to find prey, ability to build good nests. We aren't talking about a male's final reproductive success, since this will be influenced by whether females choose him. To talk about that -at this point would be to beg the whole question; it is something that may or may not emerge from the model.

2. Females cannot perceive male quality directly but have to rely upon male advertising. At this stage we make no assumption about whether the advertisements are honest. Honesty is something else that may or may not emerge from the model; again that is what the model is for. A male could grow padded shoulders, for instance, to fake an illusion of size and strength. It is the business of the model to tell us whether such a fake signal will be evolutionarily stable, or whether natural selection will enforce decent, honest, and truthful advertising standards.

3. Unlike the females eyeing them, males do in a sense 'know' their own quality; and they adopt a 'strategy' for advertising, a rule for advertising conditionally in the light of their quality. As usual, by 'know' I don't mean cognitively know. But males are assumed to have genes that are switched on conditionally upon the male's own quality (and privileged access to this information is a not unreasonable assumption; a male's genes, after all, are immersed in his internal biochemistry and far better placed than female genes to respond to his quality). Different males adopt different rules. For instance, one male might follow the rule 'Display a tail whose size is proportional to my true quality'; another might follow the opposite rule. This gives natural selection a chance to adjust the rules by selecting among males that are genetically programmed to adopt different ones. The advertising level doesn't have to be directly proportional to the true quality; indeed a male could adopt an inverse rule. All that we require is that males should be programmed to adopt *some* kind of rule for 'looking at' their true quality and on the basis of this choosing an advertising level—size of tail, say, or of antlers. As to which of the possible rules will end up being evolutionarily stable, that again is something that the model aims to find out.

4. Females have a parallel freedom to evolve rules of their own. In their case the rules are about choosing males on the basis of the strength of the males' advertisement (remember that they, or rather their genes, lack the males' privileged view of the quality itself). For example, one female might adopt the rule: 'Believe the males totally.' Another female might adopt the

rule: 'Ignore male advertising totally.' Yet another, the rule: 'Assume the opposite of what the advertisement says.'

So, we have the idea of males varying in their rules for relating quality to advertising level; and females varying in their rules for relating mate choice to advertising level. In both cases the rules vary continuously and under genetic influence. So far in our discussion, males can choose any rule relating quality to advertisement, and females can choose any rule relating male advertisement to what they choose. Out of this spectrum of possible male and female rules, what we seek is an evolutionarily stable pair of rules. This is a bit like the 'faithful/philanderer and coy/fast' model in that we are looking for an evolutionarily stable male rule and an evolutionarily stable female rule, stability meaning mutual stability, each rule being stable in the presence of itself and the other. If we can find such an evolutionarily stable pair of rules we can examine them to see what life would be like in a society consisting of males and females playing by these rules. Specifically, would it be a Zahavian-handicap world?

Grafen set himself the task of finding such a mutually stable pair of rules. If I were to undertake this task, I should probably slog through a laborious computer simulation. I'd put into the computer a range of males, varying in their rule for relating quality to advertisement. And I'd also put in a range of females, varying in their rule for choosing males on the basis of the males' advertising levels. I'd then let the males and females rush around inside the computer, bumping into one another, mating if the female choice criterion is met, passing on their male and female rules to their sons and daughters. And of course individuals would survive or fail to survive as a result of their inherited 'quality'. As the generations go by, the changing fortunes of each of the male rules and each of the female rules would appear as changes in frequencies in the population. At intervals I'd look inside the computer to see if some kind of stable mix was brewing.

That method would work in principle, but it raises difficulties in practice. Fortunately, mathematicians can get to the same conclusion as a simulation would by setting up a couple of equations and solving them. This is what Grafen did. I shall not reproduce his mathematical reasoning nor spell out his further, more detailed, assumptions. Instead I shall go directly to the conclusion. He did indeed find an evolutionarily stable pair of rules.

So, to the big question. Does Grafen's ESS constitute the kind of world that Zahavi would recognize as a world of handicaps and honesty? The answer is yes. Grafen found that there can indeed be an evolutionarily stable world that combines the following Zahavian properties:

1. Despite having a free strategic choice of advertising level, males choose a level that correctly displays their true quality, even if this amounts to betraying that their true quality is low. At ESS, in other words, males are honest.

2. Despite having a free strategic choice of response to male advertisement, females end up choosing the strategy 'Believe the males'. At ESS, females are justifiably 'trusting'.

3. Advertising is costly. In other words, if we could somehow ignore the effects of quality and attractiveness, a male would be better off not advertising (thereby saving energy or being less conspicuous to predators). Not only is advertising costly; it is because of its costliness that a given advertising system is chosen. An advertising system is chosen precisely because it actually has the effect of reducing the success of the advertiser—all other things being held equal.

4. Advertising is more costly to worse males. The same level of advertising increases the risk for a puny male more than for a strong male. Low-quality males incur a more serious risk from costly advertising than high-quality males.

These properties, especially 3, are full-bloodedly Zahavian. Grafen's demonstration that they are evolutionarily stable under plausible conditions seems very convincing. But so also did the reasoning of Zahavi's critics who influenced the first edition of this book and who concluded that Zahavi's ideas could not work in evolution. We should not be happy with Grafen's conclusions until we have satisfied ourselves that we understand where—if anywhere—those earlier critics went wrong. What did they assume that led them to a different conclusion? Part of the answer seems to be that they did not allow their hypothetical animals a choice from a continuous range of strategies. This often meant that they were interpreting Zahavi's verbal ideas in one or other of the first three kinds of interpretation listed by Grafen—the Qualifying Handicap, the Revealing Handicap, or the Conditional Handicap. They did not consider any version of the fourth interpretation, the Strategic Choice Handicap. The result was either that they couldn't make the handicap principle work at all or that it worked but only under special, mathematically abstract conditions, which did not have the full Zahavian paradoxical feel to them. Moreover, an essential feature of the Strategic Choice interpretation of the handicap principle is that at ESS high-quality individuals and low-quality individuals are all playing the same strategy: 'Advertise honestly'. Earlier modellers assumed that high-quality males played different strategies from low-quality males, and hence developed different advertisements. Grafen, in contrast, assumes that, at ESS, differences between high- and low-quality signallers emerge because they are all playing the same strategy—and the differences in their advertisements emerge because their differences in quality are being faithfully rendered by the signalling rule.

We always admitted that signals as a matter of fact can be handicaps. We always understood that extreme handicaps could evolve, especially as a result of sexual selection, *in spite of* the fact that they were handicaps. The part of the Zahavi theory that we all objected to was the idea that signals might be favoured by selection precisely *because* they were handicaps to the signallers. It is this that Alan Grafen has apparently vindicated.

If Grafen is correct—and I think he is—it is a result of considerable importance for the whole study of animal signals. It might even necessitate a radical change in our entire outlook on the evolution of behaviour, a radical change in our view of many of the issues discussed in this book. Sexual advertisement is only one kind of advertisement. The Zahavi–Grafen theory, if true, will turn topsy-turvy biologists' ideas of relations between rivals of the same sex, between parents and offspring, between enemies of different species. I find the prospect rather worrying, because it means that theories of almost limitless craziness can no longer be ruled out on commonsense grounds. If we observe an animal doing something really silly, like standing on its head instead of running away from a lion, it may be doing it in order to show off to a female. It may even be showing off to the lion: 'I am such a high-quality animal you would be wasting your time trying to catch me' (see p. 222).

But, no matter how crazy I think something is, natural selection may have other ideas. An animal will turn back-somersaults in front of a slavering pack of predators if the risks enhance the advertisement more than they endanger the advertiser. It is its very dangerousness that gives the gesture showing-off power. Of course, natural selection won't favour infinite danger. At the point where exhibitionism becomes downright foolhardy, it will be penalized. A risky or costly performance may look crazy to us. But it really isn't any of our business. Natural selection alone is entitled to judge.

CHAPTER 10

You scratch my back, I'll ride on yours

p. 225 ... it seems to be only in the social insects that [the evolution of sterile workers] has actually happened.

That is what we all thought. We had reckoned without naked mole rats. Naked mole rats are a species of hairless, nearly blind little rodents that live in large underground colonies in dry areas of Kenya, Somalia, and Ethiopia. They appear to be truly 'social insects' of the mammal world. Jennifer Jarvis's pioneering studies of captive colonies at the University of Capetown have now been extended by the field observations of Robert Brett in Kenya; further studies of captive colonies are being made in America by Richard Alexander and Paul Sherman. These four workers have promised a joint book, and I, for one, eagerly await it. Meanwhile, this account is based upon reading what few papers have been published and listening to research lectures by Paul Sherman and Robert Brett. I was also privileged to be shown the London Zoo's colony of naked mole rats by the then Curator of Mammals, Brian Bertram.

Naked mole rats live in extensive networks of underground burrows. Colonies typically number 70 or 80 individuals, but they can increase into the hundreds. The network of burrows occupied by one colony can be two or three miles in total length, and one colony may excavate three or four tons of soil annually. Tunnelling is a communal activity. A face worker digs at the front with its teeth, passing the soil back through a living conveyor belt, a seething, scuffling line of half a dozen little pink animals. From time to time the face-worker is relieved by one of the workers behind.

Only one female in the colony breeds, over a period of several years. Jarvis, in my opinion legitimately, adopts social insect terminology and calls her the queen. The queen is mated by two or three males only.

All the other individuals of both sexes are nonbreeding, like insect workers. And, as in many social insect species, if the queen is removed some previously sterile females start to come into breeding condition and then fight each other for the position of queen.

The sterile individuals are called 'workers', and again this is fair enough. Workers are of both sexes, as in termites (but not ants, bees, and wasps, among which they are females only). What mole rat workers actually do depends on their size. The smallest ones, whom Jarvis calls 'frequent workers', dig and transport soil, feed the young, and presumably free the queen to concentrate on childbearing. She has larger litters than is normal for rodents of her size, again reminiscent of social insect queens. The largest nonbreeders seem to do little except sleep and eat, while intermediate-sized nonbreeders behave in an intermediate manner: there is a continuum as in bees, rather than discrete castes as in many ants.

Jarvis originally called the largest nonbreeders nonworkers. But could they really be doing nothing? There is now some suggestion, both from laboratory and field observations, that they are soldiers, defending the colony if it is threatened; snakes are the main predators. There is also a possibility that they act as 'food vats' like 'honey-pot' ants (see p. 223). Mole rats are homocoprophagous, which is a polite way of saying that they eat one another's faeces (not exclusively: that would run foul of the laws of the universe). Perhaps the large individuals perform a valuable role by storing up their faeces in the body when food is plentiful, so that they can act as an emergency larder when food is scarce—a sort of constipated commissariat.

To me, the most puzzling feature of naked mole rats is that, although they are like social insects in so many ways, they seem to have no equivalent caste to the young winged reproductives of ants and termites. They have reproductive individuals, of course, but these don't start their careers by taking wing and dispersing their genes to new lands. As far as anyone knows, naked mole rat colonies just grow at the margins by expanding the subterranean burrow system. Apparently they don't throw off long-distance dispersing individuals, the equivalent of winged reproductives. This is so surprising to my Darwinian intuition that it is tempting to speculate. My hunch is that one day we shall discover a dispersal phase which has hitherto, for some reason, been overlooked. It is too much to hope that the dispersing individuals will literally sprout wings! But they might in various ways be equipped for life above ground rather than underground. They could be hairy instead of naked, for instance. Naked mole rats don't regulate their individual body temperatures in the way that normal mammals do; they are more like 'cold-blooded' reptiles. Perhaps they control temperature socially—another resemblance to termites and bees. Or could they be exploiting the well-known constant temperature of any good cellar? At all events, my hypothetical dispersing individuals might well, unlike the underground workers, be conventionally 'warm-blooded'. Is it conceivable that some already known hairy rodent, hitherto classified as an entirely different species, might turn out to be the lost caste of the naked mole rat?

There are, after all, precedents for this kind of thing. Locusts, for instance. Locusts are modified grasshoppers, and they normally live the solitary, cryptic, retiring life typical of a grasshopper. But under certain special conditions they change utterly—and terribly. They lose their camouflage and become vividly striped. One could almost fancy it a warning. If so, it is no idle one, for their behaviour changes too. They abandon their solitary ways and gang together, with menacing results. From the legendary biblical plagues to the present day, no animal has been so feared as a destroyer of human prosperity. They swarm in their millions, a combined harvester thrashing a path tens of miles wide, sometimes travelling at hundreds of miles per day, engulfing 2,000 tons of crops per day, and leaving a wake of starvation and ruin. And now we come to the possible analogy with mole rats. The difference between a solitary individual and its gregarious incarnation is as great as the difference between two ant castes. Moreover, just as we were postulating for the 'lost caste' of the mole rats, until 1921 the grasshopper Jekylls and their locust Hydes were classified as belonging to different species.

But alas, it doesn't seem terribly likely that mammal experts could have been so misled right up to the present day. I should say, incidentally, that ordinary, untransformed naked mole rats are sometimes seen above ground and perhaps travel farther than is generally thought. But before we abandon the 'transformed reproductive' speculation completely, the locust analogy does suggest another possibility. Perhaps naked mole rats do produce transformed reproductives, but only under certain conditions—conditions that have not arisen in recent decades.

In Africa and the Middle East, locust plagues are still a menace, just as they were in biblical times. But in North America, things are different. Some grasshopper species there have the potential to turn into gregarious locusts. But, apparently because conditions haven't been right, no locust plagues have occurred in North America this century (although cicadas, a totally different kind of plague insect, still erupt regularly and, confusingly, they are called 'locusts' in colloquial American speech). Nevertheless, if a true locust plague were to occur in America today, it would not be particularly surprising: the volcano is not extinct; it is merely dormant. But if we didn't have written historical records and information from other parts of the world it *would* be a nasty surprise because the animals would be, as far as anyone knew, just ordinary, solitary, harmless grasshoppers. What if naked mole rats are like American grasshoppers, primed to produce a distinct, dispersing caste, but only under conditions which, for some reason, have not been realized this century? Nineteenth-century East Africa could have suffered swarming plagues of hairy mole rats migrating like lemmings above ground, without any records surviving to us. Or perhaps they *are* recorded in the legends and sagas of local tribes?

p. 228 *… a hymenopteran female is more closely related to her sisters than she is to her offspring.*

The memorable ingenuity of Hamilton's 'relatedness' hypothesis for the special case of the Hymenoptera has proved, paradoxically, an embarrassment for the reputation of his more general and fundamental theory. The haplodiploid relatedness story is just easy enough for anyone to understand with a little effort, but just difficult enough that one is pleased with oneself for understanding it, and anxious to pass it on to others. It is a good 'meme'. If you learn about Hamilton not from reading him, but from a conversation in a pub, the chances are very high that you'll hear about nothing except haplodiploidy. Nowadays every textbook of biology, no matter how briefly it covers kin selection, is almost bound to devote a paragraph to 'relatedness'. A colleague, who is now regarded as one of the world's experts on the social behaviour of large mammals, has confessed to me that for years he thought that Hamilton's theory of kin selection *was* the relatedness hypothesis and nothing more! The upshot of all this is that if some new facts lead us to doubt the importance of the relatedness hypothesis, people are apt to think that this is evidence against the whole theory of kin selection. It is as if a great composer were to write a long and profoundly original symphony, in which one particular tune, briefly tossed out in the middle, is so immediately catchy that every barrow-boy whistles it down the streets. The symphony becomes identified with this one tune. If people then become disenchanted with the tune, they think that they dislike the whole symphony.

Take, for example, an otherwise useful article by Linda Gamlin on naked mole rats recently published in the magazine *New Scientist*. It is seriously marred by the innuendo that naked mole rats and termites are in some way embarrassing for Hamilton's hypothesis, simply because they are not haplodiploid! It is hard to believe that the author could possibly have even seen Hamilton's classic pair of papers, since haplodiploidy occupies a mere four of the fifty pages. She must have relied on secondary sources—I hope not *The Selfish Gene*.

Another revealing example concerns the soldier aphids that I described in the notes to Chapter 6. As explained there, since aphids form clones of identical twins, altruistic self-sacrifice is very much to be expected among them. Hamilton noted this in 1964 and went to some trouble to explain away the awkward fact that—as far as was then known—clonal animals did not show any special tendency towards altruistic behaviour. The discovery of soldier aphids, when it came, could hardly have been more perfectly in tune with Hamilton's theory. Yet the original paper announcing that discovery treats soldier aphids as though they constituted a difficulty for Hamilton's theory, aphids not being haplodiploid! A nice irony.

When we turn to termites—also frequently regarded as an embarrassment for the Hamilton theory—the irony continues, for Hamilton himself, in 1972, was responsible for suggesting one of the most ingenious theories about why they became social, and it can be regarded as a clever analogy to the haplodiploidy hypothesis. This theory, the cyclic inbreeding theory, is commonly attributed to S. Bartz, who developed it seven years after Hamilton originally published it. Characteristically, Hamilton himself forgot that he had thought of the 'Bartz theory' first, and I had to thrust his own paper under his nose before he would believe it! Priority matters aside, the theory itself is so interesting that I am sorry I did not discuss it in the first edition. I shall correct the omission now.

I said that the theory was a clever analogue of the haplodiploidy hypothesis. What I meant was this. The essential feature of haplodiploid animals, from the point of view of social evolution, is that an individual can be genetically closer to her sibling than to her offspring. This predisposes her to stay behind in the parental nest and rear siblings rather than leaving the nest to bear and rear her own offspring. Hamilton thought of a reason why, in termites too, siblings might be genetically closer to each other than parents are to offspring. Inbreeding provides the clue. When animals mate with their siblings, the offspring that they produce become more genetically uniform. White rats, within any one laboratory strain, are genetically almost equivalent to identical twins. This is because they are born of a long line of brother–sister matings. Their genomes become highly homozygous, to use the technical term: at almost every one of their genetic loci the two genes are identical, and also identical to the genes at the same locus in all the other individuals in the strain. We don't often see long lines of incestuous matings in nature, but there is one significant exception—the termites!

A typical termite nest is founded by a royal pair, the king and queen, who then mate with each other exclusively until one of them dies. His or her place is then taken by one of their offspring who mates incestuously with the surviving parent. If both of the original royal couple die, they are replaced by an incestuous brother–sister couple. And so on. A mature colony is likely to have lost several kings and queens, and the progeny being turned out after some years are likely to be very inbred indeed, just like laboratory rats. The average homozygosity, and the average coefficient of relatedness, within a termite nest creeps up and up as the years go by and royal reproductives are successively replaced by their offspring or their siblings. But this is only the first step in Hamilton's argument. The ingenious part comes next.

The end-product product of any social insect colony is new, winged reproductives who fly out of the parent colony, mate, and found a new colony. When these new young kings and queens mate, the chances are that these matings will *not* be incestuous. Indeed, it looks as though there are special synchronizing conventions designed to see to it that different termite nests in an area all produce winged reproductives on the same day, presumably in order to foster outbreeding. So, consider the genetic consequences of a mating between a young king from colony A and a young queen from colony B. Both are highly inbred themselves. Both are the equivalent of inbred laboratory rats. But, since they are the product of different, *independent* programs of incestuous breeding, they will be genetically different from one another. They will be like inbred white rats belonging to different laboratory strains. When they mate with each other, their offspring will be highly *heterozygous*, but *uniformly* so. Heterozygous means that at many of the genetic loci the two genes are different from each other. Uniformly heterozygous means that almost every one of the offspring will be heterozygous in exactly the same way. They will be genetically almost identical to their siblings, but at the same time they will be highly heterozygous.

Now jump forward in time. The new colony with its founding royal pair has grown. It has become peopled by a large number of identically heterozygous young termites. Think about what will happen when one or both of the founding royal pair dies. That old incest cycle will begin again, with remarkable consequences. The first incestuously produced generation will be dramatically more variable than the previous generation. It doesn't matter whether we consider a brother–sister mating, a father–daughter mating or a mother–son mating. The principle is the same for all, but it is simplest to consider a brother–sister mating. If both brother and sister are identically heterozygous their offspring will be a highly variable mish-mash of genetic recombinations. This follows from elementary Mendelian genetics and would apply, in principle, to all animals and plants, not just termites. If you take uniformly heterozygous individuals and cross them, either with each other or with one of the homozygous parental strains, all hell breaks loose, genetically speaking. The reason can be looked up in any elementary textbook of genetics and I won't spell it out. From our present point of view, the important consequence is that during this stage of the development of a termite colony, an individual is typically closer, genetically, to its siblings than to its potential offspring. And this, as we saw in the case of the haplodiploid hymenoptera, is a likely precondition for the evolution of altruistically sterile worker castes.

But even where there is no special reason to expect individuals to be closer to their siblings than to their offspring, there is often good reason to expect individuals to be *as close* to their siblings as to their offspring. The only condition necessary for this to be true is some degree of monogamy. In a way, the surprising thing from Hamilton's point of view is that there are not more species in which sterile workers look after their younger brothers and sisters. What is widespread, as we are increasingly realizing, is a kind of watered-down version of the sterile worker phenomenon, known as 'helping at the nest'. In many species of birds and mammals, young adults, before moving out to start families of their own, remain with their parents for a season or two and help to rear their younger brothers and sisters. Copies of genes for doing this are passed on in the bodies of the brothers and sisters. Assuming that the beneficiaries are full (rather than half) brothers and sisters, each ounce of food invested in a sibling brings back just the same return on investment, genetically speaking, as it would if invested in a child. But that is only if all other things are equal. We must look to the inequalities if we are to explain why helping at the nest occurs in some species and not others.

Think, for instance, about a species of birds that nest in hollow trees. These trees are precious, for only a limited supply is available. If you are a young adult whose parents are still alive, they are probably in possession of one of the few available hollow trees (they must have possessed one at least until recently; otherwise, you wouldn't exist). So you are probably living in a hollow tree that is a thriving going concern, and the new baby occupants of this productive hatchery are your full brothers and sisters, genetically as close to you as your own offspring would be. If you leave and try to go it alone, your chances of obtaining a hollow tree are low. Even if you succeed, the offspring that you rear will be no closer to you, genetically, than brothers and sisters. A given quantity of effort invested in your parents' hollow tree is better value than the same quantity of effort invested in trying to set up on your own. These conditions, then, might favour sibling care—'helping at the nest'.

In spite of all this, it remains true that some individuals—or all individuals some of the time—must go out and seek new hollow trees, or whatever the equivalent is for their species. To use the 'bearing and caring' terminology of Chapter 7, *somebody* has to do some bearing; otherwise, there would be no young to care for! The point here is not that 'otherwise, the species would go extinct'. Rather, in any population dominated by genes for pure caring, genes for bearing will tend to have an advantage. In social insects the bearer role is filled by the queens and males. They are the ones that go out into the world, looking for new 'hollow trees', and that is why they are winged, even in ants, whose workers are wingless. These reproductive castes are specialized for their whole lifetime. Birds and mammals that help at the nest do it the other way. Each individual spends part of its life (usually its first adult season or two) as a 'worker', helping to rear younger brothers and sisters, while for the remaining part of its life it aspires to be a 'reproductive'.

What about the naked mole rats described in the previous note? They exemplify the going concern or 'hollow tree' principle to perfection, though their going concern does not literally involve a hollow tree. The key to their story is probably the patchy distribution of their food supply underneath the savannah. They feed mainly on underground tubers. These tubers can be very large and very deeply buried. A single tuber of one such species can outweigh 1,000 mole rats and, once found, can last the colony for months or even years. But the problem is finding the tubers, for they are scattered randomly and sporadically throughout the savannah. For mole rats, a food source is difficult to find but well worth it once found. Robert Brett has calculated that a single mole rat, working on its own, would have to search so long to find a single tuber that it would wear its teeth out with digging. A large social colony, with its miles of busily patrolled burrows, is an efficient tuber-mine. Each individual is economically better off as part of a union of fellow miners.

A large burrow system, then, manned by dozens of cooperating workers, is a going concern just like our hypothetical 'hollow tree', only more so! Given that you live in a flourishing communal labyrinth, and given that your mother is still producing full brothers and sisters inside it, the inducement to leave and start a family of your own becomes very low indeed. Even if some of the young produced are only half-siblings, the 'going concern' argument can still be powerful enough to keep young adults at home.

p. 230 *They found a rather convincingly close fit to the 3: 1 female to male ratio predicted...*

Richard Alexander and Paul Sherman wrote a paper criticizing Trivers and Hare's methods and conclusion. They agreed that female-biased sex ratios are normal among social insects, but disputed the claim that there is a good fit to 3:1. They preferred an alternative explanation for the female-biased sex ratios, an explanation that, like Trivers and Hare's, was first suggested by Hamilton. I find Alexander and Sherman's reasoning quite persuasive, but confess to a gut feeling that a piece of work as beautiful as Trivers and Hare's cannot be all wrong.

Alan Grafen pointed out to me another and more worrying problem with the account of hymenopteran sex ratios given in the first edition of this book. I have explained his point in *The Extended Phenotype* (pp. 75–6). Here is a brief extract:

> The potential worker is *still* indifferent between rearing siblings and rearing offspring at any conceivable population sex ratio. Thus suppose the population sex ratio is female-biased, even suppose it conforms to Trivers and Hare's predicted 3:1. Since the worker is more closely related to her sister than to her brother or her offspring of either sex, it might seem that she would 'prefer' to rear siblings over offspring given such a female-biased sex ratio: is she not gaining mostly valuable sisters (plus only a few relatively worthless brothers) when she opts for siblings? But this reasoning neglects the relatively great reproductive value of males in such a population as a consequence of their rarity. The worker may not be closely related to each of her brothers, but if males are rare in the population as a whole each one of those brothers is correspondingly highly likely to be an ancestor of future generations.

p. 242 *If a population arrives at an ESS that drives it extinct, then it goes extinct, and that is just too bad.*

The distinguished philosopher the late J. L. Mackie has drawn attention to an interesting consequence of the fact that populations of my 'cheats' and 'grudgers' can be simultaneously stable. 'Just too bad' it may be if a population arrives at an ESS that drives it extinct; Mackie makes the additional point that some kinds of ESS are more likely to drive a population extinct than others. In this particular example, both Cheat and Grudger are evolutionarily stable: a population may stabilize at the Cheat equilibrium or at the Grudger equilibrium. Mackie's point is that populations that happen to stabilize at the Cheat equilibrium will be more likely subsequently to go extinct. There can therefore be a kind of higher-level, 'between ESS', selection in favour of reciprocal altruism. This can be developed into an argument in favour of a kind of group selection that, unlike most theories of group selection, might actually work. I have spelled out the argument in my paper 'In Defence of Selfish Genes'.

CHAPTER 11

Memes: the new replicators

p. 248 *I would put my money on one fundamental principle... all life evolves by the differential survival of replicating entities.*

My wager that all life, everywhere in the universe, would turn out to have evolved by Darwinian means has now been spelled out and justified more fully in my paper 'Universal Darwinism' and in the last chapter of *The Blind Watchmaker*. I show that all the alternatives to Darwinism that have ever been suggested are in principle incapable of doing the job of explaining the organized complexity of life. The argument is a general one, not based upon particular facts about life as we know it. As such it has been criticized by scientists pedestrian enough to think that slaving over a hot test tube (or cold muddy boot) is the only method of discovery in science. One critic complained that my argument was 'philosophical', as though that was sufficient condemnation. Philosophical or not, the fact is that neither he nor anybody else has found any flaw in what I said. And 'in principle' arguments such as mine, far from being irrelevant to the real world, can be *more* powerful than arguments based on particular factual research. My reasoning, if it is correct, tells us something important about life everywhere in the universe. Laboratory and field research can tell us only about life as we have sampled it here.

p. 249 Meme

The word meme seems to be turning out to be a good meme. It is now quite widely used and in 1988 it joined the official list of words being considered for future editions of Oxford English Dictionaries. This makes me the more anxious to repeat that my designs on human culture were modest almost to vanishing point. My true ambitions—and they are admittedly large—lead in another direction entirely. I want to claim almost limitless power for slightly inaccurate self-replicating entities, once they arise anywhere in the universe. This is because they tend to become the basis for Darwinian selection which, given enough generations, cumulatively builds systems of great complexity. I believe that, given the right conditions, replicators automatically band together to create systems, or machines, that carry them around and work to favour their continued replication. The first ten chapters of *The Selfish Gene* had concentrated exclusively on one kind of replicator, the gene. In discussing memes in the final chapter I was trying to make the case for replicators in general, and to show that genes were not the only members of that important class. Whether the milieu of human culture really does have what it takes to get a form of Darwinism going, I am not sure. But in any case that question is subsidiary to my concern. Chapter 11 will have succeeded if the reader closes the book with the feeling that DNA molecules are not the only entities that might form the basis for Darwinian evolution. My purpose was to cut the gene down to size, rather than to sculpt a grand theory of human culture.

p. 249 ...*memes should be regarded as living structures, not just metaphorically but technically.*

DNA is a self-replicating piece of hardware. Each piece has a particular structure, which is different from rival pieces of DNA. If memes in brains are analogous to genes they must be self-replicating brain structures, actual patterns of neuronal wiring-up that reconstitute themselves in one brain after another. I had always felt uneasy spelling this out aloud, because we know far less about brains than about genes, and are therefore necessarily vague about what such a brain structure might actually be. So I was relieved to receive recently a very interesting paper by Juan Delius of the University of Konstanz in Germany. Unlike me, Delius doesn't have to feel apologetic, because he is a distinguished brain scientist whereas I am not a brain scientist at all. I am delighted, therefore, that he is bold enough to ram home the point by actually publishing a detailed picture of what the neuronal hardware of a meme might look like. Among the other interesting things he does is to explore, far more searchingly than I had done, the analogy of memes with parasites; to be more precise, with the spectrum of which malignant parasites are one extreme, benign 'symbionts' the other extreme. I am particularly keen on this approach because of my own interest in 'extended phenotypic' effects of parasite genes on host behaviour (see

Chapter 13 of this book and especially Chapter 12 of *The Extended Phenotype*). Delius, by the way, emphasizes the clear separation between memes and their ('phenotypic') effects. And he reiterates the importance of coadapted meme complexes, in which memes are selected for their mutual compatibility.

p. 252 'Auld Lang Syne'

'Auld Lang Syne' was, unwittingly, a revealingly fortunate example for me to have chosen. This is because, almost universally, it is rendered with an error, a mutation. The refrain is, essentially always nowadays, sung as 'For the sake of auld lang syne', whereas Burns actually wrote 'For auld lang syne'. A memically minded Darwinian immediately wonders what has been the 'survival value' of the interpolated phrase, 'the sake of'. Remember that we are not looking for ways in which *people* might have survived better through singing the song in altered form. We are looking for ways in which the alteration *itself* might have been good at surviving in the meme pool. Everybody learns the song in childhood, not through reading Burns but through hearing it sung on New Year's Eve. Once upon a time, presumably, everybody sang the correct words. 'For the sake of' must have arisen as a rare mutation. Our question is: Why has the initially rare mutation spread so insidiously that it has become the norm in the meme pool?

I don't think the answer is far to seek. The sibilant 's' is notoriously obtrusive. Church choirs are drilled to pronounce 's' sounds as lightly as possible, otherwise the whole church echoes with hissing. A murmuring priest at the altar of a great cathedral can sometimes be heard, from the back of the nave, only as a sporadic sussuration of 's's. The other consonant in 'sake', 'k', is almost as penetrating. Imagine that nineteen people are correctly singing 'For auld lang syne' and one person, somewhere in the room, slips in the erroneous 'For the sake of auld lang syne'. A child, hearing the song for the first time, is eager to join in but uncertain of the words. Although almost everybody is singing 'For auld lang syne', the hiss of an 's' and the cut of a 'k' force their way into the child's ears, and when the refrain comes round again he too sings 'For the sake of auld lang syne'. The mutant meme has taken over another vehicle. If there are any other children there, or adults unconfident of the words, they will be more likely to switch to the mutant form next time the refrain comes round. It is not that they 'prefer' the mutant form. They genuinely don't know the words and are honestly eager to learn them. Even if those who know better indignantly bellow 'For auld lang syne' at the top of their voice (as I do!), the correct words happen to have no conspicuous consonants, and the mutant form, even if quietly and diffidently sung, is far easier to hear.

A similar case is 'Rule Britannia'. The correct second line of the chorus is 'Britannia, rule the waves.' It is frequently, though not quite universally, sung as 'Britannia rules the waves.' Here the insistently hissing 's' of the meme is aided by an additional factor. The intended meaning of the poet (James Thompson) was presumably imperative (Britannia, go out and rule the waves!) or possibly subjunctive (let Britannia rule the waves). But it is superficially easier to misunderstand the sentence as indicative (Britannia, as a matter of fact, does rule the waves). This mutant meme, then, has two separate survival values over the original form that it replaced: it sounds more conspicuous and it is easier to understand.

The final test of a hypothesis should be experimental. It should be possible to inject the hissing meme, deliberately, into the meme pool at a very low frequency, and then watch it spread because of its own survival value. What if just a few of us were to start singing 'God saves our gracious Queen'?

p. 252 If the meme is a scientific idea, its spread will depend on how acceptable it is to the population of individual scientists; a rough measure of its survival value could be obtained by counting the number of times it is referred to in successive years in scientific journals.

I'd hate it if this were taken to mean that 'catchiness' was the only criterion for acceptance of a scientific idea. After all, some scientific ideas are actually right, others wrong! Their rightness and wrongness can be tested; their logic can be dissected. They are really not like pop-tunes, religions, or punk hairdos. Nevertheless there is a sociology as well as a logic to science. Some bad scientific ideas can spread widely, at least for a while. And some good ideas lie dormant for years before finally catching on and colonizing scientific imaginations.

We can find a prime example of this dormancy followed by rampant propagation in one of the main ideas in this book, Hamilton's theory of

Figure 1. Yearly citations of Hamilton (1964) in the *Science Citation Index*

kin selection. I thought it would be a fitting case for trying out the idea of measuring meme spread by counting journal references. In the first edition I noted (p. 116) that 'His two papers of 1964 are among the most important contributions to social ethology ever written, and I have never been able to understand why they have been so neglected by ethologists (his name does not even appear in the index of two major text-books of ethology, both published in 1970). Fortunately there are recent signs of a revival of interest in his ideas.' I wrote that in 1976. Let us trace the course of that memic revival over the subsequent decade.

Science Citation Index is a rather strange publication in which one may look up any published paper and see tabulated, for a given year, the number of subsequent publications that have quoted it. It is intended as an aid to tracking down the literature on a given topic. University appointments committees have picked up the habit of using it as a rough and ready (too rough and too ready) way of comparing the scientific achievements of applicants for jobs. By counting the citations of Hamilton's papers, in each year since 1964, we can approximately track the progress of his ideas into the consciousness of biologists (Figure 1). The initial dormancy is very evident. Then it looks as though there is a dramatic upturn in interest in kin selection during the 1970s. If there is any point where the upward trend begins, it seems to be between 1973 and 1974. The upturn then gathers pace up to a peak in 1981, after which the annual rate of citation fluctuates irregularly about a plateau.

A memic myth has grown up that the upsurge of interest in kin selection was all triggered by books published in 1975 and 1976. The graph, with its upturn in 1974, seems to give the lie to this idea. On the contrary, the evidence could be used to support a very different hypothesis, namely that we are dealing with one of those ideas that was 'in the air', 'whose time had come'. Those mid-seventies books would, on this view, be symptoms of the bandwagon effect rather than prime causes of it.

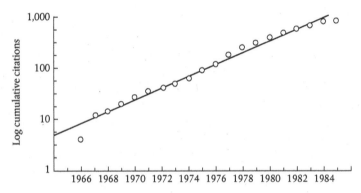

Figure 2. Log cumulative citations of Hamilton (1964)

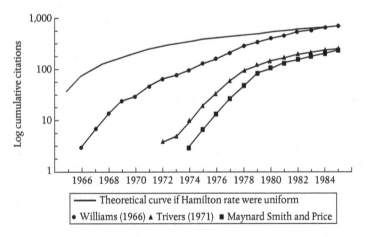

Figure 3. Log cumulative citations of three works not by Hamilton, compared with 'theoretical' curve for Hamilton (details explained in text)

Perhaps, indeed, we are dealing with a longer-term, slow-starting, exponentially accelerating bandwagon that began much earlier. One way of testing this simple, exponential hypothesis is to plot the citations cumulatively on a *logarithmic* scale. Any growth process, where rate of growth is proportional to size already attained, is called exponential growth. A typical exponential process is an epidemic: each person breathes the virus on several other people, each of whom in turn breathes on the same number again, so the number of victims grows at an ever increasing rate. It is diagnostic of an exponential curve that it becomes a straight line when plotted on a logarithmic scale. It is not necessary, but it is convenient and conventional, to plot such logarithmic graphs cumulatively. If the spread of Hamilton's meme was really like a gathering epidemic, the points on a cumulative logarithmic graph should fall on a single straight line. Do they?

The particular line drawn in Figure 2 is the straight line that, statistically speaking, best fits all the points. The apparent sharp rise between 1966 and 1967 should probably be ignored as an unreliable small-numbers effect of the kind that logarithmic plotting would tend to exaggerate. Thereafter, the graph is not a bad approximation to a single straight line, although minor overlying patterns can also be discerned. If my exponential interpretation is accepted, what we are dealing with is a single slow-burning explosion of interest, running right through from 1967 to the late 1980s. Individual books and papers should be seen both as symptoms and as causes of this long-term trend.

Do not think, by the way, that this pattern of increase is somehow trivial, in the sense of being inevitable. Any cumulative curve would, of course, rise even if the rate of citations per year were constant. But on the logarithmic scale it would rise at a steadily slower rate: it would tail off. The thick line at the top of Figure 3 shows the *theoretical* curve that we would get if every year had a constant citation rate (equal to the actual average rate of Hamilton citations, of about 37 per year). This dying away *curve* can be compared directly with the observed *straight* line in Figure 2, which indicates an exponential rate of increase. We really do have a case of increase upon increase, not a steady rate of citation.

Secondly, one might be tempted to think that there is something, if not inevitable, at least trivially expected about an exponential increase. Isn't the whole rate of publication of scientific papers, and therefore opportunities to cite other papers, itself increasing exponentially? Perhaps the size of the scientific community is increasing exponentially. To show that there is something special about the Hamilton meme, the easiest way is to plot the same kind of graph for some other papers. Figure 3 also shows the log cumulative citation frequencies of three other works (which incidentally were also highly influential on the first edition of this book). These are Williams's 1966 book, *Adaptation and Natural Selection*; Trivers's 1971 paper on reciprocal altruism; and Maynard Smith and Price's 1973 paper introducing the ESS idea. All three of them show curves that clearly are not exponential over the whole time-span. For these works too, however, the annual citation rates are far from uniform, and over part of their range they may even be exponential. The Williams curve, for instance, is approximately a straight line on the log scale from about 1970 onwards, suggesting that it, too entered an explosive phase of influence.

I have been downplaying the influence of particular books in spreading the Hamilton meme. Nevertheless, there is one apparently suggestive postscript to this little piece of memic analysis. As in the case of 'Auld Lang Syne' and 'Rule Britannia', we have an illuminating mutant error. The correct title of Hamilton's 1964 pair of papers was 'The Genetical Evolution of Social Behaviour'. In the mid to late 1970s, a rash of publications, *Sociobiology* and *The Selfish Gene* among them, mistakenly cited it as 'The Genetical Theory of Social Behaviour'. Jon Seger and Paul Harvey looked for the earliest occurrence of this mutant meme, thinking that it would be a neat marker, almost like a radioactive label, for tracing scientific influence. They traced it back to E. O. Wilson's influential book, *Sociobiology*, published in 1975, and even found some indirect evidence for this suggested pedigree.

Much as I admire Wilson's *tour de force*—I wish people would read it more and read about it less—my hackles have always risen at the entirely false suggestion that his book influenced mine. Yet, since my book also contained the mutant citation—the 'radioactive label'—it began to look alarmingly as though at least one meme had travelled from Wilson to me! This would not have been particularly surprising, since *Sociobiology* arrived in Britain just as I was completing *The Selfish Gene*, the very time when I would have been working on my bibliography. Wilson's massive bibliography would have seemed a godsend, saving hours in the library. My chagrin turned to glee, therefore, when I chanced upon an old stencilled bibliography that I had handed to the students at an Oxford lecture in 1970. Large as life, there was 'The Genetical *Theory* of Social Behaviour', a whole five years earlier than Wilson's publication. Wilson couldn't possibly have seen my 1970 bibliography. There was no doubt about it: Wilson and I had independently introduced the same mutant meme!

How could such a coincidence have happened? Once again, as in the case of 'Auld Lang Syne', a plausible explanation is not far to seek. R. A. Fisher's most famous book is called *The Genetical Theory of Natural Selection*. Such a household name has this title become in the world of evolutionary biologists, it is hard for us to hear its first two words without automatically adding the third. I suspect that both Wilson and I must have done just that. This is a happy conclusion for all concerned, since nobody minds admitting to being influenced by Fisher!

p. 255 *The computers in which memes live are human brains.*

It was obviously predictable that manufactured electronic computers, too, would eventually play host to self-replicating patterns of information—memes. Computers are increasingly tied together in intricate networks of shared information. Many of them are literally wired up together in electronic mail exchange. Others share information when their owners pass floppy discs around. It is a perfect milieu for self-replicating programs to flourish and spread. When I wrote the first edition of this book I was naïve enough to suppose that an undesirable computer meme would have to arise by a spontaneous error in the copying of a legitimate program, and I thought this an unlikely event. Alas, that was a time of innocence. Epidemics of 'viruses' and 'worms', deliberately released by malicious programmers, are now familiar hazards to computer-users all over the world. My own hard disc has to my knowledge been infected in two different virus epidemics during the past year, and that is a fairly typical experience among heavy computer-users. I shall not mention the names of particular viruses for fear of giving any nasty little satisfaction to their nasty little perpetrators. I say 'nasty', because their behaviour seems to me morally indistinguishable from that of a technician in a microbiology laboratory, who deliberately infects the drinking water and seeds epidemics in order to snigger at people getting ill. I say 'little', because these people are mentally little. There is nothing clever about designing a computer virus. Any half-way competent programmer could do it, and half-way competent programmers are two-a-penny in the modern world. I'm one myself. I shan't even bother to explain how computer viruses work. It's only too obvious.

What is less easy is to know how to combat them. Unfortunately some very expert programmers have had to waste their valuable time writing virus-detector programs, immunization programs, and so on (the analogy with medical vaccination, by the way, is astonishingly close, even down to the injection of a 'weakened strain' of the virus). The danger is that an arms race will develop, with each advance in virus-prevention being matched by counter-advances in new virus programs. So far, most anti-virus programs are written by altruists and supplied free of charge as a service. But I foresee the growth of a whole new profession—splitting into lucrative specialisms just like any other profession—of 'software doctors', on call with black bags full of diagnostic and curative floppy discs. I use the name 'doctors', but real doctors are solving natural problems that are not deliberately engineered by human malice. My software doctors, on the other hand, will be, like lawyers, solving man-made problems that should never have existed in the first place. In so far as virus-makers have any discernible motive, they presumably feel vaguely anarchistic. I appeal to them: do you really want to pave the way for a new fat-cat profession? If not, stop playing at silly memes, and put your modest programming talents to better use.

p. 257 *Blind faith can justify anything.*

I have had the predictable spate of letters from faith's victims, protesting about my criticisms of it. Faith is such a successful brainwasher in its own favour, especially a brainwasher of children, that it is hard to break its hold. But what, after all, is faith? It is a state of mind that leads people to believe something—it doesn't matter what—in the total absence of supporting evidence. If there were good supporting evidence then faith would be superfluous, for the evidence would compel us to believe it anyway. It is this that makes the often-parroted claim that 'evolution itself is a matter of faith' so silly. People believe in evolution not because they arbitrarily want to believe it but because of overwhelming, publicly available evidence.

I said 'it doesn't matter what' the faithful believe, which suggests that people have faith in entirely daft, arbitrary things, like the electric monk in Douglas Adams's delightful *Dirk Gently's Holistic Detective Agency*. He was purpose-built to do your believing for you, and very successful at it. On the day that we meet him he unshakeably believes, against all the evidence, that everything in the world is pink. I don't want to argue that the things in which a particular individual has faith are necessarily daft. They may or may not be. The point is that there is no way of deciding whether they are, and no way of preferring one article of faith over another, because evidence is explicitly eschewed. Indeed the fact that true faith doesn't need evidence is held up as its greatest virtue; this was the point of my quoting the story of Doubting Thomas, the only really admirable member of the twelve apostles.

Faith cannot move mountains (though generations of children are solemnly told the contrary and believe it). But it is capable of driving people to such dangerous folly that faith seems to me to qualify as a kind of mental illness. It leads people to believe in whatever it is so strongly that in extreme cases they are prepared to kill and to die for it without the need for further justification. Keith Henson has coined the name 'memeoids' for 'victims that have been taken over by a meme to the extent that their own survival becomes inconsequential...You see lots of these people on the evening news from such places as Belfast or Beirut.' Faith is powerful enough to immunize people against all appeals to pity, to forgiveness, to decent human feelings. It even immunizes them against fear, if they honestly believe that a martyr's death will send them straight to heaven. What a weapon! Religious faith deserves a chapter to itself in the annals of war technology, on an even footing with the longbow, the warhorse, the tank, and the hydrogen bomb.

p. 260 We, alone on earth, can rebel against the tyranny of the selfish replicators.

The optimistic tone of my conclusion has provoked scepticism among critics who feel that it is inconsistent with the rest of the book. In some cases the criticism comes from doctrinaire sociobiologists jealously protective of the importance of genetic influence. In other cases the criticism comes from a paradoxically opposite quarter, high priests of the left jealously protective of a favourite demonological icon! Rose, Kamin,

434 · ENDNOTES TO CHAPTER 11

and Lewontin in *Not in Our Genes* have a private bogey called 'reductionism'; and all the best reductionists are also supposed to be 'determinists', preferably 'genetic determinists'.

> Brains, for reductionists, are determinate biological objects whose properties produce the behaviors we observe and the states of thought or intention we infer from that behavior…Such a position is, or ought to be, completely in accord with the principles of sociobiology offered by Wilson and Dawkins. However, to adopt it would involve them in the dilemma of first arguing the innateness of much human behavior that, being liberal men, they clearly find unattractive (spite, indoctrination, etc.) and then to become entangled in liberal ethical concerns about responsibility for criminal acts, if these, like all other acts, are biologically determined. To avoid this problem, Wilson and Dawkins invoke a free will that enables us to go against the dictates of our genes if we so wish…This is essentially a return to unabashed Cartesianism, a dualistic *deus ex machina*.

I *think* that Rose and his colleagues are accusing us of eating our cake and having it. Either we must be 'genetic determinists' or we believe in 'free will'; we cannot have it both ways. But—and here I presume to speak for Professor Wilson as well as for myself—it is only in the eyes of Rose and his colleagues that we are 'genetic determinists'. What they don't understand (apparently, though it is hard to credit) is that it is perfectly possible to hold that genes exert a statistical influence on human behaviour while at the same time believing that this influence can be modified, overridden, or reversed by other influences. Genes must exert a stastical influence on any behaviour pattern that evolves by natural selection. Presumably Rose and his colleagues agree that human-sexual desire has evolved by natural selection, in the same sense as anything ever evolves by natural selection. They therefore must agree that there have been genes influencing sexual desire—in the same sense as genes ever influence anything. Yet they presumably have no trouble with curbing their sexual desires when it is socially necessary to do so. What is dualist about that? Obviously nothing. And no more is it dualist for me to advocate rebelling 'against the tyranny of the selfish replicators'. We, that is our brains, are separate and independent enough from our genes to rebel against them. As already noted, we do so in a small way every time we use contraception. There is no reason why we should not rebel in a large way, too.

BIBLIOGRAPHY

Not all the works listed here are mentioned by name in the book, but all of them are referred to by number in the index.

1. ALEXANDER, R. D. (1961) Aggressiveness, territoriality, and sexual behavior in field crickets. *Behaviour* **17**, 130–223.

2. ALEXANDER, R. D. (1974) The evolution of social behavior. *Annual Review of Ecology and Systematics* **5**, 325–83.

3. ALEXANDER, R. D. (1980) *Darwinism and Human Affairs*. London: Pitman.

4. ALEXANDER, R. D. (1987) *The Biology of Moral Systems*. New York: Aldine de Gruyter.

5. ALEXANDER, R. D. and SHERMAN, P. W. (1977) Local mate competition and parental investment in social insects. *Science* **96**, 494–500.

6. ALLEE, W. C. (1938) *The Social Life of Animals*. London: Heinemann.

7. ALTMANN, S. A. (1979) Altruistic behaviour: the fallacy of kin deployment. *Animal Behaviour* **27**, 958–9.

8. ALVAREZ, F., DE REYNA, A., and SEGURA, H. (1976) Experimental brood-parasitism of the magpie *(Pica pica)*. *Animal Behaviour* **24**, 907–16.

9. ANON. (1989) Hormones and brain structure explain behaviour. *New Scientist* **121** (1649), 35.

10. AOKI, S. (1987) Evolution of sterile soldiers in aphids. In *Animal Societies: Theories and facts* (eds. Y. ITO, J. L. BROWN, and J. KIKKAWA). Tokyo: Japan Scientific Societies Press. pp. 53–65.

11. ARDREY, R. (1970) *The Social Contract*. London: Collins.

12. AXELROD, R. (1984) *The Evolution of Cooperation*. New York: Basic Books.

13. AXELROD, R. and HAMILTON, W. D. (1981) The evolution of cooperation. *Science* **211**, 1390–6.

14. BALDWIN, B. A. and MEESE, G. B. (1979) Social behaviour in pigs studied by means of operant conditioning. *Animal Behaviour* **27**, 947–57.

15. BARTZ, S. H. (1979) Evolution of eusociality in termites. *Proceedings of the National Academy of Sciences, USA* **76** (11), 5764–8.

16. BASTOCK, M. (1967) *Courtship: A Zoological Study*. London: Heinemann.

17. BATESON, P. (1983) Optimal outbreeding. In *Mate Choice* (ed. P. BATESON). Cambridge: Cambridge University Press. pp. 257–77.

18. BELL, G. (1982) *The Masterpiece of Nature*. London: Croom Helm.

19. BERTRAM, B. C. R. (1976) Kin selection in lions and in evolution. In *Growing Points in Ethology* (eds. P. P. G. BATESON and R. A. HINDE). Cambridge: Cambridge University Press. pp. 281–301.

20. BONNER, J. T. (1980) *The Evolution of Culture in Animals*. Princeton: Princeton University Press.

21. BOYD, R. and LORBERBAUM, J. P. (1987) No pure strategy is evolution-arily stable in the repeated Prisoner's Dilemma game. *Nature* **327,** 58–9.

22. BRETT, R. A. (1986) The ecology and behaviour of the naked mole rat (*Heterocephalus glaber*). Ph.D. thesis, University of London.

23. BROADBENT, D. E. (1961) *Behaviour*. London: Eyre and Spottiswoode.

24. BROCKMANN, H. J. and DAWKINS, R. (1979) Joint nesting in a digger wasp as an evolutionarily stable preadaptation to social life. *Behaviour* **71,** 203–45.

25. BROCKMANN, H. J., GRAFEN, A., and DAWKINS, R. (1979) Evolutionarily stable nesting strategy in a digger wasp. *Journal of Theoretical Biology* **77,** 473–96.

26. BROOKE, M. DE L. and DAVIES, N. B. (1988) Egg mimicry by cuckoos *Cuculus canorus* in relation to discrimination by hosts. *Nature* **335,** 630–2.

27. BURGESS, J. W. (1976) Social spiders. *Scientific American* **234** (3), 101–6.

28. BURK, T. E. (1980) An analysis of social behaviour in crickets. D.Phil. thesis, University of Oxford.

29. CAIRNS-SMITH, A. G. (1971) *The Life Puzzle*. Edinburgh: Oliver and Boyd.

30. CAIRNS-SMITH, A. G. (1982) *Genetic Takeover*. Cambridge: Cambridge University Press.

31. CAIRNS-SMITH, A. G. (1985) *Seven Clues to the Origin of Life*. Cambridge: Cambridge University Press.

32. CAVALLI-SFORZA, L. L. (1971) Similarities and dissimilarities of soci-ocultural and biological evolution. In *Mathematics in the Archaeological and Historical Sciences* (eds. F. R. HODSON, D. G. KENDALL, and P. TAUTU). Edinburgh: Edinburgh University Press. pp. 535–41.

33. CAVALLI-SFORZA, L. L. and FELDMAN, M. W. (1981) *Cultural Transmission and Evolution: A Quantitative Approach*. Princeton: Princeton University Press.

34. CHARNOV, E. L. (1978) Evolution of eusocial behavior: offspring choice or parental parasitism? *Journal of Theoretical Biology* **75,** 451–65.

35. CHARNOV, E. L. and KREBS, J. R. (1975) The evolution of alarm calls: altruism or manipulation? *American Naturalist* **109,** 107–12.

36. CHERFAS, J. and GRIBBIN, J. (1985) *The Redundant Male.* London: Bodley Head.

37. CLOAK, F. T. (1975) Is a cultural ethology possible? *Human Ecology* **3,** 161–82.

38. CROW, J. F. (1979) Genes that violate Mendel's rules. *Scientific American* **240** (2), 104–13.

39. CULLEN, J. M. (1972) Some principles of animal communication. In *Non-verbal Communication* (ed. R. A. HINDE). Cambridge: Cambridge University Press. pp. 101–22.

40. DALY, M. and WILSON, M. (1982) *Sex, Evolution and Behavior,* 2nd edition. Boston: Willard Grant.

41. DARWIN, C. R. (1859) *The Origin of Species.* London: John Murray.

42. DAVIES, N. B. (1978) Territorial defence in the speckled wood butterfly (*Pararge aegeria*): the resident always wins. *Animal Behaviour* **26,** 138–47.

43. DAWKINS, M. S. (1986) *Unravelling Animal Behaviour.* Harlow: Longman.

44. DAWKINS, R. (1979) In defence of selfish genes. *Philosophy* **56,** 556–73.

45. DAWKINS, R. (1979) Twelve misunderstandings of kin selection. *Zeitschrift für Tierpsychologie* **51,** 184–200.

46. DAWKINS, R. (1980) Good strategy or evolutionarily stable strategy? In *Sociobiology: Beyond Nature/Nurture* (eds. G. W. BARLOW and J. SILVERBERG). Boulder, Colorado: Westview Press. pp. 331–67.

47. DAWKINS, R. (1982) *The Extended Phenotype.* Oxford: W. H. Freeman.

48. DAWKINS, R. (1982) Replicators and vehicles. In *Current Problems in Sociobiology* (eds. KING'S COLLEGE SOCIOBIOLOGY GROUP). Cambridge: Cambridge University Press. pp. 45–64.

49. DAWKINS, R. (1983) Universal Darwinism. In *Evolution from Molecules to Men* (ed. D. S. BENDALL). Cambridge: Cambridge University Press. pp. 403–25.

50. DAWKINS, R. (1986) *The Blind Watchmaker.* Harlow: Longman.

51. DAWKINS, R. (1986) Sociobiology: the new storm in a teacup. In *Science and Beyond* (eds. S. ROSE and L. APPIGNANESI). Oxford: Basil Blackwell. pp. 61–78.

52. DAWKINS, R. (1989) The evolution of evolvability. In *Artificial Life* (ed. C. LANGTON). Santa Fe: Addison-Wesley. pp. 201–20.

53. DAWKINS, R. (1993) Worlds in microcosm. In *Humanity, Environment and God* (ed. N. SPURWAY). Oxford: Basil Blackwell.

54. DAWKINS, R. and CARLISLE, T. R. (1976) Parental investment, mate desertion and a fallacy. *Nature* **262**, 131–2.

55. DAWKINS, R. and KREBS, J. R. (1978) Animal signals: information or manipulation? In *Behavioural Ecology: An Evolutionary Approach* (eds. J. R. KREBS and N. B. DAVIES). Oxford: Blackwell Scientific Publications. pp. 282–309.

56. DAWKINS, R. and KREBS, J. R. (1979) Arms races between and within species. *Proceedings of the Royal Society of London B* **205**, 489–511.

57. DE VRIES, P. J. (1988) The larval ant-organs of *Thisbe irenea* (Lepidoptera: Riodinidae) and their effects upon attending ants. *Zoological Journal of the Linnean Society* **94**, 379–93.

58. DELIUS, J. D. (1991) The nature of culture. In *The Tinbergen Legacy* (eds. M. S. DAWKINS, T. R. HALLIDAY, and R. DAWKINS). London: Chapman and Hall.

59. DENNETT, D. C. (1989) The evolution of consciousness. In *Reality Club* **3** (ed. J. BROCKMAN). New York: Lynx Publications.

60. DEWSBURY, D. A. (1982) Ejaculate cost and male choice. *American Naturalist* **119**, 601–10.

61. DIXSON, A. F. (1987) Baculum length and copulatory behavior in primates. *American Journal of Primatology* **13**, 51–60.

62. DOBZHANSKY, T. (1962) *Mankind Evolving*. New Haven: Yale University Press.

63. DOOLITTLE, W. F. and SAPIENZA, C. (1980) Selfish genes, the phenotype paradigm and genome evolution. *Nature* **284**, 601–3.

64. EHRLICH, P. R., EHRLICH, A. H., and HOLDREN, J. P. (1973) *Human Ecology*. San Francisco: Freeman.

65. EIBL-EIBESFELDT, I. (1971) *Love and Hate*. London: Methuen.

66. EIGEN, M., GARDINER, W., SCHUSTER, P., and WINKLER-OSWATITSCH, R. (1981) The origin of genetic information. *Scientific American* **244** (4), 88–118.

67. ELDREDGE, N. and GOULD, S. J. (1972) Punctuated equilibrium: an alternative to phyletic gradualism. In *Models in Paleobiology* (ed. J. M. SCHOPF). San Francisco: Freeman Cooper. pp. 82–115.

68. FISCHER, E. A. (1980) The relationship between mating system and simultaneous hermaphroditism in the coral reef fish, *Hypoplectrus nigricans* (Serranidae). *Animal Behaviour* **28**, 620–33.

69. FISHER, R. A. (1930) *The Genetical Theory of Natural Selection.* Oxford: Clarendon Press.

70. FLETCHER, D. J. C. and MICHENER, C. D. (1987) *Kin Recognition in Humans.* New York: Wiley.

71. FOX, R. (1980) *The Red Lamp of Incest.* London: Hutchinson.

72. GALE, J. S. and EAVES, L. J. (1975) Logic of animal conflict. *Nature* **254**, 463–4.

73. GAMLIN, L. (1987) Rodents join the commune. *New Scientist* **115** (1571), 40–7.

74. GARDNER, B. T. and GARDNER, R. A. (1971) Two-way communication with an infant chimpanzee. In *Behavior of Non-human Primates* **4** (eds. A. M. SCHRIER and F. STOLLNITZ). New York: Academic Press. pp. 117–84.

75. GHISELIN, M. T. (1974) *The Economy of Nature and the Evolution of Sex.* Berkeley: University of California Press.

76. GOULD, S. J. (1980) *The Panda's Thumb.* New York: W. W. Norton.

77. GOULD, S. J. (1983) *Hen's Teeth and Horse's Toes.* New York: W. W. Norton.

78. GRAFEN, A. (1984) Natural selection, kin selection and group selection. In *Behavioural Ecology: An Evolutionary Approach* (eds. J. R. KREBS and N. B. DAVIES). Oxford: Blackwell Scientific Publications. pp. 62–84.

79. GRAFEN, A. (1985) A geometric view of relatedness. In *Oxford Surveys in Evolutionary Biology* (eds. R. DAWKINS and M. RIDLEY), **2**, pp. 28–89. Oxford: Oxford University Press.

80. GRAFEN, A. (1990) Sexual selection unhandicapped by the Fisher process. *Journal of Theoretical Biology*, **144**, 473–516.

81. GRAFEN, A. and SIBLY, R. M. (1978) A model of mate desertion. *Animal Behaviour* **26**, 645–52.

82. HALDANE, J. B. S. (1955) Population genetics. *New Biology* **18**, 34–51.

83. HAMILTON, W. D. (1964) The genetical evolution of social behaviour (I and II). *Journal of Theoretical Biology* **7**, 1–16; 17–52.

84. HAMILTON, W. D. (1966) The moulding of senescence by natural selection. *Journal of Theoretical Biology* **12**, 12–45.

85. HAMILTON, W. D. (1967) Extraordinary sex ratios. *Science* **156,** 477–88.

86. HAMILTON, W. D. (1971) Geometry for the selfish herd. *Journal of Theoretical Biology* **31,** 295–311.

87. HAMILTON, W. D. (1972) Altruism and related phenomena, mainly in social insects. *Annual Review of Ecology and Systematics* **3,** 193–232.

88. HAMILTON, W. D. (1975) Gamblers since life began: barnacles, aphids, elms. *Quarterly Review of Biology* **50,** 175–80.

89. HAMILTON, W. D. (1980) Sex versus non-sex versus parasite. *Oikos* **35,** 282–90.

90. HAMILTON, W. D. and ZUK, M. (1982) Heritable true fitness and bright birds: a role for parasites? *Science* **218,** 384–7.

91. HAMPE, M. and MORGAN, S. R. (1987) Two consequences of Richard Dawkins' view of genes and organisms. *Studies in the History and Philosophy of Science* **19,** 119–38.

92. HANSELL, M. H. (1984) *Animal Architecture and Building Behaviour.* Harlow: Longman.

93. HARDIN, G. (1978) Nice guys finish last. In *Sociobiology and Human Nature* (eds. M. S. GREGORY, A. SILVERS, and D. SUTCH). San Francisco: Jossey Bass. pp. 183–94.

94. HENSON, H. K. (1985) Memes, L_5 and the religion of the space colonies. L_5 *News,* September 1985, pp. 5–8.

95. HINDE, R. A. (1974) *Biological Bases of Human Social Behaviour.* New York: McGraw-Hill.

96. HOYLE, F. and ELLIOT, J. (1962) *A for Andromeda.* London: Souvenir Press.

97. HULL, D. L. (1980) Individuality and selection. *Annual Review of Ecology and Systematics* **11,** 311–32.

98. HULL, D. L. (1981) Units of evolution: a metaphysical essay. In *The Philosophy of Evolution* (eds. U. L. JENSEN and R. HARRÉ). Brighton: Harvester. pp. 23–44.

99. HUMPHREY, N. (1986) *The Inner Eye.* London: Faber and Faber.

100. JARVIS, J. U. M. (1981) Eusociality in a mammal: cooperative breeding in naked mole-rat colonies. *Science* **212,** 571–3.

101. JENKINS, P. F. (1978) Cultural transmission of song patterns and dialect development in a free-living bird population. *Animal Behaviour* **26,** 50–78.

102. KALMUS, H. (1969) Animal behaviour and theories of games and of language. *Animal Behaviour* **17,** 607–17.

103. KREBS, J. R. (1977) The significance of song repertoires—the Beau Geste hypothesis. *Animal Behaviour* **25,** 475–8.
104. KREBS, J. R. and DAWKINS, R. (1984) Animal signals: mind-reading and manipulation. In *Behavioural Ecology: An Evolutionary Approach* (eds. J. R. KREBS and N. B. DAVIES), 2nd edition. Oxford: Blackwell Scientific Publications. pp. 380–402.
105. KRUUK, H. (1972) *The Spotted Hyena: A Study of Predation and Social Behavior.* Chicago: Chicago University Press.
106. LACK, D. (1954) *The Natural Regulation of Animal Numbers.* Oxford: Clarendon Press.
107. LACK, D. (1966) *Population Studies of Birds.* Oxford: Clarendon Press.
108. LE BOEUF, B. J. (1974) Male–male competition and reproductive success in elephant seals. *American Zoologist* **14,** 163–76.
109. LEWIN, B. (1974) *Gene Expression,* volume 2. London: Wiley.
110. LEWONTIN, R. C. (1983) The organism as the subject and object of evolution. *Scientia* **118,** 65–82.
111. LIDICKER, W. Z. (1965) Comparative study of density regulation in confined populations of four species of rodents. *Researches on Population Ecology* **7** (27), 57–72.
112. LOMBARDO, M. P. (1985) Mutual restraint in tree swallows: a test of the Tit for Tat model of reciprocity. *Science* **227,** 1363–5.
113. LORENZ, K. Z. (1966) *Evolution and Modification of Behavior.* London: Methuen.
114. LORENZ, K. Z. (1966) *On Aggression.* London: Methuen.
115. LURIA, S. E. (1973) *Life—The Unfinished Experiment.* London: Souvenir Press.
116. MACARTHUR, R. H. (1965) Ecological consequences of natural selection. In *Theoretical and Mathematical Biology* (eds. T. H. WATERMAN and H. J. MOROWITZ). New York: Blaisdell. pp. 388–97.
117. MACKIE, J. L. (1978) The law of the jungle: moral alternatives and principles of evolution. *Philosophy* **53,** 455–64. Reprinted in *Persons and Values* (eds. J. Mackie and P. Mackie, 1985). Oxford: Oxford University Press. pp. 120–31.
118. MARGULIS, L. (1981) *Symbiosis in Cell Evolution.* San Francisco: W. H. Freeman.
119. MARLER, P. R. (1959) Developments in the study of animal communication. In *Darwin's Biological Work* (ed. P. R. BELL). Cambridge: Cambridge University Press. pp. 150–206.

120. MAYNARD SMITH, J. (1972) Game theory and the evolution of fighting. In J. MAYNARD SMITH, *On Evolution*. Edinburgh: Edinburgh University Press. pp. 8–28.

121. MAYNARD SMITH, J. (1974) The theory of games and the evolution of animal conflict. *Journal of Theoretical Biology* **47**, 209–21.

122. MAYNARD SMITH, J. (1976) Group selection. *Quarterly Review of Biology* **51**, 277–83.

123. MAYNARD SMITH, J. (1976) Evolution and the theory of games. *American Scientist* **64**, 41–5.

124. MAYNARD SMITH, J. (1976) Sexual selection and the handicap principle. *Journal of Theoretical Biology* **57**, 239–42.

125. MAYNARD SMITH, J. (1977) Parental investment: a prospective analysis. *Animal Behaviour* **25**, 1–9.

126. MAYNARD SMITH, J. (1978) *The Evolution of Sex*. Cambridge: Cambridge University Press.

127. MAYNARD SMITH, J. (1982) *Evolution and the Theory of Games*. Cambridge: Cambridge University Press.

128. MAYNARD SMITH, J. (1988) *Games, Sex and Evolution*. New York: Harvester Wheatsheaf.

129. MAYNARD SMITH, J. (1989) *Evolutionary Genetics*. Oxford: Oxford University Press.

130. MAYNARD SMITH, J. and PARKER, G. A. (1976) The logic of asymmetric contests. *Animal Behaviour* **24**, 159–75.

131. MAYNARD SMITH, J. and PRICE, G. R. (1973) The logic of animal conflicts. *Nature* **246**, 15–18.

132. MCFARLAND, D. J. (1971) *Feedback Mechanisms in Animal Behaviour*. London: Academic Press.

133. MEAD, M. (1950) *Male and Female*. London: Gollancz.

134. MEDAWAR, P. B. (1952) *An Unsolved Problem in Biology*. London: H. K. Lewis.

135. MEDAWAR, P. B. (1957) *The Uniqueness of the Individual*. London: Methuen.

136. MEDAWAR, P. B. (1961) Review of P. Teilhard de Chardin, *The Phenomenon of Man*. Reprinted in P. B. MEDAWAR (1982) *Pluto's Republic*. Oxford: Oxford University Press.

137. MICHOD, R. E. and LEVIN, B. R. (1988) *The Evolution of Sex*. Sunderland, Massachusetts: Sinauer.

138. MIDGLEY, M. (1979) Gene-juggling. *Philosophy* **54,** 439–58.

139. MONOD, J. L. (1974) On the molecular theory of evolution. In *Problems of Scientific Revolution* (ed. R. HARRÉ). Oxford: Clarendon Press. pp. 11–24.

140. MONTAGU, A. (1976) *The Nature of Human Aggression.* New York: Oxford University Press.

141. MORAVEC, H. (1988) *Mind Children.* Cambridge, Massachusetts: Harvard University Press.

142. MORRIS, D. (1957) 'Typical Intensity' and its relation to the problem of ritualization. *Behaviour* **11,** 1–21.

143. *Nuffield Biology Teachers Guide IV* (1966) London: Longmans. p. 96.

144. ORGEL, L. E. (1973) *The Origins of Life.* London: Chapman and Hall.

145. ORGEL, L. E. and CRICK, F. H. C. (1980) Selfish DNA: the ultimate parasite. *Nature* **284,** 604–7.

146. PACKER, C. and PUSEY, A. E. (1982) Cooperation and competition within coalitions of male lions: kin-selection or game theory? *Nature* **296,** 740–2.

147. PARKER, G. A. (1984) Evolutionarily stable strategies. In *Behavioural Ecology: An Evolutionary Approach* (eds. J. R. KREBS and N. B. DAVIES), 2nd edition. Oxford: Blackwell Scientific Publications. pp. 62–84.

148. PARKER, G. A., BAKER, R. R., and SMITH, V. G. F. (1972) The origin and evolution of gametic dimorphism and the male–female phenomenon. *Journal of Theoretical Biology* **36,** 529–53.

149. PAYNE, R. S. and McVAY, S. (1971) Songs of humpback whales. *Science* **173,** 583–97.

150. POPPER, K. (1974) The rationality of scientific revolutions. In *Problems of Scientific Revolution* (ed. R. HARRÉ). Oxford: Clarendon Press. pp. 72–101.

151. POPPER, K. (1978) Natural selection and the emergence of mind. *Dialectica* **32,** 339–55.

152. RIDLEY, M. (1978) Paternal care. *Animal Behaviour* **26,** 904–32.

153. RIDLEY, M. (1985) *The Problems of Evolution.* Oxford: Oxford University Press.

154. ROSE, S., KAMIN, L. J., and LEWONTIN, R. C. (1984) *Not In Our Genes.* London: Penguin.

155. ROTHENBUHLER, W. C. (1964) Behavior genetics of nest cleaning in honey bees. IV. Responses of F_1 and backcross generations to disease-killed brood. *American Zoologist* **4,** 111–23.

156. RYDER, R. (1975) *Victims of Science*. London: Davis-Poynter.
157. SAGAN, L. (1967) On the origin of mitosing cells. *Journal of Theoretical Biology* **14**, 225–74.
158. SAHLINS, M. (1977) *The Use and Abuse of Biology*. Ann Arbor: University of Michigan Press.
159. SCHUSTER, P. and SIGMUND, K. (1981) Coyness, philandering and stable strategies. *Animal Behaviour* **29**, 186–92.
160. SEGER, J. and HAMILTON, W. D. (1988) Parasites and sex. In *The Evolution of Sex* (eds. R. E. MICHOD and B. R. LEVIN). Sunderland, Massachusetts: Sinauer. pp. 176–93.
161. SEGER, J. and HARVEY, P. (1980) The evolution of the genetical theory of social behaviour. *New Scientist* **87** (1208), 50–1.
162. SHEPPARD, P. M. (1958) *Natural Selection and Heredity*. London: Hutchinson.
163. SIMPSON, G. G. (1966) The biological nature of man. *Science* **152**, 472–8.
164. SINGER, P. (1976) *Animal Liberation*. London: Jonathan Cape.
165. SMYTHE, N. (1970) On the existence of 'pursuit invitation' signals in mammals. *American Naturalist* **104**, 491–4.
166. STERELNY, K. and KITCHER, P. (1988) The return of the gene. *Journal of Philosophy* **85**, 339–61.
167. SYMONS, D. (1979) *The Evolution of Human Sexuality*. New York: Oxford University Press.
168. TINBERGEN, N. (1953) *Social Behaviour in Animals*. London: Methuen.
169. TREISMAN, M. and DAWKINS, R. (1976) The cost of meiosis—is there any? *Journal of Theoretical Biology* **63**, 479–84.
170. TRIVERS, R. L. (1971) The evolution of reciprocal altruism. *Quarterly Review of Biology* **46**, 35–57.
171. TRIVERS, R. L. (1972) Parental investment and sexual selection. In *Sexual Selection and the Descent of Man* (ed. B. CAMPBELL). Chicago: Aldine. pp. 136–79.
172. TRIVERS, R. L. (1974) Parent–offspring conflict. *American Zoologist* **14**, 249–64.
173. TRIVERS, R. L. (1985) *Social Evolution*. Menlo Park: Benjamin/Cummings.
174. TRIVERS, R. L. and HARE, H. (1976) Haplodiploidy and the evolution of the social insects. *Science* **191**, 249–63.
175. TURNBULL, C. (1972) *The Mountain People*. London: Jonathan Cape.

176. WASHBURN, S. L. (1978) Human behavior and the behavior of other animals. *American Psychologist* **33**, 405–18.

177. WELLS, P. A. (1987) Kin recognition in humans. In *Kin Recognition in Animals* (eds. D. J. C. FLETCHER and C. D. MICHENER). New York: Wiley. pp. 395–415.

178. WICKLER, W. (1968) *Mimicry*. London: World University Library.

179. WILKINSON, G. S. (1984) Reciprocal food-sharing in the vampire bat. *Nature* **308**, 181–4.

180. WILLIAMS, G. C. (1957) Pleiotropy, natural selection, and the evolution of senescence. *Evolution* **11**, 398–411.

181. WILLIAMS, G. C. (1966) *Adaptation and Natural Selection*. Princeton: Princeton University Press.

182. WILLIAMS, G. C. (1975) *Sex and Evolution*. Princeton: Princeton University Press.

183. WILLIAMS, G. C. (1985) A defense of reductionism in evolutionary biology. In *Oxford Surveys in Evolutionary Biology* (eds. R. DAWKINS and M. RIDLEY), **2**, pp. 1–27. Oxford: Oxford University Press.

184. WILSON, E. O. (1971) *The Insect Societies*. Cambridge, Massachusetts: Harvard University Press.

185. WILSON, E. O. (1975) *Sociobiology: The New Synthesis*. Cambridge, Massachusetts: Harvard University Press.

186. WILSON, E. O. (1978) *On Human Nature*. Cambridge, Massachusetts: Harvard University Press.

187. WRIGHT, S. (1980) Genic and organismic selection. *Evolution* **34**, 825–43.

188. WYNNE-EDWARDS, V. C. (1962) *Animal Dispersion in Relation to Social Behaviour*. Edinburgh: Oliver and Boyd.

189. WYNNE-EDWARDS, V. C. (1978) Intrinsic population control: an introduction. In *Population Control by Social Behaviour* (eds. F. J. EBLING and D. M. STODDART). London: Institute of Biology. pp. 1–22.

190. WYNNE-EDWARDS, V. C. (1986) *Evolution through Group Selection*. Oxford: Blackwell Scientific Publications.

191. YOM-TOV, Y. (1980) Intraspecific nest parasitism in birds. *Biological Reviews* **55**, 93–108.

192. YOUNG, J. Z. (1975) *The Life of Mammals*, 2nd edition. Oxford: Clarendon Press.

193. ZAHAVI, A. (1975) Mate selection—a selection for a handicap. *Journal of Theoretical Biology* **53**, 205–14.

194. ZAHAVI, A. (1977) Reliability in communication systems and the evolution of altruism. In *Evolutionary Ecology* (eds. B. STONEHOUSE and C. M. PERRINS). London: Macmillan. pp. 253–9.

195. ZAHAVI, A. (1978) Decorative patterns and the evolution of art. *New Scientist* **80** (1125), 182–4.

196. ZAHAVI, A. (1987) The theory of signal selection and some of its implications. In *International Symposium on Biological Evolution, Bari, 9–14 April 1985* (ed. V. P. DELFINO). Bari: Adriatici Editrici. pp. 305–27.

197. ZAHAVI, A. Personal communication, quoted by permission.

INDEX AND KEY TO BIBLIOGRAPHY

I chose not to break the flow of the book with literature citations. This index should enable readers to follow up references on particular topics. The numbers in brackets refer to the numbered references in the bibliography. Other numbers refer to pages in the book, as in a normal index. Terms that are often used are not indexed every time they occur, but only in special places such as where they are defined.

EXTRACTS FROM REVIEWS

Pro bono publico

Peter Medawar in The Spectator, *15 January 1977*

When confronted by what is ostensibly altruistic or anyhow non-selfish behaviour in animals, amateurs of biology, a class that includes an increasing number of sociologists, are very easily tempted to say that it has evolved 'for the benefit of the species.'

There is a well known myth, for example, that lemmings–evidently more conscious of the need for it than we are—regulate population size by plunging over cliffs by the thousand to perish in the sea. Surely even the most gullible naturalist must have asked himself how such altruism could have become part of the behavioural repertoire of the species, having regard to the fact that the genetic make-ups conducive to it must have perished with their possessors in this grand demographic auto-da-fe. To dismiss this as a myth is not to deny, however, that genetically selfish actions may sometimes 'present' (as clinicians say) as disinterested or altruistic actions. Genetic factors conducive to grandmotherly indulgence, as opposed to callous indifference, may prevail in evolution because kindly grandmothers are selfishly promoting the survival and propagation of the fraction of their own genes that are present in their grandchildren.

Richard Dawkins, one of the most brilliant of the rising generation of biologists, gently and expertly debunks some of the favourite illusions of social biology about the evolution of altruism, but this is on no account to be thought of as a debunking kind of book: it is, on the contrary, a most skilful reformulation of the central problems of social biology in terms of the genetical theory of natural selection. Beyond this, it is learned, witty and very well written. One of the things that attracted Richard Dawkins to the study of zoology was the 'general likeableness' of animals—a point of view shared by all good biologists that shines throughout this book.

Although *The Selfish Gene* is not disputative in character, it was a very necessary part of Dawkins's programme to deflate the pretensions of such books as Lorenz's *On Aggression*, Ardrey's *The Social Contract*, and Eibl-Eibesfeldt's *Love and Hate*: 'the trouble with these books is that their authors got it totally and utterly wrong...because they misunderstood how evolution works. They made the erroneous assumption that the important thing in evolution is the good of the species (or the group) rather than the good of the individual (or the gene).'

There is indeed truth enough for a dozen sermons in the schoolboy aphorism that 'a chicken is the egg's way of making another egg.' Richard Dawkins puts it thus:

> The argument of this book is that we, and all other animals, are machines created by our genes…I shall argue that a predominant quality to be expected in a successful gene is ruthless selfishness. This gene selfishness will usually give rise to selfishness in individual behaviour. However, as we shall see, there are special circumstances in which a gene can achieve its own selfish goals best by fostering a limited form of altruism at the level of individual animals. 'Special' and 'limited' are important words in the last sentence. Much as we might wish to believe otherwise, universal love and the welfare of the species as a whole are concepts which simply do not make evolutionary sense.

We may deplore these truths, Dawkins says, but that does not make them any less true. The more clearly we understand the selfishness of the genetic process, however, the better qualified we shall be to *teach* the merits of generosity and co-operativeness and all else that works for the common good, and Dawkins expounds more clearly than most the special importance in mankind of cultural or 'exogenetic' evolution.

In his last and most important chapter, Dawkins challenges himself to formulate one fundamental principle that would certainly apply to all evolutionary systems—even perhaps to organisms in which silicon atoms took the place of carbon atoms, and to organisms like human beings in which so much of evolution is mediated through non-genetic channels. The principle is that of evolution through the net reproductive advantage of replicating entities. For ordinary organisms under ordinary circumstances these entities are the singularities in DNA molecules known as 'genes.' For Dawkins the unit of cultural transmission is that which he calls the 'meme' and in his last chapter he expounds what is in effect a Darwinian theory of memes.

To Dawkins's exhilaratingly good book I will add one footnote: the idea that the possession of a memory function is a fundamental attribute of all living things was first propounded by an Austrian physiologist Ewald Hering in 1870. He spoke of his unit as the 'mneme,' a word of conscious etymological rectitude. Richard Semon's exposition of the subject (1921) is naturally enough completely non-Darwinian, and cannot now be regarded as anything except a period piece. One of Hering's ideas was held up to ridicule by a rival nature philosopher, Professor J. S. Haldane: the idea that a compound must exist having exactly the properties we now know to be possessed by deoxyribonucleic acid, DNA.

The Play by Nature

W. D. Hamilton in Science, *13 May 1977 (extract)*

This book should be read, can be read, by almost everyone. It describes with great skill a new face of the theory of evolution. With much of the light, unencumbered style that has lately sold new and sometimes erroneous biology to the public, it is, in my opinion, a more serious achievement. It succeeds in the seemingly impossible task of using simple, untechnical English to present some rather recondite and quasi-mathematical themes of recent evolutionary thought. Seen through this book in their broad perspective at last, these will surprise and refresh even many research biologists who might have supposed themselves already in the know. At least, so they surprised this reviewer. Yet, to repeat, the book remains easily readable by anyone with the least grounding in science.

Even without intention to be snobbish, reading a popular book in a field close to one's research interests almost forces one to tally errors: this example misapplied, that point left ambiguous, that idea wrong, abandoned years ago. This book had an almost clean sheet from me. This is not to say that there are no probable errors—that could hardly be the case in a work where speculation is, in a sense, the stock in trade—but its biology as a whole is firmly the right way up and its questionable statements are at least undogmatic. The author's modest assessment of his own ideas tends to disarm criticism, and here and there the reader finds himself flattered by a suggestion that he should work out a better model if he doesn't like the one given. That such an invitation can be made seriously in a popular book vividly reflects the newness of the subject matter. Strangely, there are indeed possibilities that simple ideas as yet untested may shortly resolve some old puzzles of evolution.

What, then, is this new face of evolution? To a certain extent it is like a new interpretation of Shakespeare: it was all in the script but somehow it passed unseen. I should add, however, that the new view in question was latent not so much in Darwin's script of evolution as in nature's and that our lapse of attention is more on the scale of 20 years than of a hundred. Dawkins starts, for example, from those variable helical molecules that we now know fairly well; Darwin knew not even about chromosomes or their strange dance in the sexual process. But even 20 years is quite long enough to cause surprise.

The first chapter broadly characterizes the phenomena the book seeks to explain and shows their philosophical and practical importance to human life. Some intriguing and alarming animal examples catch our attention. The second chapter goes back to the first replicators in their primeval soup. We see these multiply and elaborate. They begin to compete for substrates, to fight, even to lyse and eat one another; they hide themselves and their gains and weapons in defensive stockades; these come to be used for shelter not only from the tactics of rivals and predators but from the physical hardships of the environments that the replicators are increasingly enabled to invade. Thus they mobilize, settle, throw up bizarre forms, pour over the beaches, across land, and right on to deserts and eternal snows. Between such frontiers, beyond which, for long, life cannot go, the soup is poured and repoured millions of times over into an ever-stranger diversity of molds; at length it is poured into ant and elephant, mandrill and man. This second chapter concludes, concerning some ultimate descendant coalitions of these ancient replicators: 'Their preservation is the ultimate rationale of our existence...Now they go by the name of genes and we are their survival machines.'

Forceful and provocative, the reader may think, but is it very new? Well, so far perhaps not, but of course evolution has not ended with our bodies. More important still, the techniques of survival in a crowded world turn out to be unexpectedly subtle, much more subtle than biologists were prepared to envisage under the old, departing paradigm of adaptation for the benefit of the species. It is this subtlety, roughly, that is the theme of the rest of the book. Take a simple example, birdsong. It seems a very inefficient arrangement: a naïve materialist looking for the techniques by which a species of *Turdus* survives hard winters, food shortages, and the like might well find the flamboyant singing of its males as improbable as ectoplasm at a séance. (On further thought he might find the fact that the species has males at all equally improbable, and this indeed is another major topic of the book: as with that of birdsong, the function of sex has been rationalized much too facilely in the past.) Yet within any bird species a whole team of replicators has concerned itself to lay down an elaborate outline for this performance. Somewhere Dawkins cites the even more extraordinary song of the humpback whale, which may make itself heard over a whole ocean; but of this song we know even less than with *Turdus* what it is about and to whom directed. So far as the evidence goes it might actually be an anthem for cetacean unity against mankind—perhaps well for whales if it were. Of course, it is other teams of teams of replicators that now turn out symphony concerts. And these certainly do sometimes cross oceans—by reflection from bodies in space which themselves were made and orbited according to plans from even more complex teams. What conjurers do with mirrors is nothing to what nature, if Dawkins is right, does with no more promising a starting material than congealed primeval soup. It will serve to characterize the new look that biology has in this and some other recent books (such as E. O. Wilson's *Sociobiology*) to say that it shines with a hope that these farthest extensions of life may soon fit more comprehensibly, in essence if not in some details (religious persons and Neo-Marxists may reverse that phrase if it suits them better), into a general pattern that includes the simplest cell wall, the simplest multicell body, and the blackbird's song.

The impression should be avoided, however, that this book is some sort of layman's or poor man's *Sociobiology*. First, it has many original ideas, and second, it counterweights a certain imbalance in Wilson's massive tome by strongly emphasizing the game-theoretic aspect of social behavior, which Wilson hardly mentioned. 'Game-theoretic' is not quite the right word, especially in the context of lower levels of social evolution, since the genes themselves don't rationalize about their methods of operation; nevertheless, it has become clear that at all levels there are useful similarities between the conceptual structures of game theory and those of social evolution. The cross-fertilization implied here is new and is still in progress: only recently, for example, I learned that game theory had already given a name ("Nash equilibrium") to a concept that corresponds roughly to the "evolutionarily stable strategy." Dawkins rightly treats the idea of evolutionary stability as all-important for his new overview of social biology. The game-like element in social behavior and social adaptation comes from the dependence, in any social situation, of the success of one individual's strategy on the strategies used by his or her interactants. The pursuit of adaptation that gets the most out of a given situation regardless of the overall good can lead to some very surprising results. Who would have supposed, for example, that the weighty matter of why in fish, contrary to the case in most other animals, it is the male that usually guards the eggs and young if either sex does, might depend on such a trivial detail as which sex is constrained to release its gametes into the water first? Yet Dawkins and a co-worker, pursuing an idea of R. L. Trivers's, have made a fair case that such a detail of timing, even if a matter of seconds, could be crucial for the whole phenomenon. Again, would we not expect that females of monogamous birds, blessed with the help of a mate, would lay larger clutches than females of polygamous species? Actually the reverse is true. Dawkins, in his somewhat alarming chapter on the "battle of the sexes," applies once more the idea of stability against exploitation (by the male in this case) and suddenly makes this odd correlation seem natural. His idea, like most of his others, remains unproven, and there may well be other, more weighty reasons; but the ones he gives, which are seen so easily from his new vantage point, demand notice.

In a textbook of game theory one sees no more of games than one sees of circles and triangles in a textbook of modern geometry. At a glance all is just algebra: game theory is a technical subject from the start. Thus it is certainly a literary feat to convey as much as this book does of even the outward feel and quality, let alone inward details, of game-theoretic situations without recourse to formulas. R. A. Fisher in his introduction to his great book on evolution wrote, "No efforts of mine could avail to make the book easy reading." In that book, under a rain of formulas and of sentences as profound as terse, the reader is soon battered into acquiescence. Having read *The Selfish Gene* I now feel that Fisher could have done better, although, admittedly, he would have had to write a different kind of book. It looks as though even the formative ideas of classic population genetics could have been made much more interesting in ordinary prose than they ever were. (Indeed, Haldane did manage somewhat better than Fisher in this, but was less profound.) But what is really remarkable is how much of the rather tedious mathematics that comes in the mainstream of population genetics following the lead of Wright, Fisher, and Haldane can be bypassed in the new, more social approach to the facts of life. I was rather surprised to find Dawkins sharing my assessment of Fisher as "the greatest biologist of the twentieth century" (a rare view, as I thought); but I was also surprised to note how little he had to reiterate Fisher's book.

Finally, in his last chapter, Dawkins comes to the fascinating subject of the evolution of culture. He floats the term "meme" (short for "mimeme") for the cultural equivalent of "gene." Hard as this term may be to delimit— it surely must be harder than gene, which is bad enough—I suspect that it will soon be in common use by biologists and, one hopes, by philosophers, linguists, and others as well and that it may become absorbed as far as the word "gene" has been into everyday speech.

Excerpted with permission from W. D. Hamilton, SCIENCE 196:757–59 (1977).

© 1977 AAAS

Genes and Memes

John Maynard Smith in The London Review of Books, *4–18 February 1982. (Extract from review of* The Extended Phenotype.*)*

The Selfish Gene was unusual in that, although written as a popular account, it made an original contribution to biology. Further, the contribution itself was of an unusual kind. Unlike David Lack's classic *The Life of the Robin*—also an original contribution in popular form—*The Selfish Gene* reports no new facts. Nor does it contain any new mathematical models—indeed it contains no mathematics at all. What it does offer is a new world view.

Although the book has been widely read and enjoyed, it has also aroused strong hostility. Much of this hostility arises, I believe, from misunderstanding, or rather, from several misunderstandings. Of these, the most fundamental is a failure to understand what the book is about. It is a book about the evolutionary process—it is not about morals, or about politics, or about the human sciences. If you are not interested in how evolution came about, and cannot conceive how anyone could be seriously concerned about anything other than human affairs, then do not read it: it will only make you needlessly angry.

Assuming, however, that you are interested in evolution, a good way to understand what Dawkins is up to is to grasp the nature of the debates which were going on between evolutionary biologists during the 1960s and 1970s. These concerned two related topics, 'group selection' and 'kin selection'. The 'group selection' debate was sparked off by Wynne-Edwards [who suggested that behavioural adaptations] had evolved by 'group selection'—i.e. through the survival of some groups and the extinction of others...

At almost the same time, W.D. Hamilton raised another question about how natural selection acts. He pointed out that if a gene were to cause its possessor to sacrifice its life in order to save the lives of several relatives, there might be more copies of the gene present afterwards than if the sacrifice had not been made... To model the process quantitatively, Hamilton introduced the concept of 'inclusive fitness'... which includes, not only an individual's own offspring, but any additional offspring raised by relatives with the help of that individual, appropriately scaled by the degree of relationship....

Dawkins, while acknowledging the debt we owe to Hamilton, suggests that he erred in making a last-ditch attempt to retain the concept of fitness, and that he would have been wiser to adopt a full-blooded 'gene's eye' view of evolution. He urges us to recognise the fundamental distinction between 'replicators'—entities whose precise structure is replicated in the process of reproduction—and 'vehicles': entities which are mortal and which are not replicated, but whose properties are influenced by replicators. The main replicators with which we are familiar are nucleic acid molecules—typically DNA molecules—of which genes and chromosomes are composed. Typical vehicles are the bodies of dogs, fruitflies and people. Suppose, then, that we observe a structure such as the eye, which is manifestly adapted for seeing. We might reasonably ask for whose benefit the eye has evolved. The only reasonable answer, Dawkins suggests, is that it has evolved for the benefit of the replicators responsible for its development. Although like me, he greatly prefers individual to group advantage as an explanation, he would prefer to think only of replicator advantage.

© John Maynard Smith, 1982

TWO CHAPTERS FROM

The Extended Phenotype

The Long Reach of the Gene

RICHARD DAWKINS

2

GENETIC DETERMINISM
AND GENE SELECTIONISM

Long after his death, tenacious rumours persisted that Adolf
Hitler had been seen alive and well in South America, or in
Denmark, and for years a surprising number of people with no
love for the man only reluctantly accepted that he was dead
(Trevor-Roper 1972). In the First World War a story that a hundred thousand Russian troops had been seen landing in Scotland
'with snow on their boots' became widely current, apparently
because of the memorable vividness of that snow (Taylor 1963).
In our own time myths such as that of computers persistently
sending householders electricity bills for a million pounds (Evans
1979), or of well-heeled welfare-scroungers with two expensive
cars parked outside their government-subsidized council houses,
are familiar to the point of cliché. There are some falsehoods, or
half-truths, that seem to engender in us an active desire to believe
them and pass them on even if we find them unpleasant, maybe in
part, perversely, *because* we find them unpleasant.

Computers and electronic 'chips' provoke more than their fair
share of such myth-making, perhaps because computer technology advances at a speed which is literally frightening. I know
an old person who has it on good authority that 'chips' are usurp-
ing human functions to the extent not only of 'driving tractors'

but even of 'fertilizing women'. Genes, as I shall show, are the source of what may be an even larger mythology than computers. Imagine the result of combining these two powerful myths, the gene myth and the computer myth! I believe that I may have inadvertently achieved some such unfortunate synthesis in the minds of a few readers of my previous book, and the result was comic misunderstanding. Happily, such misunderstanding was not widespread, but it is worth trying to avoid a repeat of it here, and that is one purpose of the present chapter. I shall expose the myth of genetic determinism, and explain why it is necessary to use language that can be unfortunately misunderstood as genetic determinism.

A reviewer of Wilson's (1978) *On Human Nature*, wrote: 'although he does not go as far as Richard Dawkins (*The Selfish Gene*...) in proposing sex-linked genes for "philandering", for Wilson human males have a genetic tendency towards polygyny, females towards constancy (don't blame your mates for sleeping around, ladies, it's not their fault they are genetically programmed). Genetic determinism constantly creeps in at the back door' (Rose 1978). The reviewer's clear implication is that the authors he is criticizing believe in the existence of genes that force human males to be irremediable philanderers who cannot therefore be blamed for marital infidelity. The reader is left with the impression that those authors are protagonists in the 'nature or nurture' debate, and, moreover, died-in-the-wool hereditarians with male chauvinist leanings.

In fact my original passage about 'philanderer males' was not about humans. It was a simple mathematical model of some unspecified animal (not that it matters, I had a bird in mind). It was not explicitly (see below) a model of genes, and if it had been about genes they would have been sex-limited, not sex-linked! It was a model of 'strategies' in the sense of Maynard Smith (1974). The 'philanderer' strategy was postulated, not as *the* way males behave, but as one of two hypothetical alternatives, the other being the 'faithful' strategy. The purpose of this very simple model was to illustrate the kinds of conditions under which philandering might be favoured by natural selection, and the kinds of conditions under which faithfulness might be favoured. There was no presumption that philandering was more likely in males than faithfulness. Indeed, the particular run of the simulation that I published culminated in a mixed male population in which faithfulness slightly predominated (Dawkins 1976a, p. 165, although see Schuster & Sigmund 1981). There is not just one misunderstanding in Rose's remarks, but multiple compounded misunderstanding. There is a wanton eagerness to misunderstand. It bears the stamp of snow-covered Russian jackboots, of little black microchips marching to usurp the male role and steal our tractor-drivers' jobs. It is a manifestation of a powerful myth, in this case the great gene myth.

The gene myth is epitomized in Rose's parenthetic little joke about ladies not blaming their mates for sleeping around. It is the myth of 'genetic determinism'. Evidently, for Rose, genetic determinism is determinism in the full philosophical sense of irreversible inevitability. He assumes that the existence of a gene 'for' X implies that X cannot be escaped. In the words of another critic of 'genetic determinism', Gould (1978, p. 238), 'If we are programmed to be what we are, then these traits are ineluctable. We may, at best, channel them, but we cannot change them either by will, education, or culture.'

The validity of the determinist point of view and, separately, its bearing on an individual's moral responsibility for his actions, has been debated by philosophers and theologians for centuries past, and no doubt will be for centuries to come. I suspect that both Rose and Gould are determinists in that they believe in a physical, materialistic basis for all our actions. So am I. We would also probably all three agree that human nervous systems are so complex that in practice we can forget about determinism and behave as if we had free will. Neurones may be amplifiers of fundamentally indeterminate physical events. The only point I wish to make is that, whatever view one takes on the question of determinism, the insertion of the word 'genetic' is not going to make any difference. If you are a full-blooded determinist you will believe that all your actions are predetermined by physical causes in the past, and you may or may not also believe that you therefore cannot be held responsible for your sexual infidelities. But, be that as it may, what difference can it possibly make whether some of those physical causes are *genetic*? Why are genetic determinants thought to be any more ineluctable, or blame-absolving, than 'environmental' ones?

The belief that genes are somehow super-deterministic, in comparison with environmental causes, is a myth of extraordinary tenacity, and it can give rise to real emotional distress. I was only dimly aware of this until it was movingly brought home to me in a question session at a meeting of the American Association for the Advancement of Science in 1978. A young woman asked the lecturer, a prominent 'sociobiologist', whether there was any evidence for genetic sex differences in human psychology. I hardly heard the lecturer's answer, so astonished was I by the emotion with which the question was put. The woman seemed to set great store by the answer and was almost in tears. After a moment of genuine and innocent bafflement the explanation hit me. Something or somebody, certainly not the eminent sociobiologist himself, had misled her into thinking that genetic determination is for keeps; she seriously believed that a 'yes' answer to her question would, if correct, condemn her as a female individual to a life of feminine pursuits, chained to the nursery and the kitchen sink. But if, unlike most of us, she is a determinist in that strong Calvinistic sense, she should be equally upset whether the causal factors concerned are genetic or 'environmental'.

What does it ever mean to say that something determines something? Philosophers, possibly with justification, make heavy weather of the concept of causation, but to a working biologist causation is a rather simple statistical concept. Operationally we can never demonstrate that a particular observed event C caused a particular result R, although it will often be judged highly likely. What biologists in practice usually do is to establish *statistically* that events of class R reliably follow events of class C. They need a number of paired instances of the two classes of events in order to do so: one anecdote is not enough.

Even the observation that R events reliably tend to follow C events after a relatively fixed time interval provides only a working hypothesis that C events cause R events. The hypothesis is confirmed, within the limits of the statistical method, only if the C events are delivered by an *experimenter* rather than simply noted by an observer, and are still reliably followed by R events. It is not necessary that every C should be followed by an R, nor that every R should be preceded by a C. (Who has not had to contend with arguments such as 'smoking cannot cause lung cancer, because I knew a non-smoker who died of it, and a heavy smoker who is still going strong at ninety'?) Statistical methods are designed to help us assess, to any specified level of probabilistic confidence, whether the results we obtain really indicate a causal relationship.

If, then, it were true that the possession of a Y chromosome had a causal influence on, say, musical ability or fondness for knitting, what would this mean? It would mean that, in some specified population and in some specified environment, an observer in possession of information about an individual's sex would be able to make a statistically more accurate prediction as to the person's musical ability than an observer ignorant of the person's sex. The emphasis is on the word 'statistically', and let us throw in an 'other things being equal' for good measure. The observer might be provided with some additional information, say on the person's education or upbringing, which would lead him to revise, or even reverse, his prediction based on sex. If females are statistically more likely than males to enjoy knitting, this does not mean that all females enjoy knitting, nor even that a majority do.

It is also fully compatible with the view that the reason females enjoy knitting is that society brings them up to enjoy knitting. If society systematically trains children without penises to knit and play with dolls, and trains children with penises to play with guns and toy soldiers, any resulting differences in male and female preferences are strictly speaking genetically determined differences! They are determined, through the medium of societal custom, by the fact of possession or non-possession of a penis, and that is determined (in a normal environment and in the absence of ingenious plastic surgery or hormone therapy) by sex chromosomes.

Obviously, on this view, if we experimentally brought up a sample of boys to play with dolls and a sample of girls to play with guns, we would expect easily to reverse the normal preferences. This might be an interesting experiment to do, for the result just might turn out to be that girls *still* prefer dolls and boys still prefer guns. If so, this might tell us something about the tenacity, in the face of a *particular* environmental manipulation, of a genetic difference. But all genetic causes have to work in the context of an environment of some kind. If a genetic sex difference makes itself felt through the medium of a sex-biased education system, it is still a genetic difference. If it makes itself felt through some other medium, such that manipulations of the education system do not perturb it, it is, in principle, no more and no less a genetic difference than in the former, education-sensitive case: no doubt some other environmental manipulation could be found which *did* perturb it.

Human psychological attributes vary along almost as many dimensions as psychologists can measure. It is difficult in practice (Kempthorne 1978), but in principle we could partition this variation among such putative causal factors as age, height, years of education, type of education classified in many different ways, number of siblings, birth order, colour of mother's eyes, father's skill in shoeing horses, and, of course, sex chromosomes. We could also examine two-way and multi-way interactions between such factors. For present purposes the important point is that the variance we seek to explain will have many causes, which interact in complex ways. Undoubtedly genetic variance is a significant cause of much phenotypic variance in observed populations, but its effects may be overridden, modified, enhanced, or reversed by other causes. Genes may modify the effects of other genes, and may modify the effects of the environment. Environmental events, both internal and external, may modify the effects of genes, and may modify the effects of other environmental events.

People seem to have little difficulty in accepting the modifiability of 'environmental' effects on human development. If a child has had bad teaching in mathematics, it is accepted that the resulting deficiency can be remedied by extra good teaching the following year. But any suggestion that the child's mathematical deficiency might have a genetic origin is likely to be greeted with something approaching despair: if it is in the genes 'it is written', it is 'determined' and nothing can be done about it; you might as well give up attempting to teach the child mathematics. This is pernicious rubbish on an almost astrological scale. Genetic causes and environmental causes are in principle no different from each other. Some influences of both types may be hard to reverse; others may be easy to reverse. Some may be usually hard to reverse but easy if the right agent is applied. The important point is that there is no general reason for expecting genetic influences to be any more irreversible than environmental ones.

What did genes do to deserve their sinister, juggernaut-like reputation? Why do we not make a similar bogey out of, say, nursery education or confirmation classes? Why are genes thought to be so much more fixed and inescapable in their effects than television, nuns, or books? Don't blame your mates for sleeping around, ladies, it's not their fault they have been inflamed by pornographic literature! The alleged Jesuit boast, 'Give me the child for his first seven years, and I'll give you the man', may have some truth in it. Educational, or other cultural influences may, in some circumstances, be just as unmodifiable and irreversible as genes and 'stars' are popularly thought to be.

I suppose part of the reason genes have become deterministic bogeys is a confusion resulting from the well-known fact of the non-inheritance of acquired characteristics. Before this century it was widely believed that experience and other acquisitions of an individual's lifetime were somehow imprinted on the hereditary substance and transmitted to the children. The abandoning of this belief, and its replacement by Weismann's doctrine of the continuity of the germ plasm, and its molecular counterpart the 'central dogma', is one of the great achievements of modern biology. If we steep ourselves in the implications of Weismannian orthodoxy, there really does seem to be something juggernaut-like and inexorable about genes. They march through generations, influencing the form and behaviour of a succession of mortal bodies, but, except for rare and non-specific mutagenic effects, they are never influenced by the experience or environment of those bodies. The genes in me came from my four grandparents; they flowed straight through my parents to me, and nothing that my parents achieved, acquired, learned, or experienced had any effect on those genes as they flowed through. Perhaps there is something a little sinister about that. But, however inexorable and undeviating the genes may be as they march down the generations, the nature of their phenotypic effects on the bodies they flow through is by no means inexorable and undeviating. If I am homozygous for a gene G, nothing save mutation can prevent my passing G on to all my children. So much is inexorable. But whether or not I, or my children, show the phenotypic effect normally associated with possession of G may depend very much on how we are brought up, what diet or education we experience, and what other genes we happen to possess. So, of the two effects that genes have on the world—manufacturing copies of themselves, and influencing phenotypes—the first is inflexible apart from the rare possibility of mutation; the second may be exceedingly flexible. I think a confusion between evolution and development is, then, partly responsible for the myth of genetic determinism.

But there is another myth complicating matters, and I have already mentioned it at the beginning of this chapter. The computer myth is almost as deep-seated in the modern mind as the gene myth. Notice that both passages I quoted contain the word 'programmed'. Thus Rose sarcastically absolved promiscuous men from blame because they are genetically *programmed*. Gould says that if we are *programmed* to be what we are then these traits are ineluctable. And it is true that we ordinarily use the word programmed to indicate unthinking inflexibility, the antithesis of freedom of action. Computers and 'robots' are, by repute, notoriously inflexible, carrying out instructions to the letter, even if the consequences are obviously absurd. Why else would they send out those famous million pound bills that everybody's friend's friend's cousin's acquaintance keeps receiving? I had forgotten the great computer myth, as well as the great gene myth, or I would have been more careful when I myself wrote of genes swarming 'inside gigantic lumbering robots', and of ourselves as 'survival machines—robot vehicles blindly programmed to preserve the selfish molecules known as genes' (Dawkins 1976a). These passages have been triumphantly quoted, and requoted apparently from secondary and even tertiary sources, as examples of rabid genetic determinism (e.g. 'Nabi' 1981). I am not apologizing for using the language of robotics. I would use it again without hesitation. But I now realize that it is necessary to give more explanation.

From 13 years' experience of teaching it, I know that a main problem with the 'selfish-gene survival machine' way of looking at natural selection is a particular risk of misunderstanding. The metaphor of the intelligent gene reckoning up how best to ensure its own survival (Hamilton 1972) is a powerful and illuminating one. But it is all too easy to get carried away, and allow hypothetical genes cognitive wisdom and foresight in planning their 'strategy'. At least three out of twelve misunderstandings of kin selection (Dawkins 1979) are directly attributable to this basic error. Time and again, non-biologists have tried to justify a form of group selection to me by, in effect, imputing foresight to genes: 'The long-term interests of a gene require the continued existence of the species; therefore shouldn't you expect adaptations to prevent species extinction, even at the expense of short-term individual reproductive success?' It was in an attempt to forestall errors like this that I used the language of automation and robotics, and used the word 'blindly' in referring to genetic programming. But it is, of course, the genes that are blind, not the animals they program. Nervous systems, like man-made computers, can be sufficiently complex to show intelligence and foresight.

Symons (1979) makes the computer myth explicit:

> I wish to point out that Dawkins's implication—through the use of words like 'robot' and 'blindly'—that evolutionary theory favors determinism is utterly without foundation.... A robot is a mindless automaton. Perhaps some animals are robots (we have no way of knowing); however, Dawkins is not referring to *some* animals, but to all animals and in this case specifically to human beings. Now, to paraphrase Stebbing, 'robot' can be opposed to 'thinking being' or it can be used figuratively to indicate a person who seems to act mechanically, but there is no common usage of language that provides a meaning for the word 'robot' in which it would make sense to say that all living things are robots [p. 41].

The point of the passage from Stebbing which Symons paraphrased is the reasonable one that X is a useless word unless there are some things that are not X. If everything is a robot, then the word robot doesn't mean anything useful. But the word robot has other associations, and rigid inflexibility was not the association I was thinking of. A robot is a programmed machine, and an important thing about programming is that it is distinct from, and done in advance of, performance of the behaviour itself. A computer is programmed to perform the behaviour of calculating square roots, or playing chess. The relationship between a chess-playing computer and the person who programmed it is not obvious, and is open to misunderstanding. It might be thought that the programmer watches the progress of the game and gives instructions to the computer move by move. In fact, however, the programming is finished before the game begins. The programmer tries to anticipate contingencies, and builds in conditional instructions of great complexity, but once the game begins he has to keep his hands off. He is not allowed to give the computer any new hints during the course of the game. If he did he would not be programming but performing, and his entry would be disqualified from the tournament. In the work criticized by Symons, I made extensive use of the analogy of computer chess in order to explain the point that genes do not control behaviour directly in the sense of interfering in its performance. They only control behaviour in the sense of programming the machine in *advance* of performance. It was this association with the word robot that I wanted to invoke, not the association with mindless inflexibility.

As for the mindless inflexibility association itself, it could have been justified in the days when the acme of automation was the rod and cam control system of a marine engine, and Kipling wrote 'McAndrew's Hymn':

> From coupler-flange to spindle-guide I see Thy Hand, O God—
> Predestination in the stride o' yon connectin'-rod.
> John Calvin might ha' forged the same—

But that was 1893 and the heyday of steam. We are now well embarked on the golden age of electronics. If machines ever had associations with rigid inflexibility—and I accept that they had—it is high time they lived them down. Computer programs have now been written that play chess to International Master standard (Levy 1978), that converse and reason in correct and indefinitely complex grammatical English (Winograd 1972), that create elegant and aesthetically satisfying new proofs of mathematical theorems (Hofstadter 1979), that compose music and diagnose illness; and the pace of progress in the field shows no sign of slowing down (Evans 1979). The advanced programming field known as artificial intelligence is in a buoyant, confident state (Boden 1977). Few who have studied it would now bet against computer programs beating the strongest Grand Masters at chess within the next 10 years. From being synonymous in the popular mind with a moronically undeviating, jerky-limbed zombie, 'robot' will one day become a byword for flexibility and rapid intelligence.

Unfortunately I jumped the gun a little in the passage quoted. When I wrote it I had just returned from an eye-opening and mind-boggling conference on the state of the art of artificial intelligence programming, and I genuinely and innocently in my enthusiasm forgot that robots are popularly supposed to be inflexible idiots. I also have to apologize for the fact that, without my knowledge, the cover of the German edition of *The Selfish Gene* was given a picture of a human puppet jerking on the end of strings descending from the word gene, and the French edition a picture of little bowler-hatted men with clockwork wind-up keys sticking out of their backs. I have had slides of both covers made up as illustrations of what I was *not* trying to say.

So, the answer to Symons is that of course he was right to criticize what he thought I was saying, but of course I wasn't actually saying it (Ridley 1980). No doubt I was partly to blame for the original misunderstanding, but I can only urge now that we put aside the preconceptions derived from common usage ('most men don't understand computers to even the slightest degree'—Weizenbaum 1976, p. 9), and actually go and read some of the fascinating modern literature on robotics and computer intelligence (e.g. Boden 1977; Evans 1979; Hofstadter 1979).

Once again, of course, philosophers may debate the ultimate determinacy of computers programmed to behave in artificially intelligent ways, but if we are going to get into that level of philosophy many would apply the same arguments to human intelligence (Turing 1950). What is a brain, they would ask, but a computer, and what is education but a form of programming? It is very hard to give a non-supernatural account of the human brain and human emotions, feelings, and apparent free will, *without* regarding the brain as, in some sense, the equivalent of a programmed, cybernetic machine. The astronomer Sir Fred Hoyle (1964) expresses very vividly what, it seems to me, any evolutionist must think about nervous systems:

> Looking back [at evolution] I am overwhelmingly impressed by the way in which chemistry has gradually given way to electronics. It is not unreasonable to describe the first living creatures as entirely chemical in character. Although electrochemical processes are important in plants, organized electronics, in the sense of data processing, does not enter or operate in the plant world. But primitive electronics begins to assume importance as soon as we have a creature that moves around….The first electronic systems possessed by primitive animals were essentially guidance systems, analogous logically to sonar or radar. As we pass to more developed animals we find electronic systems being used not merely for guidance but for directing the animal toward food….
>
> The situation is analogous to a guided missile, the job of which is to intercept and destroy another missile. Just as in our modern world attack and defense become more and more subtle in their methods, so it was the case with animals. And with increasing subtlety, better and better systems of electronics become necessary. What happened in nature has a close parallel with the development of electronics in modern military applications….I find it a sobering thought that but for the tooth-and-claw existence of the jungle we should not possess our intellectual capabilities, we should not be able to inquire into the structure of the Universe, or be able to appreciate a symphony of Beethoven….Viewed in this light, the question that is sometimes asked—can computers think?—is somewhat ironic.

Here of course I mean the computers that we ourselves make out of inorganic materials. What on earth do those who ask such a question think they themselves are? Simply computers, but vastly more complicated ones than anything we have yet learned to make. Remember that our man-made computer industry is a mere two or three decades old, whereas we ourselves are the products of an evolution that has operated over hundreds of millions of years [pp. 24–6].

Others may disagree with this conclusion, although I suspect that the only alternatives to it are religious ones. Whatever the outcome of that debate, to return to genes and the main point of this chapter, the issue of determinism versus free will is just not affected one way or the other by whether or not you happen to be considering *genes* as causal agents rather than environmental determinants.

But, it will pardonably be said, there is no smoke without fire. Functional ethologists and 'sociobiologists' must have said something to deserve being tarred with the brush of genetic determinism. Or if it is all a misunderstanding there must be some good explanation, because misunderstandings that are so widespread do not come about for no reason, even if abetted by cultural myths as powerful as the gene myth and the computer myth in unholy alliance. Speaking for myself, I think I know the reason. It is an interesting one, and it will occupy the rest of this chapter. The misunderstanding arises from the way we talk about a quite different subject, namely natural selection. Gene selectionism, which is a way of talking about evolution, is mistaken for genetic determinism, which is a point of view about development. People like me are continually postulating genes 'for' this and genes for that. We give the impression of being obsessed with genes and with 'genetically programmed' behaviour. Take this in conjunction with the popular myths of the Calvinistic determinacy of genes, and of 'programmed' behaviour as the hallmark of jactitating Disneyland puppets, and is it any wonder that we are accused of being genetic determinists?

Why, then, do functional ethologists talk about genes so much? Because we are interested in natural selection, and natural selection is differential survival of genes. If we are to so much as discuss the *possibility* of a behaviour pattern's evolving by natural selection, we have to postulate genetic variation with respect to the tendency or capacity to perform that behaviour pattern. This is not to say that there necessarily *is* such genetic variation for any particular behaviour pattern, only that there must have been genetic variation in the past if we are to treat the behaviour pattern as a Darwinian adaptation. Of course the behaviour pattern may not be a Darwinian adaptation, in which case the argument will not apply.

Incidentally, I should defend my usage of 'Darwinian adaptation' as synonymous with 'adaptation produced by natural selection', for Gould and Lewontin (1979) have recently emphasized, with approval, the 'pluralistic' character of Darwin's own thought. It is indeed true that, especially towards the end of his life, Darwin was driven by criticisms, which we can now see to be erroneous, to make some concessions to 'pluralism': he did not regard natural selection as the only important driving force in evolution. As the historian R. M. Young (1971) has sardonically put it, 'by the sixth edition the book was mistitled and should have read *On the Origin of Species by Means of Natural Selection and All Sorts of Other Things*'. It is, therefore, arguably incorrect to use 'Darwinian evolution' as synonymous with 'evolution by natural selection'. But Darwinian *adaptation* is another matter. Adaptation cannot be produced by random drift, or by any other realistic evolutionary force that we know of save natural selection. It is true that Darwin's pluralism did fleetingly allow for one other driving force that might, in principle, lead to adaptation, but that driving force is inseparably linked with the name of Lamarck, not of Darwin. 'Darwinian adaptation' could not sensibly mean anything other than adaptation produced by natural selection, and I shall use it in this sense. In several other places in this book, we shall resolve apparent disputes by drawing a distinction between evolution in general, and adaptive evolution in particular. The fixation of neutral mutations, for instance, can be regarded as evolution, but it is not adaptive evolution. If a molecular geneticist interested in gene substitutions, or a palaeontologist interested in major trends, argues with an ecologist interested in adaptation, they are likely to find themselves at cross-purposes simply because each of them emphasizes a different aspect of what evolution means.

'Genes for conformity, xenophobia, and aggressiveness are simply postulated for humans because they are needed for the theory, not because any evidence for them exists' (Lewontin 1979b). This is a fair criticism of E. O. Wilson, but not a very damning one. Apart from possible political repercussions which might be unfortunate, there is nothing wrong with cautiously speculating about a possible Darwinian survival value of xenophobia or any other trait. And you cannot begin to speculate, however cautiously, about the survival value of anything unless you postulate a genetic basis for variation in that thing. Of course xenophobia may not vary genetically, and of course xenophobia may not be a Darwinian adaptation, but we can't even discuss the possibility of its being a Darwinian adaptation unless we postulate a genetic basis for it. Lewontin himself has expressed the point as well as anybody: 'In order for a trait to evolve by natural selection it is necessary that there be genetic variation in the population for such a trait' (Lewontin 1979b). And 'genetic variation in the population for' a trait X is exactly what we mean when we talk, for brevity, of 'a gene for' X.

Xenophobia is controversial, so consider a behaviour pattern that nobody would fear to regard as a Darwinian adaptation. Pit-digging in antlions is obviously an adaptation to catch prey. Antlions are insects, neuropteran larvae with the general appearance and demeanour of monsters from outer space. They are 'sit and wait' predators who dig pits in soft sand which trap ants and other small walking insects. The pit is a nearly perfect cone, whose sides slope so steeply that prey cannot climb out once they have fallen in. The antlion sits just under the sand at the bottom of the pit, where it lunges with its horror-film jaws at anything that falls in.

Pit-digging is a complex behaviour pattern. It costs time and energy, and satisfies the most exacting criteria for recognition as an adaptation (Williams 1966; Curio 1973). It must, then, have evolved by natural selection. How might this have happened? The details don't matter for the moral I want to draw. Probably an ancestral antlion existed which did not dig a pit but simply lurked just beneath the sand surface waiting for prey to blunder over it. Indeed some species still do this. Later, behaviour leading to the creation of a shallow depression in the sand probably was favoured by selection because the depression marginally impeded escaping prey. By gradual degrees over many generations the behaviour changed so that what was a shallow depression became deeper and wider. This not only hindered escaping prey but also increased the catchment area over which prey might stumble in the first place. Later still the digging behaviour changed again so that the resulting pit became a steep-sided cone, lined with fine, sliding sand so that prey were unable to climb out.

Nothing in the previous paragraph is contentious or controversial. It will be regarded as legitimate speculation about historical events that we cannot see directly, and it will probably be thought plausible. One reason why it will be accepted as uncontroversial historical speculation is that it makes no mention of genes. But my point is that none of that history, nor any comparable history, could possibly have been true unless there was genetic variation in the behaviour at every step of the evolutionary way. Pit-digging in antlions is only one of the thousands of examples that I could have chosen. Unless natural selection has genetic variation to act upon, it cannot give rise to evolutionary change. It follows that where you find Darwinian adaptation there must have been genetic variation in the character concerned.

Nobody has ever done a genetic study of pit-digging behaviour in antlions (J. Lucas, personal communication). There is no need to do one, if all we want to do is satisfy ourselves of the sometime existence of genetic variation in the behaviour pattern. It is sufficient that we are convinced that it is a Darwinian adaptation (if you are not convinced that pit-digging is such an adaptation, simply substitute any example of which you are convinced).

I spoke of the *sometime* existence of genetic variation. This was because it is quite likely that, were a genetic study to be mounted of antlions today, no genetic variation would be found. It is in general to be expected that, where there is strong selection in favour of some trait, the original variation on which selection acted to guide the evolution of the trait will have become used up. This is the familiar 'paradox' (it is not really very paradoxical when we think about it carefully) that traits under strong selection tend to have low heritability (Falconer 1960); 'evolution by natural selection destroys the genetic variance on which it feeds' (Lewontin 1979b). Functional hypotheses frequently concern phenotypic traits, like possession of eyes, which are all but universal in the population, and therefore without contemporary genetic variation. When we speculate about, or make models of, the evolutionary production of an adaptation, we are necessarily talking about a time when there was appropriate genetic variation. We are bound, in such discussions, to postulate, implicitly or explicitly, genes 'for' proposed adaptations.

Some may balk at treating 'a genetic contribution to variation in X' as equivalent to 'a gene or genes for X'. But this is a routine genetic practice, and one which close examination shows to be almost inevitable. Other than at the molecular level, where one gene is seen directly to produce one protein chain, geneticists never deal with units of phenotype as such. Rather, they always deal with *differences*. When a geneticist speaks of a gene 'for' red eyes in *Drosophila*, he is not speaking of the cistron which acts as template for the synthesis of the red pigment molecule. He is implicitly saying: there is variation in eye colour in the population; other things being equal, a fly with this gene is more likely to have red eyes than a fly without the gene. That is all that we ever mean by a gene 'for' red eyes. This happens to be a morphological rather than a behavioural example, but exactly the same applies to behaviour. A gene 'for' behaviour X is a gene 'for' whatever morphological and physiological states tend to produce that behaviour.

A related point is that the use of single-locus models is just a conceptual convenience, and this is true of adaptive hypotheses in exactly the same way as it is true of ordinary population genetic models. When we use single-gene language in our adaptive hypotheses, we do not intend to make a point about single-gene models as against multi-gene models. We are usually making a point about *gene* models as against non-gene models, for example as against 'good of the species' models. Since it is difficult enough convincing people that they ought to think in genetic terms *at all* rather than in terms of, say, the good of the species, there is no sense in making things even more difficult by trying to handle the complexities of many loci at the outset. What Lloyd

(1979) calls the OGAM (one gene analysis model) is, of course, not the last word in genetic accuracy. *Of course* we shall eventually have to face up to multi-locus complexity. But the OGAM is vastly preferable to modes of adaptive reasoning that forget about genes altogether, and this is the only point I am trying to make at present.

Similarly we may find ourselves aggressively challenged to substantiate our 'claims' of the existence of 'genes for' some adaptation in which we are interested. But this challenge, if it is a real challenge at all, should be directed at the whole of the neo-Darwinian 'modern synthesis' and the whole of population genetics. To phrase a functional hypothesis in terms of genes is to make no strong claims about genes at all: it is simply to make explicit an assumption which is inseparably built into the modern synthesis, albeit it is sometimes implicit rather than explicit.

A few workers have, indeed, flung just such a challenge at the whole neo-Darwinian modern synthesis, and have claimed not to be neo-Darwinians. Goodwin (1979), in a published debate with Deborah Charlesworth and others, said, 'neo-Darwinism has an incoherence in it...we are not given any way of generating phenotypes from genotypes in neo-Darwinism. Therefore the theory is in this respect defective.' Goodwin is, of course, quite right that development is terribly complicated, and we don't yet understand much about how phenotypes are generated. But *that* they are generated, and *that* genes contribute significantly to their variation are incontrovertible facts, and those facts are all we need in order to make neo-Darwinism coherent. Goodwin might just as well say that, before Hodgkin and Huxley worked out how the nerve impulse fired, we were not entitled to believe that nerve impulses controlled behaviour. *Of course* it would be nice to know how phenotypes are made but, while embryologists are busy finding out, the rest of us are entitled by the known facts of genetics to carry on being neo-Darwinians, treating embryonic development as a black box. There is no competing theory that has even a remote claim to be called coherent.

It follows from the fact that geneticists are always concerned with phenotypic *differences* that we need not be afraid of postulating genes with indefinitely complex phenotypic effects, and with phenotypic effects that show themselves only in highly complex developmental conditions. Together with Professor John Maynard Smith, I recently took part in a public debate with two radical critics of 'sociobiology', before an audience of students. At one time in the discussion we were trying to establish that to talk of a gene 'for X' is to make no outlandish claim, even where X is a complex, learned behaviour pattern. Maynard Smith reached for a hypothetical example and came up with a 'gene for skill in tying shoelaces'. Pandemonium broke loose at this rampant genetic determinism! The air was thick with the unmistakable sound of worst suspicions being gleefully confirmed. Delightedly sceptical cries drowned the quiet and patient explanation of just what a *modest* claim is being made whenever one postulates a gene for, say, skill in tying shoelaces. Let me explain the point with the aid of an even more radical-sounding yet truly innocuous thought experiment (Dawkins 1981).

Reading is a learned skill of prodigious complexity, but this provides no reason in itself for scepticism about the possible existence of a gene for reading. All we would need in order to establish the existence of a gene for reading is to discover a gene for not reading, say a gene which induced a brain lesion causing specific dyslexia. Such a dyslexic person might be normal and intelligent in all respects except that he could not read. No geneticist would be particularly surprised if this type of dyslexia turned out to breed true in some Mendelian fashion. Obviously, in this event, the gene would only exhibit its effect in an environment which included normal education. In a prehistoric environment it might have had no detectable effect, or it might have had some different effect and have been known to cave-dwelling geneticists as, say, a gene for inability to read animal footprints. In our educated environment it would properly be called a gene 'for' dyslexia, since dyslexia would be its most salient consequence. Similarly, a gene which caused total blindness would also prevent reading, but it would not usefully be regarded as a gene for not reading. This is simply because preventing reading would not be its most obvious or debilitating phenotypic effect.

Returning to our gene for specific dyslexia, it follows from the ordinary conventions of genetic terminology that the wild-type gene at the same locus, the gene that the rest of the population has in double dose, would properly be called a gene 'for reading'. If you object to that, you must also object to our speaking of a gene for tallness in Mendel's peas, because the logic of the terminology is identical in the two cases. In both cases the character of interest is a *difference*, and in both cases the difference only shows itself in some specified environment. The reason why something so simple as a one gene difference can have such a complex effect as to determine whether or not a person can learn to read, or how good he is at tying shoelaces, is basically as follows. However complex a given state of the world may be, the *difference* between that state of the world and some alternative state of the world may be caused by something extremely simple.

The point I made using antlions is a general one. I could have used any real or purported Darwinian adaptation whatsoever. For further emphasis I shall use one more example. Tinbergen *et al.* (1962) investigated the adaptive significance of a particular behaviour pattern in black-headed gulls (*Larus ridibundus*), eggshell removal. Shortly after a chick hatches, the parent bird grasps the empty eggshell in the bill and removes it from the vicinity of the nest. Tinbergen and his colleagues considered a number of possible hypotheses about the survival value of this behaviour pattern. For instance they suggested that the empty eggshells might serve as breeding grounds for harmful bacteria, or the sharp edges might cut the chicks. But the hypothesis for which they ended up finding evidence was that the empty eggshell serves as a conspicuous visual beacon summoning crows and other predators of chicks or eggs to the nest. They did ingenious experiments, laying out artificial nests with and without empty eggshells, and showed that eggs accompanied by empty eggshells were, indeed, more likely to be attacked by crows than eggs without empty eggshells by their side. They concluded that natural selection had favoured eggshell removal behaviour of adult gulls, because past adults who did not do it reared fewer children.

As in the case of antlion digging, nobody has ever done a genetic study of eggshell removal behaviour in black-headed gulls. There is no direct evidence that variation in tendency to remove empty eggshells breeds true. Yet clearly the assumption that it does, or once did, is essential for the Tinbergen hypothesis. The Tinbergen hypothesis, as normally phrased in gene-free language, is not particularly controversial. Yet it, like all the rival functional hypotheses that Tinbergen rejected, rests fundamentally upon the assumption that once upon a time there must have been gulls with a genetic tendency to remove eggshells, and other gulls with a genetic tendency not to remove them, or to be less likely to remove them. There must have been genes for removing eggshells.

Here I must enter a note of caution. Suppose we actually did a study of the genetics of eggshell removal behaviour in modern gulls. It would be a behaviour-geneticist's dream to find a simple

Mendelian mutation which radically altered the behaviour pattern, perhaps abolished the behaviour altogether. By the argument given above, this mutant would truly be a gene 'for' not removing eggshells, and, by definition, its wild-type allele would have to be called a gene for eggshell removal. But now comes the note of caution. It most definitely does not follow that this particular locus 'for' eggshell removal was one of the ones upon which natural selection worked during the evolution of the adaptation. On the contrary, it seems much more probable that a complex behaviour pattern like eggshell removal must have been built up by selection on a large number of loci, each having a small effect in interaction with the others. Once the behaviour complex had been built up, it is easy to imagine a single major mutation arising, whose effect is to destroy it. Geneticists perforce must exploit the genetic variation available for them to study. They also believe that natural selection must have worked on similar genetic variation in wreaking evolutionary change. But there is no reason for them to believe that the loci controlling modern variation in an adaptation were the very same loci at which selection acted in building up the adaptation in the first place.

Consider the most famous example of single gene control of complex behaviour, the case of Rothenbuhler's (1964) hygienic bees. The point of using this example is that it illustrates well how a highly complex behaviour difference can be due to a single gene difference. The hygienic behaviour of the Brown strain of honeybees involves the whole neuromuscular system, but the fact that they perform the behaviour whereas Van Scoy bees do not is, according to Rothenbuhler's model, due to differences at two loci only. One locus determines the uncapping of cells containing diseased brood, the other locus determines the removing of diseased brood after uncapping. It would be possible, therefore, to imagine a natural selection in favour of uncapping behaviour and a natural selection in favour of removing behaviour, meaning selection of the two genes versus their respective alleles. But the point I am making here is that, although that could happen, it is not likely to be very interesting evolutionarily. The modern uncapping gene and the modern removing gene may very well not have been involved in the original natural selection process that steered the evolutionary putting together of the behaviour.

Rothenbuhler observed that even Van Scoy bees sometimes perform hygienic behaviour. They are just quantitatively much less likely to do so than are Brown bees. It is likely, therefore, that both Brown and Van Scoy bees have hygienic ancestors, and both have in their nervous systems the machinery of uncapping and removing behaviour: it is just that Van Scoy bees have genes that prevent them from turning the machinery on. Presumably if we went back even further in time we should find an ancestor of all modern bees which not only was not hygienic itself but had never had a hygienic ancestor. There must have been an evolutionary progression building up the uncapping and removing behaviour from nothing, and this evolutionary progression involved the selection of many genes which are now fixed in both the Brown and the Van Scoy strains. So, although the uncapping and the removing genes of the Brown strain really are rightly called genes for uncapping and removing, they are defined as such only because they happen to have alleles whose effect is to prevent the behaviour from being performed. The mode of action of these alleles could be boringly destructive. They might simply cut some vital link in the neural machinery. I am reminded of Gregory's (1961) vivid illustration of the perils of making inferences from ablation experiments on the brain: 'the removal of any of several widely spaced resistors may cause a radio set to emit howls, but it does not follow that howls are immediately associated with these resistors, or indeed that the causal relation is anything but the most indirect. In particular, we should not say that the function of the resistors in the normal circuit is to inhibit howling. Neurophysiologists, when faced with a comparable situation, have postulated "suppressor regions".'

This consideration seems to me to be a reason for caution, not a reason for rejecting the whole genetic theory of natural selection! Never mind if living geneticists are debarred from studying the particular loci at which selection in the past gave rise to the original evolution of interesting adaptations. It is too bad if geneticists usually are forced to concentrate on loci that are convenient rather than evolutionarily important. It is *still* true that the evolutionary putting together of complex and interesting adaptation consisted in the replacement of genes by their alleles.

This argument can contribute tangentially to the resolution of a fashionable contemporary dispute, by helping to put the issue in perspective. It is now highly, indeed passionately, controversial whether there is significant genetic variation in human mental abilities. Are some of us genetically brainier than others? What we mean by 'brainy' is also highly contentious, and rightly so. But I suggest that, by any meaning of the term, the following propositions cannot be denied. (1) There was a time when our ancestors were less brainy than we are. (2) Therefore there has been an increase in braininess in our ancestral lineage. (3) That increase came about through evolution, probably propelled by natural selection. (4) Whether propelled by selection or not, at least part of the evolutionary change in phenotype reflected an underlying genetic change: allele replacement took place and consequently mean mental ability increased over generations. (5) By definition therefore, at least in the past, there must have been significant genetic variation in braininess within the human population. Some people were genetically clever in comparison with their contemporaries; others were genetically relatively stupid.

The last sentence may engender a *frisson* of ideological disquiet, yet none of my five propositions could be seriously doubted, nor could their logical sequence. The argument works for brain size, but it equally works for any behavioural measure of cleverness we care to dream up. It does not depend on simplistic views of human intelligence as being a one-dimensional scalar quantity. The fact that intelligence is not a simple scalar quantity, important as that fact is, is simply irrelevant. So is the difficulty of measuring intelligence in practice. The conclusion of the previous paragraph is inevitable, provided only that we are evolutionists who agree to the proposition that once upon a time our ancestors were less clever (by whatever criterion) than we are. Yet in spite of all that, it still does not follow that there is any genetic variation in mental abilities left in the human population today: the genetic variance might all have been used up by selection. On the other hand it might not, and my thought experiment shows at least the inadvisability of dogmatic and hysterical opposition to the very possibility of genetic variation in human mental abilities. My own opinion, for what it is worth, is that even if there is such genetic variation in modern human populations, to base any policy on it would be illogical and wicked.

The existence of a Darwinian adaptation, then, implies the sometime existence of genes for producing the adaptation. This is not always made explicit. It is always possible to talk about the natural selection of a behaviour pattern in two ways. We can either talk about individuals with a tendency to perform the behaviour pattern being 'fitter' than individuals with a less strongly developed tendency. This is the now fashionable phraseology, within the paradigm of the 'selfish organism' and the 'central theorem of sociobiology'. Alternatively, and equivalently, we can talk directly of genes for performing the behaviour pattern surviving better than their alleles. It is always legitimate to postulate genes in any discussion of Darwinian adaptation, and it will be one of my central points in this book that it is often positively beneficial to do so. Objections, such as I have heard made, to the 'unnecessary geneticizing' of the language of functional ethology betray a fundamental failure to face up to the reality of what Darwinian selection is all about.

Let me illustrate this failure by another anecdote. I recently attended a research seminar given by an anthropologist. He was trying to interpret the incidence among various human tribes of a particular mating system (it happened to be polyandry) in terms of a theory of kin selection. A kin selection theorist can make models to predict the conditions under which we would expect to find polyandry. Thus, on one model applied to Tasmanian native hens (Maynard Smith & Ridpath 1972), the population sex ratio would need to be male-biased, and partners would need to be close kin, before a biologist would predict polyandry. The anthropologist sought to show that his polyandrous human tribes lived under such conditions, and, by implication, that other tribes showing the more normal patterns of monogamy or polygyny lived under different conditions.

Though fascinated by the information he presented, I tried to warn him of some difficulties in his hypothesis. I pointed out that the theory of kin selection is fundamentally a genetic theory, and that kin-selected adaptations to local conditions had to come about through the replacement of alleles by other alleles, over generations. Had his polyandrous tribes been living, I asked, under their current peculiar conditions for long enough—enough generations—for the necessary genetic replacement to have taken place? Was there, indeed, any reason to believe that variations in human mating systems are under genetic control at all?

The speaker, supported by many of his anthropological colleagues in the seminar, objected to my dragging genes into the discussion. He was not talking about genes, he said, but about a social behaviour pattern. Some of his colleagues seemed uncomfortable with the very mention of the four-letter word 'gene'. I tried to persuade him that it was *he* who had 'dragged genes in' to the discussion although, to be sure, he had not mentioned the word gene in his talk. That is exactly the point I am trying to make. You cannot talk about kin selection, or any other form of Darwinian selection, *without* dragging genes in, whether you do so explicitly or not. By even speculating about kin selection as an explanation of differences in tribal mating systems, my anthropologist friend was implicitly dragging genes into the discussion. It is a pity he did not make it *explicit*, because he would then have realized what formidable difficulties lay in the path of his kin selection hypothesis: either his polyandrous tribes had to have been living, in partial genetic isolation, under their peculiar conditions for a large number of centuries, or natural selection had to have favoured the universal occurrence of genes programming some complex 'conditional strategy'. The irony is that, of all the participants in that seminar on polyandry, it was I who was advancing the least 'genetically deterministic' view of the behaviour under discussion. Yet because I insisted on making the genetic nature of the kin selection hypothesis explicit, I expect I appeared to be characteristically obsessed with genes, a 'typical genetic determinist'. The story illustrates well the main message of this chapter, that frankly facing up to the fundamental genetic nature of Darwinian *selection* is all too easily mistaken for an unhealthy preoccupation with hereditarian interpretations of ontogenetic *development*.

The same prejudice against explicit mention of genes where one can get away with an individual-level circumlocution is common among biologists. The statement 'genes for performing behaviour X are favoured over genes for not performing X' has a vaguely naive and unprofessional ring to it. What evidence is there for such genes? How dare you conjure up *ad hoc* genes simply to satisfy your hypothetical convenience! To say 'individuals that perform X are fitter than individuals that do not perform X' sounds much more respectable. Even if it is not known to be true, it will probably be accepted as a permissible speculation. But the two sentences are exactly equivalent in meaning. The second one says nothing that the first does not say more clearly. Yet if we recognize this equivalence and talk explicitly about genes 'for' adaptations, we run the risk of being accused of 'genetic determinism'. I hope I have succeeded in showing that this risk results from nothing more than misunderstanding. A sensible and unexceptionable way of thinking about natural selection—'gene selectionism'—is mistaken for a strong belief about development—'genetic determinism'. Anyone who thinks clearly about the details of how adaptations come into being is almost bound to think, implicitly if not explicitly, about genes, albeit they may be hypothetical genes. As I shall show in this book, there is much to be said for making the genetic basis of Darwinian functional speculations explicit rather than implicit. It is a good way of avoiding certain tempting errors of reasoning (Lloyd 1979). In doing this we may give the impression, entirely for the wrong reason, of being obsessed with genes and all the mythic baggage that genes carry in the contemporary journalistic consciousness. But determinism, in the sense of an inflexible, tramline-following ontogeny, is, or should be, a thousand miles from our thoughts. Of course, individual sociobiologists may or may not be genetic determinists. They may be Rastafarians, Shakers, or Marxists. But their private opinions on genetic determinism, like their private opinions on religion, have nothing to do with the fact that they use the language of 'genes for behaviour' when talking about natural selection.

A large part of this chapter has been based on the assumption that a biologist might wish to speculate on the Darwinian 'function' of behaviour patterns. This is not to say that all behaviour patterns necessarily have a Darwinian function. It may be that there is a large class of behaviour patterns which are selectively neutral or deleterious to their performers, and cannot usefully be regarded as the products of natural selection. If so, the arguments of this chapter do not apply to them. But it is legitimate to say 'I am interested in adaptation. I don't necessarily think all behaviour patterns are adaptations, but I want to study those behaviour patterns that are adaptations.' Similarly, to express a preference for studying vertebrates rather than invertebrates does not commit us to the belief that all animals are vertebrates. Given that our field of interest is adaptive behaviour, we cannot talk about the Darwinian evolution of the objects of interest without postulating a genetic basis for them. And to use 'a gene for X' as a convenient way of talking about 'the genetic basis of X' has been standard practice in population genetics for over half a century.

The question of how large is the class of behaviour patterns that we can consider to be adaptations is an entirely separate question. It is the subject of the next chapter.

3

CONSTRAINTS ON PERFECTION

In one way or another, this book is largely preoccupied with the logic of Darwinian explanations of function. Bitter experience warns that a biologist who shows a strong interest in functional explanation is likely to be accused, sometimes with a passion that startles those more accustomed to scientific than ideological debate (Lewontin 1977), of believing that all animals are perfectly optimal—accused of being an 'adaptationist' (Lewontin 1979a,b; Gould & Lewontin 1979). Adaptationism is defined as 'that approach to evolutionary studies which assumes without further proof that all aspects of the morphology, physiology and behavior of organisms are adaptive optimal solutions to problems' (Lewontin 1979b). In the first draft of this chapter I expressed doubts that anyone was truly an adaptationist in the extreme sense, but I have recently found the following quotation from, ironically enough, Lewontin himself: 'That is the one point which I think all evolutionists are agreed upon, that it is virtually impossible to do a better job than an organism is doing in its own environment' (Lewontin 1967). Lewontin has since, it seems, travelled his road to Damascus, so it would be unfair to use him as my adaptationist spokesman. Indeed together with Gould he has, in recent years, been one of the most articulate and forceful critics of adaptationism. As my representative adaptationist I take A. J. Cain, who has remained (Cain 1979) consistently true to the views expressed in his trenchant and elegant paper on 'The perfection of animals'.

Writing as a taxonomist, Cain (1964) is concerned to attack the traditional dichotomy between 'functional' characters, which by implication are not reliable taxonomic indicators, and 'ancestral' characters, which are. Cain argues forcefully that ancient 'ground-plan' characters, like the pentadactyl limb of tetrapods and the aquatic phase of amphibians, are there because they are functionally useful, rather than because they are inescapable historical legacies as is often implied. If one of two groups 'is in any way more primitive than the other, then its primitiveness must in itself be an adaptation to some less specialized mode of life which it can pursue successfully; it cannot be merely a sign of inefficiency' (p. 57). Cain makes a similar point about so-called trivial characters, criticizing Darwin for being too ready, under the at first sight surprising influence of Richard Owen, to concede functionlessness: 'No one will suppose that the stripes on the whelp of a lion, or the spots on the young blackbird, are of any use to these animals'. Darwin's remark must sound foolhardy today even to the most extreme critic of adaptationism. Indeed, history seems to be on the side of the adaptationists, in the sense that in particular instances they have confounded the scoffers again and again. Cain's own celebrated work, with Sheppard and their school, on the selection pressures maintaining the banding polymorphism in the snail *Cepaea nemoralis* may have been partly provoked by the fact that 'it had been confidently asserted that it could not matter to a snail whether it had one band on its shell or two' (Cain, p. 48). 'But perhaps the most remarkable functional interpretation of a "trivial" character is given by Manton's work on the diplopod *Polyxenus*, in which she has shown that a character formerly described as an "ornament" (and what could sound more useless?) is almost literally the pivot of the animal's life' (Cain, p. 51).

Adaptationism as a working hypothesis, almost as a faith, has undoubtedly been the inspiration for some outstanding discoveries. Von Frisch (1967), in defiance of the prestigious orthodoxy of von Hess, conclusively demonstrated colour vision in fish and in honeybees by controlled experiments. He was driven to undertake those experiments by his refusal to believe that, for example, the colours of flowers were there for no reason, or simply to delight men's eyes. This is, of course, not evidence for the validity of adaptationist faith. Each question must be tackled afresh, on its merits.

Wenner (1971) performed a valuable service in questioning von Frisch's dance language hypothesis, since he provoked J. L. Gould's (1976) brilliant confirmation of von Frisch's theory. If Wenner had been more of an adaptationist Gould's research might never have been done, but Wenner would also not have allowed himself to be so blithely wrong. Any adaptationist, while perhaps conceding that Wenner had usefully exposed lacunae in von Frisch's original experimental design, would instantly have jumped, with Lindauer (1971), on the fundamental question of why bees dance at all. Wenner never denied that they dance, nor that the dance contained all the information about the direction and distance of food that von Frisch claimed. All he denied was that other bees used the dance information. An adaptationist could not have rested happy with the idea of animals performing such a time-consuming, and above all complex and statistically improbable, activity for nothing. Adaptationism cuts both ways, however. I am now delighted that Gould did his clinching experiments, and it is entirely to my discredit that, even in the unlikely event of my having been ingenious enough to think of them, I would have been too adaptationist to have bothered. I just *knew* Wenner was wrong (Dawkins 1969)!

Adaptationist thinking, if not blind conviction, has been a valuable stimulator of testable hypotheses in physiology. Barlow's (1961) recognition of the overwhelming functional need in sensory systems to reduce redundancy in input led him to a uniquely coherent understanding of a variety of facts about sensory physiology. Analogous functional reasoning can be applied to the motor system, and to hierarchical systems of organization generally (Dawkins 1976b; Hailman 1977). Adaptationist conviction cannot tell us about physiological mechanism. Only physiological experiment can do that. But cautious adaptationist reasoning can suggest which of many possible physiological hypotheses are most promising and should be tested first.

I have tried to show that adaptationism can have virtues as well as faults. But this chapter's main purpose is to list and classify constraints on perfection, to list the main reasons why the student of adaptation should proceed with caution. Before coming to my list of six constraints on perfection, I should deal with three others that have been proposed, but which I find less persuasive. Taking, first, the modern controversy among biochemical geneticists about 'neutral mutations', repeatedly cited in critiques of adaptationism, it is simply irrelevant. If there are neutral mutations in the biochemists' sense, what this means is that any change in polypeptide structure which they induce has no effect on the enzymatic activity of the protein. This means that the neutral mutation will not change the course of embryonic development, will have no phenotypic effect *at all*, as a whole-organism biologist would understand phenotypic effect. The biochemical controversy over neutralism is concerned with the interesting and important question of whether all gene substitutions have phenotypic effects. The adaptationism controversy is quite different. It is concerned with whether, *given* that we are dealing with a phenotypic effect big enough to see and ask questions about, we should assume that it is the product of natural selection. The biochemist's 'neutral mutations' are more than neutral. As far as those of us who look at gross morphology, physiology, and behaviour are concerned, they are not mutations at all. It was in this spirit that Maynard Smith (1976) wrote: 'I interpret "rate of evolution" as a rate of adaptive change. In this sense, the substitution of a neutral allele would not constitute evolution'. If a whole-organism biologist sees a genetically determined difference among phenotypes, he already knows he cannot be dealing with neutrality in the sense of the modern controversy among biochemical geneticists.

He might, nevertheless, be dealing with a neutral character in the sense of an earlier controversy (Fisher & Ford 1950; Wright 1951). A genetic difference could show itself at the phenotypic level, yet still be selectively neutral. But mathematical calculations such as those of Fisher (1930b) and Haldane (1932a) show how unreliable human subjective judgement can be on the 'obviously trivial' nature of some biological characters. Haldane, for example, showed that, with plausible assumptions about a typical population, a selection pressure as weak as 1 in 1000 would take only a few thousand generations to push an initially rare mutation to fixation, a small time by geological standards. It appears that, in the controversy referred to above, Wright was misunderstood (see below). Wright (1980) was embarrassed at finding the idea of evolution of nonadaptive characters by genetic drift labelled the 'Sewall Wright effect', 'not only because others had previously advanced the same idea, but because I myself had strongly rejected it from the first (1929), stating that pure random drift leads "inevitably to degeneration and extinction"'. I have

attributed apparent nonadaptive taxonomic differences to pleiotropy, where not merely ignorance of an adaptive significance.'
Wright was in fact showing how a subtle mixture of drift and
selection can produce adaptations *superior* to the products of
selection alone (see pp. 39–40).

A second suggested constraint on perfection concerns allometry (Huxley 1932): 'In cervine deer, antler size increases more
than proportionately to body size...so that larger deer have
more than proportionately large antlers. It is then unnecessary
to give a specifically adaptive reason for the extremely large antlers of large deer' (Lewontin 1979b). Well, Lewontin has a point
here, but I would prefer to rephrase it. As it stands it suggests that
the allometric constant is constant in a God-given immutable
sense. But constants on one time scale can be variables on
another. The allometric constant is a parameter of embryonic
development. Like any other such parameter it may be subject to
genetic variation and therefore it may change over evolutionary
time (Clutton-Brock & Harvey 1979). Lewontin's remark turns
out to be analogous to the following: all primates have teeth; this
is just a plain fact about primates, and it is therefore unnecessary
to give a specifically adaptive reason for the presence of teeth in
primates. What he probably meant to say is something like the
following.

Deer have evolved a developmental mechanism such that
growth of antlers relative to body size is allometric with a particular constant of allometry. Very probably the evolution of this
allometric system of development occurred under the influence
of selection pressures having nothing to do with the social function of antlers: probably it was conveniently compatible with
pre-existing developmental processes in a way which we shall
not understand until we know more about the biochemical and
cellular details of embryology. Maybe ethological consequences

of the extra large antlers of large deer exert a selective effect, but this selection pressure is likely to be swamped in importance by other selection pressures concerned with concealed internal embryological details.

Williams (1966, p. 16) invoked allometry in the service of a speculation about the selection pressures leading to increased brain size in man. He suggested that the prime focus of selection was on early teachability, at an elementary level, of children. 'The resulting selection for acquiring verbal facility as early as possible might have produced, as an allometric effect on cerebral development, populations in which an occasional Leonardo might arise.' Williams, however, did not see allometry as a weapon against the use of adaptive explanations. One feels that he was rightly less loyal to his particular theory of cerebral hypertrophy than to the general principle enunciated in his concluding rhetorical question: 'Is it not reasonable to anticipate that our understanding of the human mind would be aided greatly by knowing the purpose for which it was designed?'

What has been said of allometry applies also to pleiotropy, the possession by one gene of more than one phenotypic effect. This is the third of the suggested constraints on perfection that I want to get out of the way before embarking on my main list. It has already been mentioned in my quotation from Wright. A possible source of confusion here is that pleiotropy has been used as a weapon by both sides in this debate, if indeed it is a real debate. Fisher (1930b) reasoned that it was unlikely that any one of a gene's phenotypic effects was neutral, so how much more unlikely was it that *all* of a gene's pleiotropic effects could be neutral. Lewontin (1979b), on the other hand, remarked that 'many changes in characters are the result of pleiotropic gene action, rather than the direct result of selection on the character itself. The yellow color of the Malpighian tubules of an insect cannot itself be the subject of natural selection since that color can never be seen by any organism. Rather it is the pleiotropic consequence of red eye pigment metabolism, which may be adaptive.' There is no real disagreement here. Fisher was talking of the selective effects on a genetic mutation, Lewontin of selective effects on a phenotypic character; it is the same distinction, indeed, as I was making in discussing neutrality in the biochemical geneticists' sense.

Lewontin's point about pleiotropy is related to another one which I shall come on to below, about the problem of defining what he calls the natural 'suture lines', the 'phenotypic units' of evolution. Sometimes the dual effects of a gene are in principle inseparable; they are different views of the same thing, just as Everest used to have two names depending on which side it was seen from. What a biochemist sees as an oxygen-carrying molecule may be seen by an ethologist as red coloration. But there is a more interesting kind of pleiotropy in which the two phenotypic effects of a mutation are separable. The phenotypic effect of any gene (versus its alleles) is not a property of the gene alone, but also of the embryological context in which it acts. This allows abundant opportunities for the phenotypic effects of one mutation to be modified by others, and is the basis of such respected ideas as Fisher's (1930a) theory of the evolution of dominance, the Medawar (1952) and Williams (1957) theories of senescence, and Hamilton's (1967) theory of Y-chromosome inertness. In the present connection, if a mutation has one beneficial effect and one harmful one, there is no reason why selection should not favour modifier genes that detach the two phenotypic effects, or that reduce the harmful effect while enhancing the beneficial one. As in the case of allometry, Lewontin took too static a view of gene action, treating pleiotropy as if it was a property of the gene rather than of the interaction between the gene and its (modifiable) embryological context.

This brings me to my own critique of naive adaptationism, my own list of constraints on perfection, a list which has much in common with those of Lewontin and Cain, and those of Maynard Smith (1978b), Oster and Wilson (1978), Williams (1966), Curio (1973), and others. There is, indeed, much more agreement than the polemical tone of recent critiques would suggest. I shall not be concerned with particular cases, except as examples. As Cain and Lewontin both stress, it is not of general interest to challenge our ingenuity in dreaming up possible advantages of particular strange things that animals do. Here we are interested in the more general question of what the theory of natural selection entitles us to expect. My first constraint on perfection is an obvious one, mentioned by most writers on adaptation.

Time lags

The animal we are looking at is very probably out of date, built under the influence of genes that were selected in some earlier era when conditions were different. Maynard Smith (1976) gives a quantitative measure of this effect, the 'lag load'. He (Maynard Smith 1978b) cites Nelson's demonstration that gannets, who normally lay only one egg, are quite capable of successfully incubating and rearing two if an extra one is experimentally added. Obviously an awkward case for the Lack hypothesis on optimal clutch size, and Lack himself (1966) was not slow to use the 'time-lag' escape route. He suggested, entirely plausibly, that the gannet clutch size of one egg evolved during a time when food was less plentiful, and that there had not yet been time for them to evolve to meet the changed conditions.

Such *post hoc* rescuing of a hypothesis in trouble is apt to provoke accusations of the sin of unfalsifiability, but I find such accusations rather unconstructive, almost nihilistic. We are not in Parliament or a court of law, with advocates of Darwinism scoring debating points against opponents, and vice versa. With the exception of a few genuine opponents of Darwinism, who are unlikely to be reading this, we are all in this together, all Darwinians who substantially agree on how we interpret what is, after all, the only workable theory we have to explain the organized complexity of life. We should all sincerely want to *know* why gannets lay only one egg when they could lay two, rather than treating the fact as a debating point. Lack's invoking of the 'time-lag' hypothesis may have been *post hoc*, but it is still thoroughly plausible, and it is testable. No doubt there are other possibilities which, with luck, may also be testable. Maynard Smith is surely right that we should leave aside the 'defeatist' (Tinbergen 1965) and untestable 'natural selection has bungled again' explanation as a last resort, as a matter of simple research strategy if nothing else. Lewontin (1979b) says much the same: 'In a sense, then, biologists are forced to the extreme adaptationist program because the alternatives, although they are undoubtedly operative in many cases, are untestable in particular cases.'

Returning to the time-lag effect itself, since modern man has drastically changed the environment of many animals and plants over a time-scale that is negligible by ordinary evolutionary standards, we can expect to see anachronistic adaptations rather often. The hedgehog antipredator response of rolling up into a ball is sadly inadequate against motor cars.

Lay critics frequently bring up some apparently maladaptive feature of modern human behaviour—adoption, say, or contraception—and fling down a challenge to 'explain that if you can with your selfish genes'. Obviously, as Lewontin, Gould and others have rightly stressed, it would be possible, depending on one's ingenuity, to pull a 'sociobiological' explanation out of a hat, a 'just-so story', but I agree with them and Cain that the answering of such challenges is a trivial exercise; indeed it is likely to be positively harmful. Adoption and contraception, like reading, mathematics, and stress-induced illness, are products of an animal that is living in an environment radically different from the one in which its genes were naturally selected. The question, about the adaptive significance of behaviour in an artificial world, should never have been put; and although a silly question may deserve a silly answer, it is wiser to give no answer at all and to explain why.

A useful analogy here is one that I heard from R. D. Alexander. Moths fly into candle flames, and this does nothing to help their inclusive fitness. In the world before candles were invented, small sources of bright light in darkness would either have been celestial bodies at optical infinity, or they might have been escape holes from caves or other enclosed spaces. The latter case immediately suggests a survival value for approaching light sources. The former case also suggests one, but in a more indirect sense (Fraenkel & Gunn 1940). Many insects use celestial bodies as compasses. Since these are at optical infinity, rays from them are parallel, and an insect that maintains a fixed orientation of, say, 30° to them will go in a straight line. But if the rays do not come from infinity they will not be parallel, and an insect that behaves in this way will spiral in to the light source (if steering an acute-angled course) or spiral away (if steering an obtuse-angled course) or orbit the source (if steering a course of exactly 90° to the rays). Self-immolation by insects in candle flames, then, has no survival value in itself: according to this hypothesis, it is a byproduct of the useful habit of steering by means of sources of light which are 'assumed' to be at infinity. That assumption was once safe. It now is safe no longer, and it may be that selection is even now working to modify the insects' behaviour. (Not necessarily, however. The overhead costs of making the necessary improvements may outweigh the benefits they might bring: moths that pay the costs of discriminating candles from stars may be less successful, on average, than moths that do not attempt the costly discrimination and accept the low risk of self-immolation.)

But now we have reached a problem which is more subtle than the simple time-lag hypothesis itself. This is the problem, already mentioned, about what characteristics of animals we choose to recognize as units which require explanation. As Lewontin (1979b) puts it, 'What are the "natural" suture lines for evolutionary dynamics? What is the topology of phenotype in evolution? What are the phenotypic units of evolution?' The candle flame paradox arose only because of the way in which we chose to characterize the moth's behaviour. We asked 'Why do moths fly into candle flames?' and were puzzled. If we had characterized the behaviour differently and asked 'Why do moths maintain a fixed angle to light rays (a habit which incidentally causes them to spiral into the light source if the rays happen not to be parallel)?', we should not have been so puzzled.

Consider human male homosexuality as a more serious example. On the face of it, the existence of a substantial minority of men who prefer sexual relations with their own sex rather than with the opposite sex constitutes a problem for any simple Darwinian theory. The rather discursive title of a privately circulated homosexualist pamphlet, which the author was kind enough to send me, summarizes the problem: 'Why are there "gays" at all? Why hasn't evolution eliminated "gayness" millions of years ago?' The author, incidentally, thinks the problem so important that it seriously undermines the whole Darwinian view of life. Trivers (1974), Wilson (1975, 1978), and especially Weinrich (1976) have considered various versions of the possibility that homosexuals may, at some time in history, have been functionally equivalent to sterile workers, foregoing personal reproduction the better to care for other relatives. I do not find this idea particularly plausible (Ridley & Dawkins 1981), certainly no more so than a 'sneaky male' hypothesis. According to this latter idea, homosexuality represents an 'alternative male tactic' for obtaining matings with females. In a society with harem defence by dominant males, a male who is known to be homosexual is more likely to be tolerated by a dominant male than a known heterosexual male, and an otherwise subordinate male may be able, by virtue of this, to obtain clandestine copulations with females.

But I raise the 'sneaky male' hypothesis not as a plausible possibility so much as a way of dramatizing how easy and inconclusive it is to dream up explanations of this kind (Lewontin, 1979b, used the same didactic trick in discussing apparent homosexuality in *Drosophila*). The main point I wish to make is quite different and much more important. It is again the point about how we characterize the phenotypic feature that we are trying to explain.

Homosexuality is, of course, a problem for Darwinians only if there is a genetic component to the difference between homosexual and heterosexual individuals. While the evidence is controversial (Weinrich 1976), let us assume for the sake of argument that this is the case. Now the question arises, what does it *mean* to say there is a genetic component to the difference, in common parlance that there is a gene (or genes) 'for' homosexuality? It is a fundamental truism, of logic more than of genetics, that the phenotypic 'effect' of a gene is a concept that has meaning only if the context of environmental influences is specified, environment being understood to include all the other genes in the genome. A gene 'for' A in environment X may well turn out to be a gene for B in environment Y. It is simply meaningless to speak of an absolute, context-free, phenotypic effect of a given gene.

Even if there are genes which, in today's environment, produce a homosexual phenotype, this does not mean that in another environment, say that of our Pleistocene ancestors, they would have had the same phenotypic effect. A gene for homosexuality in our modern environment might have been a gene for something utterly different in the Pleistocene. So, we have the possibility of a special kind of 'time-lag effect' here. It may be that the phenotype which we are trying to explain did not even exist in some earlier environment, even though the gene did then exist. The ordinary time-lag effect which we discussed at the beginning of this section was concerned with changes in the environment as manifested in changed selection pressures. We have now added the more subtle point that changes in the environment may change the very nature of the phenotypic character we set out to explain.

Historical constraints

The jet engine superseded the propeller engine because, for most purposes, it was superior. The designers of the first jet engine started with a clean drawing board. Imagine what they would have produced if they had been constrained to 'evolve' the first jet engine from an existing propeller engine, changing one component at a time, nut by nut, screw by screw, rivet by rivet. A jet engine so assembled would be a weird contraption indeed. It is hard to imagine that an aeroplane designed in that evolutionary way would ever get off the ground. Yet in order to complete the biological analogy we have to add yet another constraint. Not only must the end product get off the ground; so must every intermediate along the way, and each intermediate must be superior to its predecessor. When looked at in this light, far from expecting animals to be perfect we may wonder that anything about them works at all.

Examples of the Heath Robinson (or Rube Goldberg—Gould 1978) character of animals are harder to be confident of than the

previous paragraph might lead us to expect. A favourite example, suggested to me by Professor J. D. Currey, is the recurrent
laryngeal nerve. The shortest distance from the brain to the lar-
ynx in a mammal, especially a giraffe, is emphatically not via the
posterior side of the aorta, yet that is the route taken by the recurrent laryngeal. Presumably there once was a time in the remote
ancestry of the mammals when the straight line from origin to
end organ of the nerve did run posterior to the aorta. When, in
due course, the neck began to lengthen, the nerve lengthened its
detour posterior to the aorta, but the marginal cost of each step
in the lengthening of the detour was not great. A major mutation
might have re-routed the nerve completely, but only at a cost of
great upheaval in early embryonic processes. Perhaps a pro-
phetic, God-like designer back in the Devonian could have fore-
seen the giraffe and designed the original embryonic routing of
the nerve differently, but natural selection has no foresight. As
Sydney Brenner has remarked, natural selection could not be
expected to have favoured some useless mutation in the Cambrian
simply because 'it might come in handy in the Cretaceous'.

The Picasso-like face of a flatfish such as a sole, grotesquely
twisted to bring both eyes round to the same side of the head, is
another striking demonstration of a historical constraint on perfection. The evolutionary history of these fish is so clearly written into their anatomy that the example is a good one to thrust
down the throats of religious fundamentalists. Much the same
could be said of the curious fact that the retina of the vertebrate
eye appears to be installed backwards. The light-sensitive 'photo-
cells' are at the back of the retina, and light has to pass through
the connecting circuitry, with some inevitable attenuation,
before it reaches them. Presumably it would be possible to write
down a very long sequence of mutations which would eventually
lead to the production of an eye whose retina was 'the right way

round' as it is in cephalopods, and this might be, in the end, slightly more efficient. But the cost in embryological upheaval would be so great that the intermediate stages would be heavily disfavoured by natural selection in comparison with the rival, patched-up job which does, after all, work pretty well. Pittendrigh (1958) has well said of adaptive organization that it is 'a patchwork of makeshifts pieced together, as it were, from what was available when opportunity knocked, and accepted in the hindsight, not the foresight, of natural selection' (see also Jacob, 1977, on 'tinkering').

Sewall Wright's (1932) metaphor, which has become known under the name of the 'adaptive landscape', conveys the same idea that selection in favour of local optima prevents evolution in the direction of ultimately superior, more global optima. His somewhat misunderstood (Wright 1980) emphasis on the role of genetic drift in allowing lineages to escape from the pull of local optima, and thereby attain a closer approximation to what a human might recognize as 'the' optimal solution, contrasts interestingly with Lewontin's (1979b) invoking of drift as an 'alternative to adaptation'. As in the case of pleiotropy, there is no paradox here. Lewontin is right that 'the finiteness of real populations results in random changes in gene frequency so that, with a certain probability, genetic combinations with lower reproductive fitness will be fixed in a population'. But on the other hand it is also true that, to the extent that local optima constitute a limitation on the attainment of design perfection, drift will tend to provide an escape (Lande 1976). Ironically, then, a *weakness* in natural selection can theoretically *enhance* the likelihood of a lineage attaining optimal design! Because it has no foresight, unalloyed natural selection is in a sense an *anti*-perfection mechanism, hugging, as it will, the tops of the low foot-hills of Wright's landscape. A mixture of strong selection interspersed with periods of relaxation of selection and drift may be the formula for crossing the valleys to the high uplands. Clearly if 'adaptationism' is to become an issue where debating points are scored, there is scope for both sides to have it both ways!

My own feeling is that somewhere here may lie the solution to the real paradox of this section on historical constraints. The jet engine analogy suggested that animals ought to be risible monstrosities of lashed-up improvisation, top-heavy with grotesque relics of patched-over antiquity. How can we reconcile this reasonable expectation with the formidable grace of the hunting cheetah, the aerodynamic beauty of the swift, the scrupulous attention to deceptive detail of the leaf insect? Even more impressive is the detailed agreement between different convergent solutions to common problems, for instance the multiple parallels that exist between the mammal radiations of Australia, South America, and the Old World. Cain (1964) remarks that 'Up to now it has usually been assumed, by Darwin and others, that convergence will never be so good as to mislead us' but he goes on to give examples where competent taxonomists have been fooled. More and more groups which had hitherto been regarded as decently monophyletic, are now being suspected of polyphyletic origin.

The citation of example and counter-example is mere idle fact-dropping. What we need is constructive work on the relation between local and global optima in an evolutionary context. Our understanding of natural selection itself needs to be supplemented by a study of 'escapes from specialization' to use Hardy's (1954) phrase. Hardy himself was suggesting neoteny as an escape from specialization, while in this chapter, following Wright, I have emphasized drift in this role.

Müllerian mimicry in butterflies may prove to be a useful case-study here. Turner (1977) remarks that 'among the long-winged butterflies of the tropical American rainforests (ithomiids, heliconids, danaids, pierids, pericopids) there are six distinct warning patterns, and although all the warningly colored species belong to one of these mimicry "rings" the rings themselves coexist in the same habitats through most of the American tropics and remain very distinct.... Once the difference between two patterns is too great to be jumped by a single mutation, convergence becomes virtually impossible, and the mimicry rings will coexist indefinitely.' This is one of the only cases where 'historical constraints' may be close to being understood in full genetic detail. It may provide a worthwhile opportunity also for the study of the genetic details of 'valley-crossing', which in the present case would consist in the detachment of a type of butterfly from the orbit of one mimicry ring, and its eventual 'capture' by the 'pull' of another mimicry ring. Though he does not invoke drift as an explanation in this case, Turner tantalizingly indicates that 'In southern Europe *Amata phegea*…has…captured *Zygenea ephialtes* from the Müllerian mimicry ring of zygaenids, homopterans, etc. to which it still belongs outside the range of *A. phegea* in northern Europe'.

At a more general theoretical level, Lewontin (1978) notes that 'there may often be several alternative stable equilibriums of genetic composition even when the force of natural selection remains the same. Which of these adaptive peaks in the space of genetic composition is eventually reached by a population depends entirely on chance events at the beginning of the selective process…. For example, the Indian rhinoceros has one horn and the African rhinoceros has two. Horns are an adaptation for protection against predators but it is not true that one horn is specifically adaptive under Indian conditions as opposed to two horns on the African plains. Beginning with two somewhat different developmental systems, the two species responded to the same selective forces in slightly different ways.' The point is basically a good one, although it is worth adding that Lewontin's uncharacteristically 'adaptationist' blunder about the functional significance of rhinoceros horns is not trivial. If horns really *were* an adaptation against predators it would indeed be hard to imagine how a single horn could be more useful against Asian predators while two horns were of more help against African predators. However if, as seems much more likely, rhinoceros horns are an adaptation for intraspecific combat and intimidation, it could well be the case that a one-horned rhino would be at a disadvantage in one continent while a two-horned rhino would suffer in the other. Whenever the name of the game is intimidation (or sexual attraction as Fisher taught us long ago), mere conformity to the majority style, whatever that majority style may happen to be, can have advantages. The details of a threat display and its associated organs may be arbitrary, but woe betide any mutant individual that departs from established custom (Maynard Smith & Parker 1976).

Available genetic variation

No matter how strong a potential selection pressure may be, no evolution will result unless there is genetic variation for it to work on. 'Thus, although I might argue that the possession of wings in addition to arms and legs might be advantageous to some vertebrates, none has ever evolved a third pair of appendages, presumably because the genetic variation has never been available' (Lewontin 1979b). One could reasonably dissent from this opinion. It may be that the only reason pigs have no wings is that selection has never favoured their evolution. Certainly we must be careful before we assume, on human-centred common-sense grounds, that it would obviously be handy for any animal to have a pair of wings even if it didn't use them very often, and that therefore the absence of wings in a given lineage must be due to lack of available mutations. Female ants can sprout wings if they happen to be nurtured as queens, but if nurtured as workers they do not express their capacity to do so. More strikingly, the queens in many species use their wings only once, for their nuptial flight, and then take the drastic step of biting or breaking them off at the roots in preparation for the rest of their life underground. Evidently wings have costs as well as benefits.

One of the most impressive demonstrations of the subtlety of Charles Darwin's mind is given by his discussion of winglessness and the costs of having wings in the insects of oceanic islands. For present purposes, the relevant point is that winged insects may risk being blown out to sea, and Darwin (1859, p. 177) suggested that this is why many island insects have reduced wings. But he also noted that some island insects are far from wingless; they have extra large wings.

> This is quite compatible with the action of natural selection. For when a new insect first arrived on the island, the tendency of natural selection to enlarge or to reduce the wings, would depend on whether a greater number of individuals were saved by successfully battling with the winds, or by giving up the attempt and rarely or never flying. As with mariners shipwrecked near a coast, it would have been better for the good swimmers if they had been able to swim still further, whereas it would have been better for the bad swimmers if they had not been able to swim at all and had stuck to the wreck.

A neater piece of evolutionary reasoning would be hard to find, although one can almost hear the baying chorus of 'Unfalsifiable! Tautological! Just-so story!'

Returning to the question of whether pigs ever could develop wings, Lewontin is undoubtedly right that biologists interested in adaptation cannot afford to ignore the question of the availability of mutational variation. It is certainly true that many of us, with Maynard Smith (1978a) though without his and Lewontin's authoritative knowledge of genetics, tend to assume 'that genetic variance of an appropriate kind will usually exist'. Maynard Smith's grounds are that 'with rare exceptions, artificial selection has always proved effective, whatever the organism or the selected character'. A notorious case, fully conceded by Maynard Smith (1978b), where the genetic variation necessary to an optimality theory often seems to be lacking, is that of Fisher's (1930a) sex ratio theory. Cattle breeders have had no trouble in breeding for high milk yield, high beef production, large size, small size, hornlessness, resistance to various diseases, and fierceness in fighting bulls. It would obviously be of immense benefit to the dairy industry if cattle could be bred with a bias towards producing heifer calves rather than bull calves. All attempts to do this have singularly failed, apparently because the necessary genetic variation does not exist. It may be the measure of how misled is my own biological intuition that I find this fact rather astonishing, indeed worrying. I would like to think that it is an exceptional case, but Lewontin is certainly right that we need to pay more attention to the problem of the limitations of available genetic variation. From this point of view, a compilation of the amenability or resistance to artificial selection of a wide variety of characters would be of great interest.

Meanwhile, there are certain common-sense things that can be said. Firstly, it may make sense to invoke lack of available mutation to explain why animals do not have some adaptation which we think reasonable, but it is harder to apply the argument the other way round. For instance, we might indeed think that pigs would be better off with wings and suggest that they lack them only because their ancestors never produced the necessary mutations. But if we see an animal with a complex organ, or a complex and time-consuming behaviour pattern, we would seem to be on strong grounds in guessing that it must have been put together by natural selection. Habits such as dancing in bees as already discussed, 'anting' in birds, 'rocking' in stick insects, and egg-shell removal in gulls are positively time-consuming, energy-consuming, and complex. The working hypothesis that they must have a Darwinian survival value is overwhelmingly strong. In a few cases it has proved possible to find out what that survival value is (Tinbergen 1963).

The second common-sense point is that the hypothesis of 'no available mutations' loses some of its force if a related species, or the same species in other contexts, has shown itself capable of producing the necessary variation. I shall mention below a case where the known capabilities of the digger wasp *Ammophila campestris* were used to illuminate the lack of similar capabilities in the related species *Sphex ichneumoneus*. A more subtle version of the same argument can be applied within any one species. For instance, Maynard Smith (1977, see also Daly 1979) concludes a paper with an up-beat question: Why do male mammals not lactate? We need not go into the details of why he thought they ought to; he may have been wrong, his model may have been wrongly set up, and the real answer to his question may be that it would not pay male mammals to lactate. The point here is that this is a slightly different kind of question from 'Why don't pigs have wings?' We know that male mammals contain the genes necessary for lactation, because all the genes in a female mammal have passed through male ancestors and may be handed on to male descendants. Genetic male mammals treated with hormones, indeed, can develop as lactating females. This all makes it less plausible that the reason male mammals don't lactate is simply that they haven't 'thought of it' mutationally speaking.

(Indeed, I bet I could breed a race of spontaneously lactating males by selecting for increased sensitivity to progressively reduced dosages of injected hormone, an interesting practical application of the Baldwin/Waddington Effect.)

The third common-sense point is that if the variation that is being postulated consists in a simple quantitative extension of already existing variation it is more plausible than a radical qualitative innovation. It may be implausible to postulate a mutant pig with wing rudiments, but it is not implausible to postulate a mutant pig with a curlier tail than existing pigs. I have elaborated this point elsewhere (Dawkins 1980).

In any case, we need a more subtle approach to the question of what is the evolutionary impact of differing degrees of mutability. It is not good enough to ask, in an all or none way, whether there is or is not genetic variation available to respond to a given selection pressure. As Lewontin (1979a) rightly says, 'Not only is the qualitative possibility of adaptive evolution constrained by available genetic variation, but the relative rates of evolution of different characters are proportional to the amount of genetic variance for each.' I think this opens up an important line of thought when combined with the notion of historical constraints treated in the previous section. The point can be illustrated with a fanciful example.

Birds fly with wings made of feathers, bats with wings consisting of flaps of skin. Why do they not both have wings made in the same way, whichever way is 'superior'? A confirmed adaptationist might reply that birds must be better off with feathers and bats better off with skin flaps. An extreme anti-adaptationist might say that very probably feathers would actually be better than skin-flaps for both birds and bats, but bats never had the good fortune to produce the right mutations. But there is an intermediate position, one which I find more persuasive than either extreme. Let us concede to the adaptationist that, given enough time, the ancestors of bats probably could have produced the sequence of mutations necessary for them to sprout feathers. The operative phrase is 'given enough time'. We are not making an all-or-none distinction between impossible and possible mutational variation, but simply stating the undeniable fact that some mutations are quantitatively more probable than others. In this case, ancestral mammals might have produced both mutants with rudimentary feathers and mutants with rudimentary skin flaps. But the proto-feather mutants (they might have had to go through an intermediate stage of small scales) were so slow in making their appearance in comparison with the skin-flap mutants, that skin-flap wings had long ago appeared and led to the evolution of passably efficient wings.

The general point is akin to the one already made about adaptive landscapes. There we were concerned with selection preventing lineages from escaping the clutches of local optima. Here we have a lineage faced with two alternative routes of evolution, one leading to, say, feathered wings, the other to skin-flap wings. The feathered design may be not only a global optimum but the present local optimum as well. The lineage, in other words, may be sitting exactly at the foot of the slope leading to the feathered peak of the Sewall Wright landscape. If only the necessary mutations were available it would climb easily up the hill. Eventually, according to this fanciful parable, those mutations might have come, but—and this is the important point—they were too late. Skin-flap mutations had come before them, and the lineage had already climbed too far up the slopes of the skin-flap adaptive hill to turn back. As a river takes the line of least resistance downhill, thereby meandering in a route that is far from the most direct one to the sea, so a lineage will evolve according to the effects of selection on the variation available at any given moment. Once a lineage has begun to evolve in a given direction, this may in itself close options that were formerly available, sealing off access to a global optimum. My point is that lack of available variation does not have to be absolute in order to become a significant constraint on perfection. It need only be a quantitative brake to have dramatic qualitative effects. In spirit, then, I agree with Gould and Calloway (1980) when they say, citing Vermeij's (1973) stimulating paper on the mathematics of morphological versatility, that 'Some morphologies can be twisted, bent and altered in a variety of ways, and others cannot.' But I would prefer to soften 'cannot', to make it a quantitative constraint, not an absolute barrier.

McCleery (1978), in an agreeably comprehensible introduction to the McFarland school of ethological optimality theory, mentions H. A. Simon's concept of 'satisficing' as an alternative to optimizing. If optimizing systems are concerned with maximizing something, satisficing systems get away with doing just enough. In this case, doing enough means doing enough to stay alive. McCleery contents himself with complaining that such 'adequacy' concepts have not generated much experimental work. I think evolutionary theory entitles us to be a bit more negative *a priori*. Living things are not selected for their capacity simply to stay alive; they are staying alive in competition with other such living things. The trouble with satisficing as a concept is that it completely leaves out the competitive element which is fundamental to all life. In Gore Vidal's words: 'It is not enough to succeed. Others must fail.'

On the other hand 'optimizing' is also an unfortunate word because it suggests the attainment of what an engineer would recognize as the best design in a global sense. It tends to overlook the constraints on perfection which are the subject of this chapter. In many ways the word 'meliorizing' expresses a sensible middle way between optimizing and satisficing. Where *optimus* means best, *melior* means better. The points we have been considering about historical constraints, about Wright's adaptive landscapes, and about rivers following the line of immediate least resistance are all related to the fact that natural selection chooses the better of present available alternatives. Nature does not have the foresight to put together a sequence of mutations which, for all that they may entail temporary disadvantage, set a lineage on the road to ultimate global superiority. It cannot refrain from favouring slightly advantageous available mutations now, so as to take better advantage of superior mutations which may arrive later. Like a river, natural selection blindly meliorizes its way down successive lines of immediately available least resistance. The animal that results is not the most perfect design conceivable, nor is it merely good enough to scrape by. It is the product of a historical sequence of changes, each one of which represented, at best, the *better* of the alternatives that happened to be around at the time.

Constraints of costs and materials

'If there were no constraints on what is possible, the best phenotype would live for ever, would be impregnable to predators, would lay eggs at an infinite rate, and so on' (Maynard Smith 1978b). 'An engineer, given carte blanche on his drawing board could design an "ideal" wing for a bird, but he would demand to know the constraints under which he must work. Is he constrained to use feathers and bones, or may he design the skeleton in titanium alloy? How much is he allowed to spend on the wings, and how much of the available economic investment must be diverted into, say, egg production?' (Dawkins & Brockmann 1980). In practice, an engineer will normally be given a specification of minimum performance such as 'The bridge must bear a load of ten tons.... The aeroplane wing must not break until it receives a stress three times what would be expected in worst-case turbulent conditions; now go ahead and build it as cheaply as you can.' The best design is the one that satisfies ('satisfices') the criterion specification at the least cost. Any design that achieves 'better' than the specified criterion performance is likely to be rejected, because presumably the criterion could be achieved more cheaply.

The particular criterion specification is an arbitrary working rule. There is nothing magic about a safety margin of three times the expected worst-case conditions. Military aircraft may be designed with more risky safety margins than civilian ones. In effect, the engineer's optimization instructions amount to a monetary evaluation of human safety, speed, convenience, pollution of the atmosphere, etc. The price put on each of these is a matter of judgement, and is often a matter of controversy.

In the evolutionary design of animals and plants, judgement does not enter into it, nor does controversy except among the human spectators of the show. In some way, however, natural selection must provide the equivalent of such judgement: risks of predation must be evaluated against risks of starving and benefits of mating with an extra female. For a bird, resources spent on making breast muscles for powering wings are resources that could have been spent on making eggs. An enlarged brain would permit a finer tuning of behaviour to environmental details, past and present, but at a cost of an enlarged head, which means extra weight at the front end of the body, which in turn necessitates a larger tail for aerodynamic stability, which in turn....Winged aphids are less fecund than wingless ones of the same species (J. S. Kennedy, personal communication). That every evolutionary adaptation must cost something, costs being measured in lost opportunities to do other things, is as true as that gem of traditional economic wisdom, 'There is no such thing as a free lunch'.

Of course the mathematics of biological currency-conversion, of evaluating the costs of wing muscle, singing time, predator-vigilance time, etc., in some common currency such as 'gonad equivalents', are likely to be very complex. Whereas the engineer is allowed to simplify his mathematics by working to an arbitrarily chosen minimum threshold of performance, the biologist is granted no such luxury. Our sympathy and admiration must go out to those few biologists who have attempted to grapple with these problems in detail (e.g. Oster & Wilson 1978; McFarland & Houston 1981).

On the other hand, although the mathematics may be formidable, we don't need mathematics to deduce the most important point, which is that any view of biological optimization that denies the existence of costs and trade-offs is doomed. An adaptationist who looks at one aspect of an animal's body or behaviour, say the aerodynamic performance of its wings, while forgetting that efficiency in the wings can only be bought at a cost which will be felt somewhere else in the animal's economy, would deserve all the criticism he gets. It has to be admitted that too many of us, while never actually denying the importance of costs, forget to mention them, perhaps even forget to think about them, when we discuss biological function. This has probably provoked some of the criticism that has come our way. In an earlier section I quoted Pittendrigh's remark that adaptive organization was a 'patchwork of makeshifts'. We must also not forget that it is a tangle of compromises (Tinbergen 1965).

In principle, it would seem a valuable heuristic procedure to *assume* that an animal is optimizing something under a given set of constraints, and to try to work out what those constraints are. This is a restricted version of what McFarland and his colleagues call the 'reverse optimality' approach (e.g. McCleery 1978). As a case study I shall take some work with which I happen to be familiar.

Dawkins and Brockmann (1980) found that the digger wasps (*Sphex ichneumoneus*) studied by Brockmann behaved in a way that a naive human economist might have criticized as maladaptive. Individual wasps appeared to commit the 'Concorde Fallacy' of valuing a resource according to how much they had already spent on it, rather than according to how much they could get out of it in the future. Very briefly, the evidence is as follows. Solitary females provision burrows with stung and paralysed katydids which are to serve as food for their larvae. Occasionally two females find themselves provisioning the same burrow, and they usually end up fighting over it. Each fight goes on until one wasp, thereby defined as the loser, flees from the area, leaving the winner in control of the burrow and all the katydids caught by both wasps. We measured the 'real value' of a burrow as the number of katydids which it contained. The 'prior investment' by each wasp in the burrow was measured as the number of katydids which she, as an individual, had put into it. The evidence suggested that each wasp fought for a time proportional to her own investment, rather than proportional to the 'true value' of the burrow.

Such a policy has great human psychological appeal. We too tend to fight tenaciously for property which we have put great effort into acquiring. The fallacy gets its name from the fact that, at a time when sober economic judgement of future prospects counselled abandoning the developing of the Concorde airliner, one of the arguments in favour of continuing with the half completed project was retrospective: 'We have already spent so much on it that we cannot back out now.' A popular argument for prolonging wars gave rise to the other name for the fallacy, the 'Our boys shall not have died in vain' fallacy.

When Dr Brockmann and I first realized that digger wasps behaved in like manner, I was, it has to be confessed, a little disconcerted, possibly because of my own past investment of effort (Dawkins & Carlisle 1976; Dawkins 1976a) in persuading my colleagues that the psychologically appealing Concorde Fallacy was, indeed, a fallacy! But then we started to think more seriously about cost constraints. Could it be that what appeared to be maladaptive was better interpreted as an optimum, *given certain constraints*? The question then became: Is there a constraint such that the wasps' Concordian behaviour is the best they can achieve under it?

In fact the question was more complicated than that, because it was necessary to substitute Maynard Smith's (1974) concept of evolutionarily stability ('ESS') for that of simple optimality, but the principle remains that a reverse optimality approach might be heuristically valuable. If we can show that an animal's behaviour is what would be produced by an optimizing system working under constraint X, maybe we can use the approach to learn something of the constraints under which animals actually do work.

In the present case it seemed that the relevant constraint might be one of sensory capacity. If the wasps, for some reason, cannot count katydids in the burrow, but can meter some aspect of their own hunting efforts, there is an asymmetry of information possessed by the two combatants. Each one 'knows' that the burrow contains at least b katydids, where b is the number she herself has caught. She may 'estimate' that the true number in the burrow is larger than b, but she does not know how much larger. Under such conditions Grafen has shown that the expected ESS is approximately the one originally calculated by Bishop and Cannings (1978) for the so-called 'generalized war of attrition'. The mathematical details can be left aside; for present purposes what matters is that the behaviour expected by the extended war of attrition model would look very like the Concordian behaviour actually shown by the wasps.

If we were interested in testing the general hypothesis that animals optimize, this kind of *post hoc* rationalization would be suspect. By *post hoc* modification of the details of the hypothesis, one is bound to find a version which fits the facts. Maynard Smith's (1978b) reply to this kind of criticism is very relevant: 'in testing a model we are *not* testing the general proposition that nature optimizes, but the specific hypotheses about constraints, optimization criteria, and heredity'. In the present case we are making a general assumption that nature does optimize within constraints, and testing particular models of what those constraints might be.

The particular constraint suggested—inability of the wasp's sensory system to assess the contents of a burrow—is in accordance with independent evidence from the same population of wasps (Brockmann, Grafen & Dawkins 1979; Brockmann & Dawkins 1979). There is no reason to regard it as an irrevocably binding limitation for all time. Probably the wasps could evolve the capacity to assess nest contents, but only at a cost. Digger wasps of the related species *Ammophila campestris* have long been known to make an assessment of the contents of each of their nests every day (Baerends 1941). Unlike *Sphex*, which provisions one burrow at a time, lays an egg, then fills the burrow in with soil and leaves the larva to eat the provision on its own, *Ammophila campestris* is a progressive provisioner of several burrows concurrently. A female tends two or three growing larvae, each in a separate burrow, at the same time. The ages of her various larvae are staggered, and their food needs are different. Every morning she assesses the current contents of each burrow on a special early morning 'inspection round'. By experimentally changing the contents of burrows, Baerends showed that the female adjusts her whole day's provisioning of each burrow according to what it contained at the time of her morning inspection. The contents of the burrow at any other time of day have no effect on her behaviour, even though she is provisioning it all day. She appears, therefore, to use her assessment faculty sparingly, switching it off for the rest of the day after the morning inspection, almost as though it was a costly, power-consuming instrument. Fanciful as that analogy may be, it surely suggests that the assessment faculty, whatever it is, may have overhead running costs, even if (G. P. Baerends, personal communication) these consist only in the time consumed.

Sphex ichneumoneus, not being a progressive provisioner, and tending only one burrow at a time, presumably has less need than *Ammophila* for a burrow-assessment faculty. By not attempting to count prey in the burrow, it can save itself not only the running expenditure that *Ammophila* seems so careful to ration; it can also save itself the initial manufacturing costs of the necessary neural and sensory apparatus. Probably it could benefit slightly from having an ability to assess burrow contents, but only on the comparatively rare occasions when it finds itself competing for a burrow with another wasp. It is easy to believe that the costs outweigh the benefits, and that selection has therefore never favoured the evolution of assessment apparatus. I think this is a more constructive and interesting hypothesis than the alternative hypothesis that the necessary mutational variation has never arisen. Of course we have to admit that the latter might be right, but I would prefer to keep it as a hypothesis of last resort.

Imperfections at one level due to selection at another level

One of the main topics to be tackled in this book is that of the level at which natural selection acts. The kind of adaptations we should see if selection acted at the level of the group would be quite different from the adaptations we should expect if selection

acts at the level of the individual. It follows that a group selectionist might well see as imperfections features which an individual
selectionist would see as adaptations. This is the main reason
why I regard as unfair Gould and Lewontin's (1979) equating of
modern adaptationism with the naive perfectionism that Haldane
named after Voltaire's Dr Pangloss. With reservations due to the
various constraints on perfection, an adaptationist may believe
that all aspects of organisms are 'adaptive optimal solutions to
problems', or that 'it is virtually impossible to do a better job
than an organism is doing in its given environment'. Yet the same
adaptationist may be extremely fussy about the kind of meaning
he allows to words like 'optimal' and 'better'. There are many
kinds of adaptive, indeed Panglossian, explanations, for example
most group-selectionist ones, which would be utterly ruled out
by the modern adaptationist.

For the Panglossian the demonstration that something is
'beneficial' (to whom or to what is often not specified) is a sufficient explanation for its existence. The neo-Darwinian adaptationist, on the other hand, insists upon knowing the exact nature
of the selective process that has led to the evolution of the putative adaptation. In particular, he insists on precise language
about the level at which natural selection is supposed to have
acted. The Panglossian looks at a one-to-one sex ratio and sees
that it is good: does it not minimize the wastage of the population's resources? The neo-Darwinian adaptationist considers in
detail the fates of genes acting on parents to bias the sex ratio of
their offspring, and calculates the evolutionarily stable state of
the population (Fisher 1930a). The Panglossian is disconcerted by
1:1 sex ratios in polygynous species, in which a minority of males
hold harems and the rest sit about in bachelor herds consuming
almost half the population's food resources yet contributing
not at all to the population's reproduction. The neo-Darwinian

adaptationist takes this in his stride. The system may be hideously uneconomical from the population's point of view, but, from the point of view of the genes influencing the trait concerned, there is no mutant that could do better. My point is that neo-Darwinian adaptationism is not a catch-all, blanket faith in all being for the best. It rules out of court most of the adaptive explanations that readily occur to the Panglossian.

Some years ago, a colleague received an application from a prospective graduate student wishing to work on adaptation, who was brought up a religious fundamentalist and did not believe in evolution. He believed in adaptations, but thought they were designed by God, designed for the benefit of... ah, but that is just the problem! It might be thought that it did not matter whether the student believed adaptations were produced by natural selection or by God. Adaptations are 'beneficial' whether because of natural selection or because of beneficient design, and could not a fundamentalist student be usefully employed in uncovering the detailed ways in which they were beneficial? My point is that this argument will not do, because what is beneficial to one entity in the hierarchy of life is harmful to another, and creationism gives us no grounds for supposing that one entity's welfare will be preferred to another's. In passing, the fundamentalist student might pause to wonder at a God who goes to great trouble to provide predators with beautiful adaptations to catch prey, while with the other hand giving prey beautiful adaptations to thwart them. Perhaps He enjoys the spectator sport. Returning to the main point, if adaptations were designed by God, He might have designed them to benefit the individual animal (its survival or—not the same thing—its inclusive fitness), the species, some other species such as mankind (the usual view of religious fundamentalists), the 'balance of nature', or some other inscrutable purpose known only to Him. These are frequently incompatible alternatives. It really *matters* for whose benefit adaptations are designed. Facts such as the sex ratio in harem-forming mammals are inexplicable on certain hypotheses and easily explicable on others. The adaptationist working within the framework of a proper understanding of the genetical theory of natural selection countenances only a very restricted set of the possible functional hypotheses which the Panglossian might admit.

One of the main messages of this book is that, for many purposes, it is better to regard the level at which selection acts as neither the organism, nor the group or any larger unit, but the gene or small genetic fragment. This difficult topic is discussed in later chapters. Here, it is sufficient to note that selection at the level of the gene can give rise to apparent imperfections at the level of the individual. The classic example is the case of heterozygous advantage. A gene may be positively selected because of its beneficial effects when heterozygous, even though it has harmful effects when homozygous. As a consequence of this, a predictable proportion of the individual organisms in the population will have defects. The general point is this. The genome of an individual organism in a sexual population is the product of a more or less random shuffling of the genes in the population. Genes are selected over their alleles because of their phenotypic effects, averaged over all the individual bodies in which they are distributed, over the whole population, and through many generations. The effects that a given gene has will usually depend upon the other genes with which it shares a body: heterozygous advantage is just a special case of this. A certain proportion of bad bodies seems an almost inevitable consequence of selection for good genes, where good refers to the average effects of a gene on a statistical sample of bodies in which it finds itself permuted with other genes.

Inevitable, that is, as long as we accept the Mendelian shuffle as given and inescapable. Williams (1979), disappointed at finding no evidence for adaptive fine-adjustment of the sex ratio, makes the perceptive point that

> Sex is only one of many offspring characters that would seem adaptive for a parent to control. For instance, in human populations affected by sickle-cell anaemia, it would be advantageous for a heterozygous woman to have her *A* eggs fertilized only by *a*-bearing sperm, and vice versa, or even to abort all homozygous embryos. Yet if mated to another heterozygote she will reliably submit to the Mendelian lottery, even though this means markedly lowered fitness for half her children.... The really fundamental questions in evolution may be answerable only by regarding each gene as ultimately in conflict with every other gene, even those at other loci in the same cell. A really valid theory of natural selection must be based ultimately on selfish replicators, genes and all other entities capable of the biased accumulation of different variant forms.

Amen!

Mistakes due to environmental unpredictability or 'malevolence'

However well adapted an animal may be to environmental conditions, those conditions must be regarded as a statistical average. It will usually be impossible to cater for every conceivable contingency of detail, and any given animal will therefore frequently be observed to make 'mistakes', mistakes which can easily be fatal. This is not the same point as the time-lag problem already mentioned. The time-lag problem arises because of non-stationarities in the statistical properties of the environment: average conditions now are different from the average conditions experienced by the animal's ancestors. The present point is more inescapable. The modern animal may be living under identical *average* conditions to those of an ancestor, yet the detailed moment to moment occurrences facing either of them are not the same from day to day, and are too complex for precise prediction to be possible.

It is particularly in behaviour that such mistakes are seen. The more static attributes of an animal, its anatomical structure for instance, are obviously adapted only to long-term average conditions. An individual is either big or small, it cannot change size from minute to minute as the need arises. Behaviour, rapid muscular movement, is that part of an animal's adaptive repertoire which is specifically concerned with high speed adjustment. The animal can be now here, now there, now up a tree, now underground, rapidly accommodating to environmental contingencies. The number of such *possible* contingencies, when defined in all their detail, is like the number of possible chess positions, virtually infinite. Just as chess-playing computers (and chess-playing people) learn to classify chess positions into a manageable number of generalized classes, so the best that an adaptationist can hope for is that an animal will have been programmed to behave in ways appropriate to a manageable number of general contingency classes. Actual contingencies will fit these general classes only approximately, and apparent mistakes are therefore bound to be made.

The animal that we see up a tree may come from a long line of tree-dwelling ancestors. The trees in which the ancestors underwent natural selection were, in general, much the same as the trees of today. General rules of behaviour which worked then, such as 'Never go out on a limb that is too thin', still work. But the details of any one tree are inevitably different from the details of another. The leaves are in slightly different places, the breaking strain of the branches is only approximately predictable from their diameter, and so on. However strongly adaptationist our beliefs may be, we can only expect animals to be average statistical optimizers, never perfect anticipators of every detail.

So far we have considered the environment as statistically complex and therefore hard to predict. We have not reckoned on its being actively malevolent from our animal's point of view. Tree boughs surely do not deliberately snap out of spite when monkeys venture on to them. But a 'tree bough' may turn out to be a camouflaged python, and our monkey's last mistake is then no accident but is, in a sense, deliberately engineered. Part of a monkey's environment is non-living or at least indifferent to the monkey's existence, and the monkey's mistakes can be put down to statistical unpredictability. But other parts of the monkey's environment consist of living things that are themselves adapted to profit at the expense of monkeys. This portion of the monkey's environment may be called malevolent.

Malevolent environmental influences may themselves be hard to predict for the same reasons as indifferent ones, but they introduce an added hazard; an added opportunity for the victim to make 'mistakes'. The mistake made by a robin in feeding a cuckoo in its nest is presumably in some sense a maladaptive blunder. This is not an isolated, unpredictable occurrence such as arises because of the statistical unpredictability of the non-malevolent part of the environment. It is a recurrent blunder, afflicting generation after generation of robins, even the same robin several times in its life. Examples of this kind always make us wonder at the compliance, in evolutionary time, of the organisms that are manipulated against their best interests. Why doesn't selection simply eliminate the susceptibility of robins to the deception of cuckoos? This kind of problem is one of many which I believe will one day become the stock in trade of a new subdiscipline of biology—the study of manipulation, arms races, and the extended phenotype.

GLOSSARY FOR
THE EXTENDED PHENOTYPE
CHAPTERS

The Extended Phenotype was primarily intended for biologists who will have no need of a glossary, but a brief explanation of a few technical terms appearing in these two selected chapters, based on the glossary in the original work, is included here to make the discussion more widely accessible.

adaptation A technical term which has evolved somewhat away from its common usage as a near synonym of 'modification'. From sentences like 'cricket wings are adapted (modified from their primary function of flying) for singing' (and by implication are well designed for singing), 'an adaptation' has come to mean approximately an attribute of an organism that is 'good' for something. Good in what sense?, and good for what or for whom?, are difficult questions which are discussed at length in *The Extended Phenotype*. Examples of adaptations are the wing or the eye. An important point to note is that adaptations can only arise from natural selection.

adaptive landscape A powerful metaphorical multidimensional visualization, devised by Sewall Wright (1932), of the reproductive fitness of varying genotypes in terms of a landscape in which peaks represent higher fitness. The evolution of a population can be charted as a path in the landscape that tends, by natural selection, towards peaks (adaptations). Several variations of this representation are used by evolutionary biologists and population geneticists.

alleles (short for **allelomorphs**) Each gene is able to occupy only a particular region of chromosome, its *locus*. At any given locus there may exist, in the population, alternative forms of the gene. These alternatives are called alleles of one another. *The Extended Phenotype* emphasizes that there is a sense in which alleles are competitors of each other, because over evolutionary time successful alleles achieve numerical superiority over others at the same locus, in all the chromosomes of the population.

allometry A disproportionate relationship between size of a body part and size of the whole body, the comparisons being made either across individuals or across different life stages in the same individual. For example, large ants (but small humans) tend to have relatively very large heads; the head grows at a different rate from the body as a whole. Mathematically, the size of the part is usually taken as being related to the size of the whole raised to a power, which may be fractional.

Baldwin/Waddington Effect First proposed by Spalding in 1873. A largely hypothetical evolutionary process (also called *genetic assimilation*) whereby natural selection can create an illusion of the inheritance of acquired characteristics. Selection in favour of a genetic tendency to acquire a characteristic in response to environmental stimuli leads to the evolution of increased sensitivity to the same environmental stimuli, and eventual emancipation from the need for them. In the chapter 'Constraints on Perfection', I suggest that we might breed a race of spontaneously milk-producing male mammals by treating successive generations of males with female hormones and selecting for increased sensitivity to female hormones. The role of the hormones, or other environmental treatment, is to bring out into the open genetic variation which would otherwise lie dormant.

central dogma In molecular biology the dogma that nucleic acids act as templates for the synthesis of proteins, but never the reverse. More generally, the dogma that genes exert an influence over the form of a body, but the form of a body is never translated back into genetic code: acquired characteristics are not inherited.

chromosome One of the chains of genes found in cells. In addition to DNA itself, there is usually a complicated supporting structure of protein. Chromosomes become visible under the light microscope only at certain times in the cell cycle, but their number and linearity may be inferred by statistical reasoning from the facts of inheritance alone. The chromosomes are usually present in all cells in the body, even though only a minority of them will be active in any one cell. There are usually two sex chromosomes in every diploid cell as well as a number of autosomes (44 in humans).

cistron One way of defining a gene. In molecular genetics the cistron has a precise definition in terms of a specific experimental test. More loosely it is used to refer to a length of chromosome responsible for the encoding of one chain of amino acids in a protein.

evolutionarily stable strategy (ESS) [Note 'evolutionarily' *not* 'evolutionary'. The latter is a common grammatical error in this context.] A strategy that does well in a population dominated by the same strategy. This definition captures the intuitive essence of the idea, but is somewhat imprecise; for a mathematical definition, see Maynard Smith (1974).

extended phenotype All effects of a gene upon the world. As always, 'effect' of a gene is understood as meaning in comparison with its alleles. The conventional phenotype is the special case in which the effects are regarded as being confined to the individual body in which the gene sits. In practice it is convenient to limit 'extended phenotype' to cases where the effects influence the survival chances of the gene, positively or negatively.

genetic drift Changes in gene frequencies over generations, resulting from chance rather than selection.

genotype The genetic constitution of an organism at a particular locus or set of loci. Sometimes used more loosely as the whole genetic counterpart to phenotype (q.v.).

group selection A hypothetical process of natural selection among groups of organisms. Often invoked to explain the evolution of altruism. Sometimes confused with kin selection (q.v.).

heritability A term that is often misunderstood, it is a statistic that specifies the proportion of the observed variability in a phenotypic trait in a population that results from genetic variation within the population, as opposed to environmental or other causes.

heterozygous The condition of having non-identical alleles at a chromosomal locus. Is usually applied to an individual organism, in which case it refers to two different alleles at a given locus. More loosely it may refer to the overall statistical within-locus heterogeneity of alleles averaged over all loci in an individual or in a population.

homozygous The condition of having identical alleles at a chromosomal locus. Is usually applied to an individual organism, in which case it indicates that the individual has two identical alleles at the locus. More loosely it may refer to the overall statistical within-locus homogeneity of alleles averaged over all loci in an individual or in a population.

inclusive fitness A concept developed mathematically by Bill Hamilton (1964) as part of kin selection theory (q.v.), by which the fitness of an individual is estimated in terms of the fitness of its close kin, which would share the same genes.

kin selection Selection of genes causing individuals to favour close kin, owing to the high probability that kin share those genes. Strictly speaking 'kin' includes immediate offspring, but it is unfortunately undeniable that many biologists use the phrase 'kin selection' specifically when talking about kin other than offspring. Kin selection is also sometimes confused with group selection (q.v.), from which it is logically distinct, although where species happen to go around in discrete kin groups the two may incidentally amount to the same thing—'kin group selection'.

locus The position on a chromosome occupied by a gene (or a set of alternative alleles). For instance, there might be an eye-colour locus, at which the alternative alleles code for green, brown, and red. Usually applied at the level of the cistron (q.v.), the concept of the locus can be generalized to smaller or larger lengths of chromosome.

Mendelian inheritance Non-blending inheritance by means of pairs of discrete hereditary factors (now identified with genes), one member of each pair coming from each parent. The main theoretical alternative is 'blending inheritance'. In Mendelian inheritance genes may blend in their effects on a body, but they themselves do not blend, and they are passed on intact to future generations.

monophyletic A group of organisms is said to be monophyletic if all are descended from a common ancestor which would also have been classified as a member of the group. For instance, the birds are probably a monophyletic group since the most recent common ancestor of all birds would probably have been classified as a bird. The reptiles, however, are probably *polyphyletic*, in that the most recent common ancestor of all reptiles would probably not have been classified as a reptile.

Müllerian mimicry A form of mimicry in which two species of prey organisms which are both distasteful and which share a common predator come to mimic each other in their warning signals, for example warning coloration in the case of butterflies. The species need not be related to each other. Individuals in both species benefit from the mimicry, since a predator that has once caught one of the distasteful forms will in future avoid individuals from either species. Further species, related or not, may come to mimic the same warning signals with time, resulting in a 'mimicry ring'.

mutation An inherited change in the genetic material. In Darwinian theory mutations are said to be random. This does not mean that they are not lawfully caused, but only that there is no specific tendency for them to be directed towards improved adaptation. Improved adaptation comes about only through selection, but it needs mutation as the ultimate source of the variants among which it selects.

neo-Darwinism A term coined (actually re-coined, for the word was used in the 1880s for a very different group of evolutionists) in the middle part of the 20th century. Its purpose was to emphasize (and in my opinion exaggerate) the distinctness of the modern synthesis of Darwinism and Mendelian genetics, achieved in the 1920s and 1930s, from Darwin's own view of evolution. I think the need for the 'neo' is fading, and Darwin's own approach to 'the economy of nature' now looks very modern.

neoteny An evolutionary slowing down of bodily development relative to the development of sexual maturity, with the result that reproduction comes to be practised by organisms which resemble the juvenile stages of ancestral forms. It is hypothesized that some major steps in evolution, for example the origin of the vertebrates, came about through neoteny.

neutral mutation A mutation that has no selective advantage or disadvantage in comparison with its allele. Theoretically, a neutral mutation may become 'fixed' (i.e. numerically predominant in the population at its locus) after a number of generations, and this would be a form of evolutionary change. There is legitimate controversy over the importance of such random fixations in evolution, but there should be no controversy over their importance in the direct production of adaptation: it is zero.

phenotype The manifested attributes of an organism, the joint product of its genes and their environment during ontogeny. A gene may be said to have phenotypic expression in, say, eye colour. In *The Extended Phenotype* the concept of phenotype was *extended* to include functionally important consequences of gene differences, outside the bodies in which the genes sit.

pleiotropy The phenomenon whereby a change at one genetic locus can bring about a variety of apparently unconnected phenotypic changes. For instance a particular mutation might at one and the same time affect eye colour, toe length, and milk yield. Pleiotropy is probably the rule rather than the exception, and is entirely to be expected from all that we understand about the complex way in which development happens.

polyphyletic *See* monophyletic.

replicator Any entity in the universe of which copies are made.

survival value The quality for which a characteristic was favoured by natural selection.

Weismannism The doctrine of a rigid separation between an immortal germ-line and the succession of mortal bodies which house it. In particular the doctrine that the germ-line may influence the form of the body, but not the other way around. *See also* central dogma.

wild-type gene Or wild-type allele, the allele most commonly found in a population.

REFERENCES FOR
THE EXTENDED PHENOTYPE
CHAPTERS

Baerends, G. P. (1941). Fortpflanzungsverhalten und Orientierung der Grabwespe *Ammophila campestris* Jur. *Tijdschrift voor Entomologie* **84**, 68–275.

Barlow, H. B. (1961). The coding of sensory messages. In *Current Problems in Animal Behaviour* (eds W. H. Thorpe & O. L. Zangwill), pp. 331–60. Cambridge: Cambridge University Press.

Bishop, D. T. & Cannings, C. (1978). A generalized war of attrition. *Journal of Theoretical Biology* **70**, 85–124.

Boden, M. (1977). *Artificial Intelligence and Natural Man*. Brighton: Harvester Press.

Brockmann, H. J. & Dawkins, R. (1979). Joint nesting in a digger wasp as an evolutionarily stable preadaptation to social life. *Behaviour* **71**, 203–45.

Brockmann, H. J., Grafen, A. & Dawkins, R. (1979). Evolutionarily stable nesting strategy in a digger wasp. *Journal of Theoretical Biology* **77**, 473–96.

Cain, A. J. (1964). The perfection of animals. In *Viewpoints in Biology*, 3 (eds J. D. Carthy & C. L. Duddington), pp. 36–63. London: Butterworths.

Cain, A. J. (1979). Introduction to general discussion. In *The Evolution of Adaptation by Natural Selection* (eds J. Maynard Smith & R. Holliday). *Proceedings of the Royal Society of London*, B **205**, 599–604.

Clutton-Brock, T. H. & Harvey, P. H. (1979). Comparison and adaptation. *Proceedings of the Royal Society of London*, B **205**, 547–65.

Curio, E. (1973). Towards a methodology of teleonomy. *Experientia* **29**, 1045–58.

Daly, M. (1979). Why don't male mammals lactate? *Journal of Theoretical Biology* **78**, 325–45.

Darwin, C. R. (1859). *The Origin of Species*. 1st edn, reprinted 1968. Harmondsworth, Middx: Penguin.

Dawkins, R. (1969). Bees are easily distracted. *Science* **165**, 751.

Dawkins, R. (1976a). *The Selfish Gene*. Oxford: Oxford University Press.

Dawkins, R. (1976b). Hierarchical organisation: a candidate principle for ethology. In *Growing Points in Ethology* (eds P. P. G. Bateson & R. A. Hinde), pp. 7–54. Cambridge: Cambridge University Press.

Dawkins, R. (1979). Twelve misunderstandings of kin selection. *Zeitschrift für Tierpsychologie* **51**, 184–200.

Dawkins, R. (1980). Good strategy or evolutionarily stable strategy? In *Sociobiology: Beyond Nature/Nurture?* (eds G. W. Barlow & J. Silverberg), pp. 331–67. Boulder: Westview Press.

Dawkins, R. (1981). In defence of selfish genes. *Philosophy*, October.

Dawkins, R. & Brockmann, H. J. (1980). Do digger wasps commit the Concorde fallacy? *Animal Behaviour* **28**, 892–6.

Dawkins, R. & Carlisle, T. R. (1976). Parental investment, mate desertion and a fallacy. *Nature* **262**, 131–3.

Evans, C. (1979). *The Mighty Micro*. London: Gollancz.

Falconer, D. S. (1960). *Introduction to Quantitative Genetics*. London: Longman.

Fisher, R. A. (1930a). *The Genetical Theory of Natural Selection*. Oxford: Clarendon Press.

Fisher, R. A. (1930b). The distribution of gene ratios for rare mutations. *Proceedings of the Royal Society of Edinburgh* **50**, 204–19.

Fisher, R. A. & Ford, E. B. (1950). The Sewall Wright effect. *Heredity* **4**, 47–9.

Fraenkel, G. S. & Gunn, D. L. (1940). *The Orientation of Animals*. Oxford: Oxford University Press.

Frisch, K. von (1967). *A Biologist Remembers*. Oxford: Pergamon Press.

Goodwin, B. C. (1979). Spoken remark in *Theoria to Theory* **13**, 87–107.

Gould, J. L. (1976). The dance language controversy. *Quarterly Review of Biology* **51**, 211–44.

Gould, S. J. (1978). *Ever Since Darwin*. London: Burnett.

Gould, S. J. & Calloway, C. B. (1980). Clams and brachiopods—ships that pass in the night. *Paleobiology* **6**, 383–96.

Gould, S. J. & Lewontin, R. C. (1979). The spandrels of San Marco and the Panglossian paradigm: a critique of the adaptationist programme. *Proceedings of the Royal Society of London*, B **205**, 581–98.

Gregory, R. L. (1961). The brain as an engineering problem. In *Current Problems in Animal Behaviour* (eds W. H. Thorpe & O. L. Zangwill), pp. 307–30. Cambridge: Cambridge University Press.

Hailman, J. P. (1977). *Optical Signals*. Bloomington: Indiana University Press.

Haldane, J. B. S. (1932a). *The Causes of Evolution*. London: Longman's Green.

Hamilton, W. D. (1964). The genetical evolution of social behavior. I and II. *Journal of Theoretical Biology* **7**, 1–52.

Hamilton, W. D. (1967). Extraordinary sex ratios. *Science* **156**, 477–88.

Hamilton, W. D. (1972). Altruism and related phenomena, mainly in social insects. *Annual Review of Ecology and Systematics* **3**, 193–232.

Hardy, A. C. (1954). Escape from specialization. In *Evolution as a Process* (eds J. S. Huxley, A. C. Hardy & E. B. Ford), pp. 122–40. London: Allen & Unwin.

Hofstadter, D. R. (1979). *Gödel, Escher, Bach: An Eternal Golden Braid*. Brighton: Harvester Press.

Hoyle, F. (1964). *Man in the Universe*. New York: Columbia University Press, 24–26.

Huxley, J. S. (1932). *Problems of Relative Growth*. London: McVeagh.

Jacob, F. (1977). Evolution and tinkering. *Science* **196**, 1161–6.

Kempthorne, O. (1978). Logical, epistemological and statistical aspects of nature–nurture data interpretation. *Biometrics* **34**, 1–23.

Lack, D. (1966). *Population Studies of Birds*. Oxford: Oxford University Press.

Lande, R. (1976). Natural selection and random genetic drift. *Evolution* **30**, 314–34.

Levy, D. (1978). Computers are now chess masters. *New Scientist* **79**, 256–8.

Lewontin, R. C. (1967). Spoken remark in *Mathematical Challenges to the Neo-Darwinian Interpretation of Evolution* (eds P. S. Moorhead & M. Kaplan). *Wistar Institute Symposium Monograph* **5**, 79.

Lewontin, R. C. (1977). Caricature of Darwinism. *Nature* **266**, 283–4.

Lewontin, R. C. (1978). Adaptation. *Scientific American* **239** (3), 156–69.

Lewontin, R. C. (1979a). Fitness, survival and optimality. In *Analysis of Ecological Systems* (eds D. J. Horn, G. R. Stairs & R. D. Mitchell), pp. 3–21. Columbus: Ohio State University Press.

Lewontin, R. C. (1979b). Sociobiology as an adaptationist program. *Behavioral Science* **24**, 5–14.

Lindauer, M. (1971). The functional significance of the honeybee waggle dance. *American Naturalist* **105**, 89–96.

Lloyd, J. E. (1979). Mating behavior and natural selection. *Florida Entomologist* **62** (1), 17–23.

McCleery, R. H. (1978). Optimal behaviour sequences and decision making. In *Behavioural Ecology* (eds J. R. Krebs & N. B. Davies), pp. 377–410. Oxford: Blackwell Scientific.

McFarland, D. J. & Houston, A. I. (1981). *Quantitative Ethology*. London: Pitman.

Maynard Smith, J. (1974). The theory of games and the evolution of animal conflicts. *Journal of Theoretical Biology* **47**, 209–21.

Maynard Smith, J. (1976). What determines the rate of evolution? *American Naturalist* **110**, 331–8.

Maynard Smith, J. (1977). Parental investment: a prospective analysis. *Animal Behaviour* **25**, 1–9.

Maynard Smith, J. (1978a). *The Evolution of Sex*. Cambridge: Cambridge University Press.

Maynard Smith, J. (1978b). Optimization theory in evolution. *Annual Review of Ecology and Systematics* **9**, 31–56.

Maynard Smith, J. & Parker, G. A. (1976). The logic of asymmetric contests. *Animal Behaviour* **24**, 159–75.

Maynard Smith, J. & Ridpath, M. G. (1972). Wife sharing in the Tasmanian native hen, *Tribonyx mortierii*: a case of kin selection? *American Naturalist* **106**, 447–52.

Medawar, P. B. (1952). *An Unsolved Problem in Biology*. London: H. K. Lewis.

'Nabi, I.' (1981). Ethics of genes. *Nature* **290**, 183.

Oster, G. F. & Wilson, E. O. (1978). *Caste and Ecology in the Social Insects*. Princeton: Princeton University Press.

Pittendrigh, C. S. (1958). Adaptation, natural selection, and behavior. In *Behavior and Evolution* (eds A. Roe & G. G. Simpson), pp. 390–416. New Haven: Yale University Press.

Ridley, M. (1980). Konrad Lorenz and Humpty Dumpty: some ethology for Donald Symons. *Behavioral and Brain Sciences* **3**, 196.

Ridley, M. & Dawkins, R. (1981). The natural selection of altruism. In *Altruism and Helping Behavior* (eds J. P. Rushton & R. M. Sorentino), pp. 19–39. Hillsdale, N.J.: Erlbaum.

Rose, S. (1978). Pre-Copernican sociobiology? *New Scientist* **80**, 45–6.

Rothenbuhler, W. C. (1964). Behavior genetics of nest cleaning in honey bees. IV. Responses of F1 and backcross generations to disease-killed brood. *American Zoologist* **4**, 111–23.

Schuster, P. & Sigmund, K. (1981). Coyness, philandering and stable strategies. *Animal Behaviour* **29**, 186–92.

Spalding, D. A. (1873). Instinct. With original observations on young animals. *Macmillan's Magazine* **27**, 282–93.

Symons, D. (1979). *The Evolution of Human Sexuality*. New York: Oxford University Press.

Taylor, A. J. P. (1963). *The First World War*. London: Hamish Hamilton.

Tinbergen, N. (1963). On aims and methods of ethology. *Zeitschrift für Tierpsychologie* **20**, 410–33.

Tinbergen, N. (1965). Behaviour and natural selection. In *Ideas in Modern Biology* (ed. J. A. Moore), pp. 519–42. New York: Natural History Press.

Tinbergen, N., Broekhuysen, G. J., Feekes, F., Houghton, J. C. W., Kruuk, H. & Szulc, E. (1962). Egg shell removal by the black-headed gull, *Larus ridibundus*, L.; a behaviour component of camouflage. *Behaviour* **19**, 74–117.

Trevor-Roper, H. R. (1972). *The Last Days of Hitler*. London: Pan.

Trivers, R. L. (1974). Parent-offspring conflict. *American Zoologist* **14**, 249–64.

Turing, A. (1950). Computing machinery and intelligence. *Mind* **59**, 433–60.

Turner, J. R. G. (1977). Butterfly mimicry: the genetical evolution of an adaptation. In *Evolutionary Biology*, Vol. 10 (eds M. K. Hecht *et al.*), pp. 163–206. New York: Plenum Press.

Vermeij, G. J. (1973). Adaptation, versatility and evolution. *Systematic Zoology* **22**, 466–77.

Weinrich, J. D. (1976). Human reproductive strategy: the importance of income unpredictability, and the evolution of non-reproduction. PhD dissertation, Harvard University, Cambridge, Mass.

Weizenbaum, J. (1976). *Computer Power and Human Reason*. San Francisco: W. H. Freeman.

Wenner, A. M. (1971). *The Bee Language Controversy: An Experience in Science*. Boulder: Educational Programs Improvement Corporation.

Williams, G. C. (1957). Pleiotropy, natural selection, and the evolution of senescence. *Evolution* **11**, 398–411.

Williams, G. C. (1966). *Adaptation and Natural Selection*. Princeton, N.J.: Princeton University Press.

Williams, G. C. (1979). The question of adaptive sex ratio in outcrossed vertebrates. *Proceedings of the Royal Society of London*, B **205**, 567–80.

Wilson, E. O. (1975). *Sociobiology: the New Synthesis*. Cambridge, Mass.: Harvard University Press.

Wilson, E. O. (1978). *On Human Nature*. Cambridge, Mass.: Harvard University Press.

Winograd, T. (1972). *Understanding Natural Language*. Edinburgh: Edinburgh University Press.

Wright, S. (1932). The roles of mutation, inbreeding, crossbreeding and selection in evolution. *Proceedings of the 6th International Congress of Genetics* **1**, 356–68.

Wright, S. (1951). Fisher and Ford on the Sewall Wright effect. *American Science Monthly* **39**, 452–8.

Wright, S. (1980). Genic and organismic selection. *Evolution* **34**, 825–43.

Young, R. M. (1971). Darwin's metaphor: does nature select? *The Monist* **55**, 442–503.